D1727466

Nanowire Energy Storage Devices

Nanowire Energy Storage Devices

Synthesis, Characterization, and Applications

Edited by Liqiang Mai

WILEY-VCH

The Editor

Prof. Liqiang Mai
Wuhan University of Technology
122 Luoshi Road
Wuhan
China
430070

Cover Image: © MF3d/Getty Images

Library of Congress Card No.: applied for

British Library Cataloguing-in-Publication Data
A catalogue record for this book is available from the British Library.

Bibliographic information published by the Deutsche Nationalbibliothek
The Deutsche Nationalbibliothek lists this publication in the Deutsche Nationalbibliografie; detailed bibliographic data are available on the Internet at <http://dnb.d-nb.de>.

Print ISBN: 978-3-527-34917-3
ePDF ISBN: 978-3-527-83245-3
ePub ISBN: 978-3-527-83247-7
oBook ISBN: 978-3-527-83246-0

Typesetting Straive, Chennai, India

Contents

Preface

With the increasing prominence of worldwide energy crisis, materials science and nanotechnologies have become the key to the fields of new energy storage and utilization. With the in-depth study of nanotechnologies, nanoenergy storage materials are in the rapid development stage. Especially, owing to their excellent chemistry, electricity, thermal, mechanical, and other properties, nanowire materials exhibit unique electron and ion transport characterization, making them distinctive among nanomaterials. From this perspective, nanowire energy storage devices (ESDs) show outstanding electrochemical and energy storage characteristics. And nanowire materials also have more unique advantages in the construction of micro and nanoenergy storage devices and the assembly of flexible ESDs. To this end, nanowire materials offer various opportunities to address the challenges of advanced ESD, thus attracting the great attention and strong interest of the majority of scientific researchers.

In this book, we present a comprehensive discourse on nanowire energy storage devices (ESDs) from the perspective of synthesis, characterization, and applications. This publication is structured into 10 chapters, each with coherent connections and distinct focal points. Readers can read systematically or selectively, according to their needs.

The **Chapter 1** is written by *Prof. Liqiang Mai and Wen Luo*. **Chapter 1** intends to offer an all-sided introduction to one-dimensional nanomaterials, energy storage science and technology, and review the nanowire energy storage materials and devices.

The **Chapter 2** is written by *Prof. Liqiang Mai and Chaojiang Niu*. **Chapter 2** summarizes the physicochemical characteristics of nanowire, the unique advantages of nanowires as electrode materials, the basic performance parameters of nanowire electrochemical ESDs, the optimization mechanism of electrochemical performance, and briefly introduces the theoretical calculations related to nanowire electrode materials.

The **Chapter 3** is written by *Prof. Liqiang Mai and Lin Xu*. **Chapter 3** focuses on the common synthesis methods and growth mechanisms of various types of nanowire materials based on the classification of nanowires with different morphologies.

The **Chapter 4** is written by *Prof. Liqiang Mai and Xiaocong Tian*. **Chapter 4** briefly introduces various characterization methods and summarizes the progress in the characterization of nanowires up to now. Chapter 4 will also discuss the future of characterization for nanowires and give perspectives on the opportunities of nanowires in post-lithium energy storage systems.

The **Chapter 5** is written by *Prof. Liqiang Mai and Ya You*. **Chapter 5** elaborates the application of nanowires in lithium-ion batteries, discusses the unique characteristic of nanowires for energy storage, and gives perspectives on the opportunities of nanowires in post-lithium energy storage systems.

The **Chapter 6** is written by *Prof. Liqiang Mai and Ting Zhu*. **Chapter 6** summarizes the application of nanowire cathode and anode materials in sodium-ion batteries. The synthesis, characterization, and reaction mechanism of the nanowires are completely demonstrated. The perspectives on the development of nanowires in sodium-ion batteries are also discussed in this chapter.

The **Chapter 7** is written by *Prof. Liqiang Mai and Xu Xu*. **Chapter 7** expounds the principles and applications of various new batteries from the perspective of nanowire electrode materials and explains the characteristics and advantages of nanowire electrode materials in new batteries.

The **Chapter 8** is written by *Prof. Liqiang Mai and Liang Zhou*. **Chapter 8** focuses on one-dimensional nanomaterials applied to three typical capacitors, including Electric double-layer capacitors (EDLCs), pseudocapacitive supercapacitors, and hybrid supercapacitors. The relationship between one-dimensional structure and performance of capacitors is explored, which could provide theoretical guidance for the deeper improvement of nanowire-based supercapacitors.

The **Chapter 9** is written by *Prof. Liqiang Mai and Qinyou An*. **Chapter 9** introduces a broader application of nanowire ESDs in the various fields of multivalent-ion batteries, such as magnesium-ion battery, calcium-ion battery, zinc-ion battery, and aluminum-ion batteries. In the chapter, we also give a simple summary and outlook on nanowires for multivalent-ion batteries.

The last chapter is written by *Prof. Liqiang Mai and Wen Luo*. **Chapter 10** provides a systematic conclusion and outlook on nanowire ESD and gives perspectives on the structure design, performance optimization, and advanced characterization methods of 1D nanomaterials. Moreover, in this chapter, we also intend to present a broad picture of micro-nano device structure optimization and fabrication process improvement for different applications.

This book is available for institutions of higher learning and scientific research units engaged in the development of nanomaterials and ESDs research related researchers and practitioners to use, also can be used as institutions of higher learning materials, chemical energy, physical and related professional teachers, graduate, and senior undergraduate professional reference book. Overall, despite the best efforts of the author, due to the limitations of level and time, it is inevitable that there will be some mistakes and omissions in the book. We sincerely hope that the readers will contact us with valuable suggestions for its continued improvement.

Finally, I would like to thank all the contributors to this book. We thank the support of the National Key Research and Development Program of China (2020YFA0715000), the National Natural Science Foundation of China (51832004, 52127816, 51802239, 52172233).

February 2023 *Liqiang Mai*
Wuhan, China

1

Nanowire Energy Storage Devices: Synthesis, Characterization, and Applications

1.1 Introduction

1.1.1 One-Dimensional Nanomaterials

As an important member of the large family of nanomaterials, one-dimensional nanowire (NW) materials, including nanorods, nanotubes, nanobelts, and nanocables, have gradually received attention from researchers (Figure 1.1). A NW can be defined as a one-dimensional structure that is less than 100 nm in the lateral direction (there is no limitation in the longitudinal direction). The aspect ratio of a part of a NW is more than 1000 [1]. According to their different compositions, NWs can be divided into different types, including metal NWs (such as Ni, Pt, and Au), semiconductor NWs (such as InP, Si, and GaN), and insulator NWs (such as SiO_2 and TiO_2). NW materials have important implications for theoretical research and technical applications. This kind of material has peculiar physical and chemical properties, such as the transition from metal to insulator, super mechanical strength, high luminous efficiency, lower laser threshold, and enhanced thermoelectric coefficient [2].

The research on NWs can be traced to the early 1960s. In 1964, Wagner and Ellis [3] used vapor–liquid–solid (VLS) growth to epitaxially grow Si single crystal whiskers on a single crystal Si(111) substrate, creating a precedent for Si NW research. In 1975, Givargizov [4] conducted a systematic study on the process of VLS growth and gave a reasonable VLS growth mechanism for NWs. In 1998, Morales [5] and Zhang et al. [6] used laser ablation deposition (LAD) to successfully grow Si NWs. NW materials exhibit unique optical, electrical, and magnetic properties that many bulk materials do not have. Therefore, they have an extremely important position in the field of producing nanodevices, various sensors, microtools, microelectrodes, device-integrated connection lines, and next-generation EL display devices. Current research results show that many one-dimensional nanomaterials have demonstrated essential applications. For example, Huang et al. [7] grew ZnO NW array with a diameter of 20–150 nm and a length of about 10 μm through a vapor-phase transport process on a sapphire substrate, successfully preparing

Nanowire Energy Storage Devices: Synthesis, Characterization, and Applications, First Edition.
Edited by Liqiang Mai.

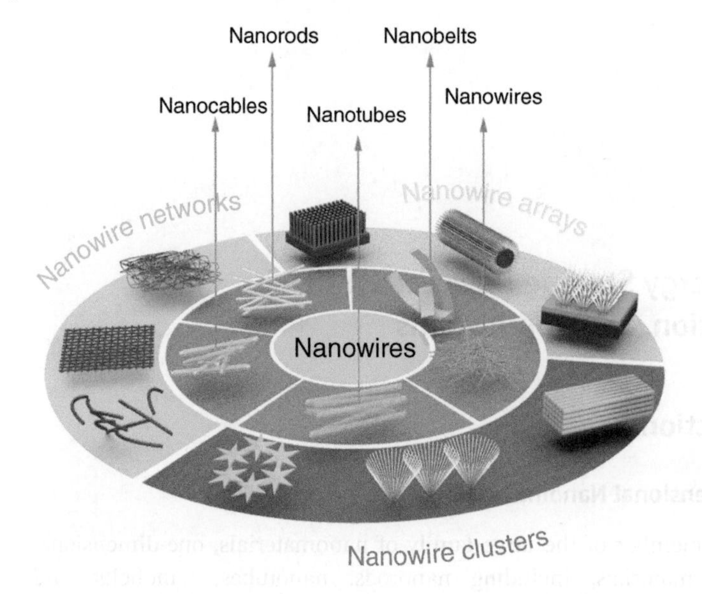

Figure 1.1 Schematic diagram of different types of nanowires and their secondary structures.

nanolasers in 2001. In 2007, Prof. Lieber from Harvard University and his team developed the first single NW solar cell [8]. They reported a solar cell prepared by a single coaxial silicon nanowire (SiNW), which adopts a p–i–n coaxial structure. The synthesis and application of NWs have become hot spots for scientists in recent years because of their unique advantages. Prof. Mai from Wuhan University of Technology and his team designed and assembled the first single NW electrochemical energy storage device to reveal the intrinsic mechanism of electrochemical performance degradation in 2010 [9]. In the same year, Huang et al. [10] used in situ TEM to observe the lithiation phenomenon of a single NW of SnO_2 for the first time. In 2012, Kouwenhoven from Delft University of Technology and his team verified the hypothesis that Majorana fermions in NWs are coupled with superconductors [11]. In 2014, Prof. Cheng from Monash University and his team built a wearable pressure sensor using gold nanowires [12]. In 2019, scientists from Cambridge University, King's College London, Peking University, Zhejiang University, Shanghai Jiao Tong University, and other universities developed the first single NW spectrometer in the world [13]. In 2020, researchers from Chinese Academy of Sciences demonstrated a new flexible dual- NW structure consisting of a GeSn layer with Sn content of 10% heteroepitaxially grown on the sidewall of a Ge NW by molecular beam epitaxy (MBE), which effectively suppressed the formation of defects at the GeSn/Ge interface and greatly reduced the dark current and static power consumption of the photodetector with GeSn/Ge dual-NW structure [14]. NWs and devices with various structures have been continuously developed, and their applications in scientific research and industrialization have become increasingly extensive with the continuous progress of nanotechnology.

The important development process is shown in Figure 1.2. This book focuses on the introduction of NW materials in electrochemical energy storage.

1.1.1.1 Nanorods

Nanorods and NWs are well-known one-dimensional nanostructures, which are frequently used not only in nano-electromechanical systems but also in biomedical treatments, dentistry, production of energy from solar cells, and humidity-sensitive analysis [15]. Nanorods consist of rod-like nanoparticles. Its length is usually much larger than its size in the two-dimensional direction and can achieve a macroscopic magnitude. It should be noted that there is no strict distinction between NWs and nanorods. Generally, those with a large slenderness ratio are called nanofibers, and those with a small aspect ratio are called nanorods. The dividing line is located at 1 μm. The preparation methods for nanorods mainly adopt VLS growth and template-based synthesis. Nanorods can also be formed by using LAD. Panasonic, the U.S. Bureau of Standards and Metrology, and the University of Florida have done a large quantity of work in this area, using the pulsed laser method to successfully prepare one-dimensional SiNWs and boron nitride nanotubes. The one-dimensional nanomaterial synthesized by this method has the advantages of high purity, large output, and uniform diameter. Researchers at the University of Minnesota and Princeton University successfully prepared a quantum disk in 1998. This disk is a nano-array system composed of magnetic nanorods with a density of 109 bit cm^{-2}.

1.1.1.2 Carbon Nanofibers

Carbon nanofibers (CNFs) are sp^2-based linear, noncontinuous filaments that are different from carbon fibers, which are continuous with diameter of several micrometers [16]. They are a new type of quasi-one-dimensional carbon material that has attracted more attention in recent years. Its diameter is generally 50–200 nm, the length is 50–100 μm, and the aspect ratio is 100–500. Its structure and performance are in the transition state between ordinary carbon fiber and carbon nanotubes, and are formed by stacking nano-sized graphite sheets at different angles to the axial direction of the fiber in space [17]. One-dimensional CNFs have many superior properties, so their application prospects are extensive. CNFs have pores on molecular level over the surface and also have pores inside, as well as a large specific surface area [18]. Therefore, it can absorb a large amount of gas and is a potential hydrogen storage material. It can also be used as a high-efficiency adsorbent, catalyst, and catalyst carrier [19]. CNFs also have high electrical conductivity and can be used as cathode materials for lithium-ion secondary batteries and electrodes for electric double-layer capacitors [20].

1.1.1.3 Nanotubes

Carbon nanotubes, the one-dimensional allotropes of carbon, have attracted significant research interest ever since their discovery due to their outstanding material properties. Carbon nanotubes are a typical representative of nanotubes. In the 1970s,

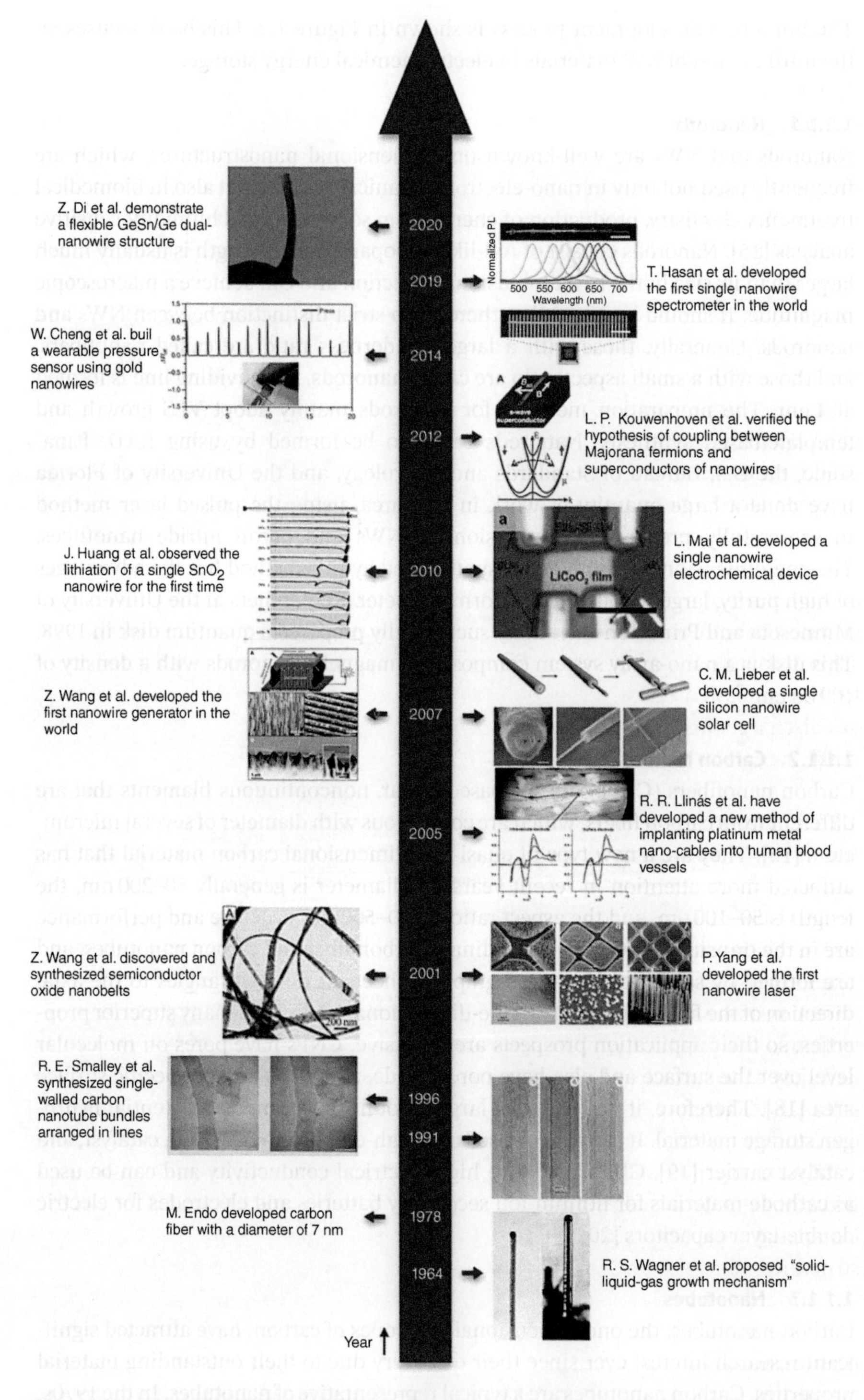

Figure 1.2 Important development history of nanowire materials. Source: Refs. [3–14].

M. Endo from the University of Orléans in France successfully synthesized carbon fibers with a diameter of 7 nm, but he did not carefully evaluate and characterize the structure of these carbon fibers [21]. Until 1991, Iijima [22] of the Japanese NEC company discovered carbon nanotubes for the first time with a high-resolution electron microscope. Since then, nanomaterials have attracted widespread attention in the global scientific research field. In 1996, the famous American Nobel Prize winner Smalley and others synthesized single-walled carbon nanotube (SWNT) bundles arranged in rows, which signifies that scientists can prepare batches of NWs in a controlled manner [23]. Xie and coworkers [24] from Institute of Physics, Chinese Academy of Sciences, have achieved the directional growth of carbon nanotubes and successfully synthesized ultra-long (millimeter-level) carbon nanotubes, reported the synthesis of strong, highly conducting, and transparent SWNT films in 2007.

Because carbon nanotubes have the advantages of large specific surface area, good electrical conductivity, and high chemical stability, they have also been widely used for electric double-layer supercapacitor electrodes or as conductive additives in other electrochemical energy storage systems, such as pseudocapacitors and metal ion batteries [25]. At the same time, the internal and surface bond states of carbon nanotubes are different, and there is a phenomenon of incomplete coordination of surface atoms, which leads to an increase in active sites on the surface. These characteristics make carbon nanotubes one of the ideal catalyst support materials [26]. In addition, carbon nanotubes are also expected to be used for field emission tubes, field emission flat panel displays, microwave generators, gas discharge tubes, and fluorescent lamps, and their nano-scale hollow pipes also make them have great potential for hydrogen storage [27].

1.1.1.4 Nanobelts

It is difficult to maintain stability during large-scale industrial production of carbon nanotubes, slowing the research of nanomaterials. In 2001, Prof. Zhonglin Wang and his team used the high-temperature solid gas phase method to discover and synthesize semiconductor oxide nanobelt structures for the first time in the world [28]. Then, they successfully synthesized the ribbonlike structure of broadband semiconductor systems such as zinc oxide, tin oxide, indium oxide, cadmium oxide, and gallium oxide. The purity of these ribbonlike structure materials is as high as 95%, and they have the advantages of large output, perfect structure, clean surface, no internal defects, and no dislocations. In 2007, Zhonglin Wang and his team developed the first NW generator in the world [29].

The cross section of the nanobelt is a narrow rectangular structure with a bandwidth of 30–300 nm, a thickness of 5–10 nm, and length of up to several millimeters. Compared with carbon nanotubes, silicon, and compound semiconductor linear structures, nanobelts are the only broadband semiconductor one-dimensional ribbon structures that have been found to have a controllable structure, be defect-free, and have more unique and superior features of structure and physical properties than carbon nanotubes. Although nanobelts lack the high structural strength of cylindrical nanotubes, they can ensure the necessary stability of electronic equipment and the uniformity of the material structure during mass production because of their superior performance, which is very important in nanophysics research and nanodevice applications. This structure is an ideal system for researching the transport processes of light, electricity, and heat in one-dimensional functional and smart materials. It can enable scientists to use single oxide nanobelts to make nanometer-sized gas and liquid sensors, nanofunctions, and smart optoelectronic components, laying a solid foundation for nanooptoelectronics [30].

1.1.1.5 Nanocables

In 1997, French scientists C. Colliex and his team found a sandwich-geometry C–BN–C tube in the product obtained by analyzing the arc discharge. Because its geometry is similar to that of a coaxial cable and the diameter is at the nanometer level, it is called a coaxial nanocable [31]. In 1998, Zhang et al. [32] used laser ablation to synthesize a coaxial nanocable with a diameter of tens of nanometers and a length of 50 μm. Their experiment shows that if the mixed powder of BN, C, and SiO_2 is used as the raw material, a single-core nanocable can be formed with β-SiC core wire inside and amorphous SiO_2 single wire outside. If Li_3N is added to the raw materials, a coaxial nanocable with another structure can be formed, that is, the core is SiC, the middle layer is amorphous SiO_2, and the outermost layer is BNC with a graphite structure. At the same time, Meng et al. [33] developed a new coaxial nanocable preparation method, which is sol–gel and carbothermal reduction with evaporation–condensation, and successfully synthesized a coaxial nanocable with SiC semiconductor core and SiO_2 insulator on the outer layer. After 2000, people have successfully prepared hundreds of coaxial nanocables using different synthesis methods and different kinds of materials, such as SiGaN/SiO_xN_y [34], Fe/C [35], Ag/polypyrrole [36], GaP/ZnS [37], CdSe/TiO_2 [38], and three-layer structure of Ag/SiO_2/ppy [39] and Nd/FM (FM = Fe, Co, Ni)/PA66 [40].

Coaxial nanocables play a crucial role in the fields of biomedicine, nano-electronic devices, nano-microprocessing, and testing technology. In terms of practical applications, the components in the ultrahigh-density-integrated circuit are mainly connected by nanocable coupling; hence, the nanocable also plays an important role in the connection line. Considering the research on solar cells, the US National Renewable Energy Laboratory has developed a kind of solar cell with high-energy conversion using coaxial nanocable. Respecting medical research, in 2005, scientists in the Unites States and Japan successfully developed a new method of implanting platinum metal nanocables with a diameter of 1% of human hair into human blood vessels and hoped that one day it would help doctors treat certain human diseases [41]. Nanocables can also be used as probes for miniature detectors on important nano-resolution detectors such as field emission detectors, biological detectors, and atomic force microscopes [42]. It is believed that with the continuous discovery of new special properties of nanocables, coaxial nanocables will also become a new force in the field of NW materials.

1.1.2 Energy Storage Science and Technology

The widespread application of fossil fuels (oil, coal, and natural gas) has greatly promoted the rapid development of the world economy. Nowadays, the improvement of the world economy is still largely based on the use of fossil fuels. However, with the continuous consumption of nonrenewable energy, traditional fossil fuels such as oil, coal, and natural gas will eventually face exhaustion. It is predicted that the life span of coal, natural gas, and oil will be less than 100 years on average if calculated at the current rate of fossil fuel consumption. The global demand for energy will reach 28 TW by 2050 [43], which is equivalent to the energy produced by consuming 1010 tons of oil each year. Obviously, one of the biggest issues of the twenty-first century is the energy crisis. The deterioration of the energy crisis will lead to a world economic crisis, which in turn leads to an intensification of economic conflicts. At the same time, major problems such as environmental pollution and global warming caused by the growing consumption of fossil fuels pose a severe threat to the sustainable development of mankind. Therefore, energy and environmental issues have become international issues that threaten human survival and development. In summary, the development of renewable energy is the key to solving energy and environmental problems and is the only choice for human beings to achieve sustainable development. Renewable energy mainly includes solar energy, biomass energy, wind energy, hydropower, geothermal energy, and tidal energy. The most important way to solve energy and environmental problems is to vigorously develop renewable energy. However, renewable energy sources such as solar and wind energy are discontinuous and have large, unpredictable, and variable characteristics. Their acquisition and output are unstable, and their energy density is low, which has a great impact on the reliability of the power grid. It is difficult to integrate into the grid. Hence, how to achieve effective conversion and storage between various energy forms is particularly important. The development of energy storage technology can effectively solve this problem enabling renewable energy to be stored and applied in a stable form. In addition, with energy storage technology as the development direction of the future power grid, smart grids can adjust grid peak using energy storage devices, increasing the capacity of the transmission and distribution system and optimizing efficiency. Energy storage technology can be widely used for the generation, transmission, distribution, and usage of the entire power industry. In this book, energy storage mainly refers to electrical energy storage. It is a technology in which electrical energy is stored in another form and released when needed, which can effectively solve the above problems.

The existing energy storage methods are mainly divided into the following categories: electrochemical energy storage (lead-acid batteries, flow batteries, alkaline metal secondary batteries, multivalent ion batteries, metal-chalcogenide batteries, supercapacitors, etc.); mechanical energy storage (pumped hydro storage, flywheel energy storage, compressed air energy storage, etc.); electromagnetic energy storage (superconducting energy storage); and latent heat storage.

1.1.2.1 Mechanical Energy Storage

The essence of mechanical energy storage is that electrical energy is converted into mechanical energy for storage. There are three main types of mechanical energy storage technologies that have been applied in industry: pumped hydro storage, flywheel energy storage, and compressed air energy storage.

Pumped Hydro Storage

Pumped hydro storage is currently the most mature and widely used energy storage technology in the world, which is mainly used for system backup and peak load and frequency modulation. The energy storage system is equipped with two storage reservoirs: the upper reservoir and the lower reservoir. Water is pumped from the lower reservoir to the upper reservoir when storing electricity. Water flows from the upper reservoir to the pump location when using electricity, and the potential energy of the water flow is used to drive the turbine to generate electricity. Pumped hydro energy storage has the longest life cycle (30–60 years), the largest number of cycles (10 000–30 000 times) and the largest capacity (500–8000 MWh) compared with other energy storage [44]. However, the disadvantage is that the site selection is greatly affected by the geographical location and the construction period is long as well as the investment cost is high.

Compressed Air Energy Storage

Compressed air energy storage system (CAES) uses gas turbine energy to generate electricity. The air compression system consumes excess electricity to compress the air and store it in underground salt mines, abandoned quarries, or large ground storage tanks when storing electricity. The stored air is released from the air storage chamber and burned with fuel when using electricity. The produced gas, with its high temperature and pressure, drives the gas turbine to generate electricity. The system construction investment and power generation cost are lower than pumped hydro storage. Besides, it has the following advantages: long service life (20–40 years), large capacity, and large quantities of charge–discharge cycles. It has wide applications in the fields of power production, transportation, and consumption, including peak shaving and valley filling, power load balancing, renewable energy access, and backup power supplies [45]. It has broad application prospects in conventional power systems, renewable energy, distributed function systems, and smart grids. However, the negative aspects of CAES are obvious. To begin with, its energy density is low. To make matters worse, it depends on large gas storage chambers and causes fossil fuel consumption and environmental pollution.

Flywheel Energy Storage

Flywheel energy storage and power generation technology uses the conversion between electrical and kinetic energy to store and generate energy. The flywheel is connected to the power grid, and the electrical energy provided by the power grid drives the flywheel to rotate at a high speed, converting the electrical energy into kinetic energy and storing it when storing energy. The rotating flywheel drives the motor to generate electricity, converting kinetic energy into electrical energy when using energy. Compared with other energy storage technologies, flywheel energy storage has the advantages of high energy conversion efficiency (85–95%), non-polluting, simple maintenance, free from geographical environment restrictions, and continuous operation [46]. Mainly used for uninterruptible power supply (UPS), grid peak shaving, and frequency control. However, short power generation times and high equipment costs are important factors restricting its development.

1.1.2.2 Electromagnetic Energy Storage

Superconducting magnetic energy storage (SMES) is the most common energy storage technology for electromagnetic energy storage. SMES system directly stores electromagnetic energy in the superconducting coil. The stored electromagnetic energy is sent back to the grid or other loads when electricity is needed. SMES system has unique benefits that other energy storage technologies do not have. SMES systems can store energy almost lossless for a long time (conversion efficiency exceeds 90%) [47], release energy rapidly, have a small size, and are environmentally friendly. The low-frequency power oscillation of the grid can be reduced or eliminated, and the active and reactive power can be adjusted more easily with the existence of SMES system. Therefore, SMES can improve the stability of the power system and enhance the controllability of power generation with new energy. However, its short board is high in cost, complicated in process, and operating conditions.

1.1.2.3 Electrochemical Energy Storage

The principle of electrochemical energy storage is that electrical energy is converted into chemical energy for storage. It includes various types of batteries (lead–acid batteries, nickel-metal hydride (NiMH) batteries, lithium-ion batteries, etc.) and supercapacitors. Electrochemical energy storage is convenient and safe, is not restricted by region, has less environmental pollution, is not restricted by Carnot cycle in energy conversion, and has high conversion efficiency compared with other methods [48]. Electrochemical energy storage is the most important component in the field of energy storage. Since Le Clancy invented lead-acid batteries in 1859, various chemical batteries involving different energy storage systems have been developing in the direction of high capacity, high power, low pollution, long service life, and high safety to meet the needs of different fields. At present, lithium-ion battery has the advantages of high energy density, high conversion efficiency (close to 100%), long cycle life, and low self-discharge. They have been widely used in various electronics and electric vehicles and are gradually being deployed in the field of large-scale energy storage. Many large-scale lithium battery energy storage demonstration systems and bases have been built in most countries. However, in 2018, the U.S. Geological Survey Mineral and Commodity Summaries showed that the global lithium reserves could not meet the rapid development of global electric vehicles [49]. Thus, in order to meet the rapid development of electric vehicles and the increasing demand for energy storage devices in other fields, other types of secondary batteries and energy storage devices have also been significantly developed, such as new alkaline metal secondary batteries (sodium ion battery, potassium ion battery), multivalent ion batteries (magnesium ion battery, zinc ion battery, aluminum ion battery, etc.), metal-chalcogenide batteries, metal-oxygen batteries, and supercapacitors. The common types of electrochemical energy storage involved and their main characteristics are shown in Table 1.1. Electrode materials are the core component of energy storage devices. Improving the performance of chemical energy storage equipment by exploring and developing new energy storage materials has become a global issue, which has received great attention and support from all countries.

Table 1.1 Common types of electrochemical energy storage and their main advantages and disadvantages.

Type of battery	Mechanism of energy storage	Analysis of advantages and disadvantages
Lead-acid batteries	$PbO_2 + 2H_2SO_4 + Pb \leftrightarrow 2H_2O + 2PbSO_4$	Pros: mature technology and low cost Cons: short cycle life and pollution problem
NiMH batteries	$Ni(OH)_2 + M \leftrightarrow NiOOH + MH$	Pros: withstand overcharge and over discharge, strong capability of high rate discharge, safety and high power density Cons: low voltage and low energy density
Monovalent ion batteries	Embedding reaction, alloying reaction, conversion reaction	Pros: mature technology for lithium-ion batteries, small self-discharge, light weight, small size, high working voltage, environmentally friendly, no memory effect, charging rapidly and high energy capacity Cons: internal impedance is high, the working voltage changes greatly and the capacity decays quickly
Multivalent ion batteries	Embedding reaction, alloying reaction, conversion reaction	Pros: the theoretical volume ratio is high, and the anode has rich metal resources Cons: multivalent ions have a large radius, the polarization phenomenon is serious, and configuring a suitable electrolyte is difficult
Metal-chalcogenide batteries	The anode (M) is mainly the deposition and dissolution reaction of metal ions, and the oxidation–reduction reaction of the chalcogen element (X) occurs in the cathode. The reaction of the entire battery system can be summarized as: $M + X \leftrightarrow MX$	Pros: high energy density Cons: low power density, short cycle life, immature system, and difficult to achieve commercialization
Metal-air batteries	Catalytic materials are used to catalyze the oxygen or pure oxygen in the air as the cathode active material, the metal is the anode, and the ammonium chloride or caustic solution is used as the electrolyte to participate in the electrochemical reaction.	Pros: relatively low cost and stable output voltage Cons: low power density, short cycle life, and low output voltage

(Continued)

Table 1.1 (Continued)

Type of battery	Mechanism of energy storage	Analysis of advantages and disadvantages
Supercapacitors	Electric double layer, pseudocapacitance (underpotential deposition, redox reaction, and embedded pseudocapacitance)	Pros: high power density and good cycle stability Cons: low energy density and low output voltage
Flow batteries	A high-performance battery that uses positive and negative electrolytes is separated and circulate, respectively	Pros: long cycle life Cons: low energy storage density

Lead-Acid Batteries

The basic chemical composition of all types of lead-acid battery is the same. In the charged state, the cathode is lead dioxide, the anode is metallic lead, and the electrolyte is sulfuric acid solution. In the discharge state, the main components of the positive and negative electrodes are lead sulfate. The global lead-acid battery industry has always been larger than other batteries because of a series of advantages such as low cost, high working voltage, safety, and reliability that other batteries cannot replace [50]. Lead-acid batteries have a long history of development, and the technology is mature and is now widely used in many fields. In the automotive industry, lead-acid batteries are used for the starting, ignition, and lighting of various cars, motorcycles, and ships. In the electric vehicle industry, lead-acid batteries provide power for electric cars and electric bicycles. In addition, the market demand for lead-acid batteries in the communications industry is huge. In addition, when new energy sources generate electricity, it also needs to be supplied with lead-acid batteries [51]. Lead-acid batteries can replace expensive gas and oil turbine generators to meet load balancing requirements as an alternative power supply method during peak electricity consumption.

NiMH Batteries

NiMH batteries are a new generation of high-energy secondary batteries that replace Ni–Cd batteries [52]. NiMH batteries are a kind of green battery compared with lead-acid batteries. The cathode of NiMH battery is $Ni(OH)_2$. The anode material is metal hydride (MH). The electrolyte is an aqueous potassium hydroxide solution. In the field of small batteries, although the energy density of NiMH batteries is higher than Ni–Cd batteries, its energy density is lower than that of lithium-ion batteries. Therefore, at present, the market utilization rate of NiMH batteries is lower than that of lithium-ion batteries. There are also some electric tools that retain NiMH batteries, like sweeping robots. In the automotive industry, NiMH batteries are used as power batteries. NiMH batteries have outstanding advantages compared with lithium-ion batteries and hydrogen fuel cells, which mainly show in their fast charging and discharging speeds, high mass-specific power, and good stability.

In addition, the energy storage batteries currently used in smart grids mainly include lead-acid batteries, lithium-ion batteries, sodium–sulfur batteries, all-vanadium flow batteries, and NiMH batteries. Among them, lead-acid batteries will produce lead pollution, and many European countries have gradually banned lead-acid batteries. Li-ion batteries are not yet mature in large-scale integration technology. The technology of NiMH batteries as a smart grid energy storage system is relatively mature, with good safety performance and long service life [53].

Monovalent Ion Batteries

Lithium-ion batteries are the most representative of monovalent ion batteries. It mainly realizes charge and discharge through repeated insertion and extraction of lithium ions between the positive and negative electrodes. Lithium ions are extracted from the cathode material and inserted into the anode material through the electrolyte, and the anode material is in a lithium-rich state when charging. The situation is opposite when discharging. Before the advent of lithium-ion batteries, NiMH batteries were widely used in the portable electronic equipment industry. After the advent of lithium-ion batteries, they are superior to NiMH batteries in terms of weight, capacity, and power. The energy density of lithium-ion batteries is greater than that of NiMH batteries, the self-discharge phenomenon is weakened, as well as the charging time is shorter. Lithium-ion battery energy storage technology currently occupies a leading position in the energy storage industry and has been widely used in electric vehicles, smart grids, cutting-edge defense equipment, biomedicine, portable devices, artificial intelligence, etc., as shown in Figure 1.3.

However, large-scale applications of lithium-ion batteries are still limited, and security and high cost are the main influencing factors. The current commercial cathode materials for lithium-ion batteries are mainly $LiCoO_2$, $LiMn_2O_4$, and $LiFePO_4$. The anode materials are mainly carbon materials and $Li_4Ti_5O_{12}$ [54]. The electrolyte is a nonaqueous organic electrolyte. In recent years, $LiFePO_4$

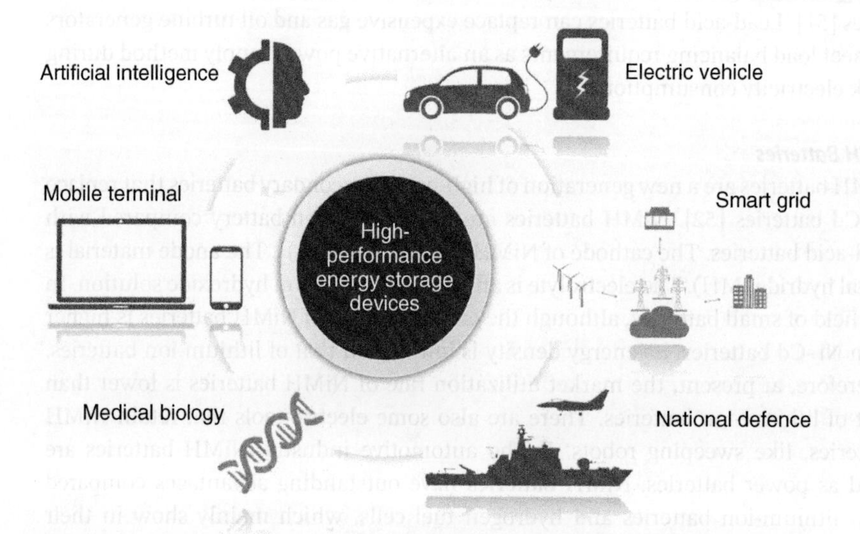

Figure 1.3 Main application areas of high-performance energy storage devices.

has received widespread attention around the world due to its better safety and high-rate discharge performance, and its application in electric vehicles and grid energy storage has been rapidly promoted [55].

Multivalent Ion Batteries

There is an increasing demand for lithium-ion batteries in various fields due to their continuous development. The shortage of lithium resources is one of the important factors hindering the development of lithium-ion batteries. In order to alleviate this problem, multivalent ion batteries, including magnesium ion batteries, zinc ion batteries, aluminum ion batteries, and calcium ion batteries, have been rapidly developed. Multivalent ion batteries have higher safety performance and lower cost than lithium ion batteries. Moreover, multivalent ion batteries directly use metals as negative electrode materials, so they have the potential to greatly increase the energy density of the batteries compared with lithium ion battery using carbon materials as negative electrodes. Therefore, it has good application value and development prospects in the fields of large-scale energy storage and civil batteries. Unfortunately, multivalent ion batteries are still in the research stage, and the strong polarization effect and high requirements for electrolyte are important factors that limit the development of multivalent ion batteries [56].

Supercapacitors

Supercapacitors are different from traditional batteries. The energy storage mechanism of supercapacitors was divided into physical energy storage and chemical energy storage. The physical mechanism is that a double layer structure formed by the interface of electrode/electrolyte with porous carbon material with high specific surface area is used for energy storage [57]. The chemical mechanism is that the rapid, reversible redox reactions between electrodes and electrolyte are used for energy storage (pseudocapacitance) [58]. Therefore, according to different energy storage mechanisms, supercapacitors can be divided into electric double layer supercapacitors (EDLC), pseudocapacitors, and hybrid capacitors (both physical energy storage and chemical energy storage). It combines the advantages of traditional capacitors and batteries, not only has the advantages of high discharge power of electrostatic capacitors, but also has a large charge storage capacity, as same as batteries. Its capacity can reach the farad level or even thousands of farad levels. At the same time, it has the advantages of high power density, good cycle stability, strong temperature adaptability, and being environmentally friendly. Supercapacitors are currently widely used in different market areas such as auxiliary peak power, backup power, storage of renewable energy, and alternative power supplies. More importantly, it also has very broad market prospects in industrial control, wind and solar power generation, transportation, military industry, and other directions [59].

1.1.3 Overview of Nanowire Energy Storage Materials and Devices

Nanomaterials have a high specific surface area and excellent activity. When used as a battery electrode material, it has a large contact area with the electrolyte and a short ion diffusion distance, which can effectively improve the electrical activity of the material. It also has significant advantages when used as a high-power battery electrode material. At the same time, a reasonable design of nanostructures is beneficial to alleviate stress release during cycling, and is conducive to improving the stability of the structure in order to obtain a longer cycle life. For this reason, research on large-capacity, high-power, long-life, and low-cost electrochemical energy storage technology based on new nano-electrode materials is one of the frontiers and areas of focus in the low-carbon economy era. Lithium-ion batteries, supercapacitors, and lithium-air batteries have been extensively studied due to their respective advantages, but traditional electrode materials are still difficult to meet the needs of high capacity and high power. Numerous researches have shown that the electrochemical properties of electrode materials are closely related to their scale, internal crystal structure, and apparent morphology. NW electrode materials, due to their unique anisotropy, rapid axial electron transmission, and radial ion diffusion characteristics, are suitable for the design, integration, and performance control of alkali metal ion batteries, supercapacitors, transparent flexible energy storage devices, and hybrid devices.

In recent years, the development of nanowires has shown diversified characteristics, including multilayer NWs, NW clusters, nanotube clusters, NW clusters, and other one-dimensional NW materials with composite structures that have been able to develop rapidly. It is used for the tip of a scanning tunneling microscope, nanodevices, ultralarge integrated circuit wires, optical fibers, micro-drills in microelectronics, electrode materials for energy hunting systems, active materials for catalytic systems such as photocatalysis and electrocatalysis, and electrode materials for energy storage systems (such as metal ion batteries and supercapacitors). This chapter focuses on the electrochemical energy storage system by discussing the physical and chemical properties, synthetic chemistry, and practical applications of NW electrode materials.

There are many ways to synthesize NW materials, such as hydrothermal methods [60], electrospinning methods [61], and template methods [62]. Various directional control methods can synthesize NW materials with different morphologies. However, the physical and chemical properties of NWs will be different, and the external properties will also be different. With the progress of technology, NW materials have also developed, from conventional NWs to ultra-long NWs [63], mesoporous NWs [64], dendritic NWs [65], etc. The uniformity, controllability, and electrochemical properties of NW materials have been continuously improved with the continuous optimization of the structure. The specific preparation methods for different NW structures will be described in detail in Chapter 3.

One-dimensional NW materials have the following advantages for electrochemical energy storage. Firstly, compared to granular nanomaterials, NW electrode materials provide transmission paths with continuous axial electron and radial ion transmission paths, which endow NW materials with better rate performance. Secondly, NW materials can directly grow on the substrate of metal or carbon material. It can be used as a framework to composite with other materials without the addition of a binder, constructing a complex and multi-element electrode structure. Thirdly, NW materials have geometric characteristics, which are tens of microns in length and tens of nanometers in diameter, making it easy to make a single NW electrochemical device for in situ electron microscopy and spectroscopy characterization [66].

At the same time, the use of NW materials for electrochemical energy storage also has the following challenges. Firstly, when semiconductor NWs are used for electrochemical energy storage, their poor conductivity requires additional conductive agents or conductive material coatings like carbon coatings, which increases the complexity of the electrode construction process. Besides, the continuous charging and discharging processes will cause the deterioration of the material structure, which will affect the axial electron conduction of the NW material, thus affecting the cycle stability of the battery. Additionally, the materials in the electrode network created by overlapping NWs are largely in point contact with one another, which reduces the conductivity of the entire electrode and raises the internal resistance of the battery [66]. In view of the advantages and challenges of using NW materials for energy storage, researchers have proposed a variety of optimization strategies and designed and constructed NW electrode materials with various structures.

Energy storage materials and devices with high energy density, good power density, and long service lives are major requirements for the development of new energy vehicles, smart grids, and cutting-edge national defense equipment. However, rapid capacity decay and difficulty balancing energy and power density are major bottlenecks, restricting the development of high-performance energy storage materials and device technologies. In order to solve the above bottlenecks, the author of this book has been committed to the research of NW energy storage materials and devices in combination with the advantages of NW materials described above. The author and his team focused on basic research and application development of NW electrode materials in basic theory, controlled growth, performance regulation, device assembly, in situ characterization, and energy storage applications. They designed and assembled the first single NW solid-state energy storage device for monitoring capacity decay in real-time and took the lead in realizing the large-scale preparation and application of high-performance NW energy storage devices and key materials, achieving many innovative research results. The main research content of the NW energy storage devices involved in this book is shown in Figure 1.4.

1.1.3.1 Si Nanowires

Silicon materials have the characteristics of high specific capacity (4200 $mAh\ g^{-1}$) that other materials cannot achieve as negative electrode materials. It has attracted the attention of researchers and is the third-generation negative electrode material that can replace graphite materials [67]. In addition, as silicon element is generally nontoxic, abundant, and widely available, new systems may have cost and recycling benefits. The application of silicon in lithium batteries originated in the 1970s. Nelson et al. [68] and Sharma et al. [69] reported that the phenomenon of reversible lithiation and delithiation of silicon at high temperature. Subsequently, researchers such as Weydanz et al. [70], Gao et al. [71], Limthongkul et al. [72], Li et al. [73],

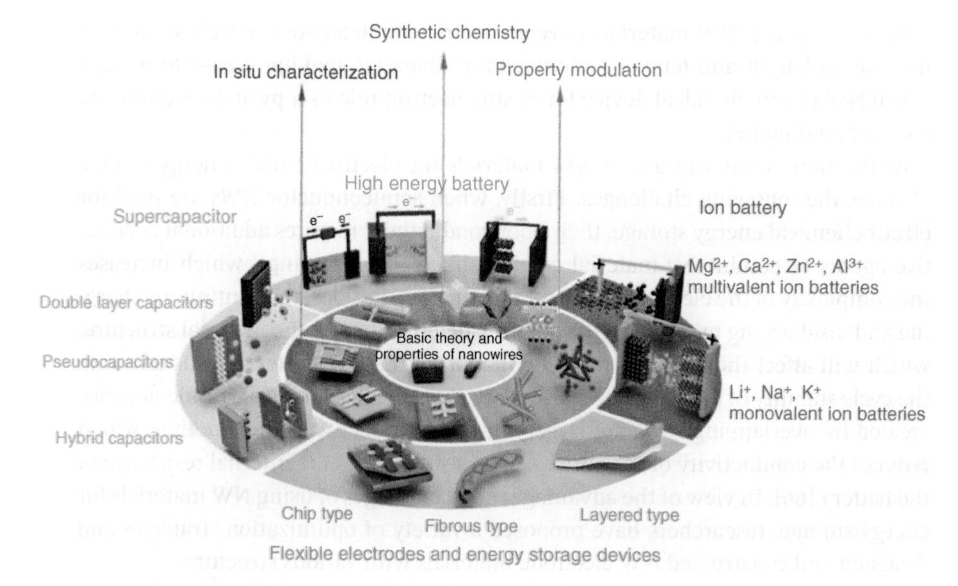

Figure 1.4 Main research contents of nanowire energy storage devices.

and Obrovac et al. [74] conducted in-depth research and gradually improved the reversible lithium storage mechanism of silicon materials.

Nanotechnology applied to silicon materials can resolve the volume expansion of silicon-based anode materials during cycle. The volume expansion phenomenon will cause structure collapse of the active materials and shedding of the current collector to lose electrical contact when the lithium ions are released, thereby affecting the cycle's performance. Yue et al. [75] used nanosilicon powders with different sizes (5, 10, and 20 nm) to prepare anode materials and found that silicon particles with a particle size of 10 nm have the highest specific capacity after electrochemical tests. One-dimensional NWs and nanotubes can also improve the problem of volume expansion, except for zero-dimensional nanoparticles. SiNWs have many excellent properties that are different from those of general silicon materials, such as electron transport, field emission characteristics, surface activity, and quantum confinement [76]. Therefore, it has a wide range of application prospects in the manufacture of low-dimensional nanodevices [77]. At present, the preparation and application of nanodevices such as SiNW transistors, sensors, and NW batteries have been realized.

The battery made with SiNWs as the anode has better storage capacity than the battery made with traditional graphite as the anode. Traditional graphite anodes need six carbon atoms to store one Li^+, while one Si atom can store four Li^+. However, a large number of moving Li atoms with high speed will break the Si in a battery made of Si thin layers or Si ions and will also destroy the bond between Si and the substrate, weakening the power storage capacity. Cui et al. of Stanford University prepared Si NWs on a stainless steel substrate and then made a lithium ion battery with SiNWs, which not only increased the anode capacity of the battery but also ensured that the SiNWs never fell off the substrate after a series of expansion and contraction experiments. This is because the shape of the SiNWs enables the Si atomic lattice to expand and contract rapidly along the NW, thereby alleviating the structural strain and making the NW firmly adhere to the metal substrate. In this way, the stored electricity of Si anodes is 10 times higher than that of traditional graphite anodes [78].

1.1.3.2 ZnO Nanowires

The band gap of ZnO nanowires at room temperature is 3.37 eV. ZnO hardly produces any light scattering under natural light. In addition, its thermal ionization energy under room temperature is low (26 meV) [79] and much lower than the exciton binding energy (60 meV), so the ZnO NWs have great exciton stability. The crystal structure of ZnO NWs has intrinsic defects that can cause energy transitions to emit light at various energy levels. ZnO NWs show unique properties in the fields of thermodynamics, electricity, optics, and magnetism [80].

Bae et al. [81] reported that ZnO NWs have a larger electron concentration and field-effect mobility than bulk ZnO. Choi et al. [82] developed a kind of composite nanofiber containing insulating $ZnGa_2O_4$ nanocrystals as components for building field-effect transistors (FET). The composite NW material has excellent electrical properties and high field effects. This is because the insulating spinel phase of $ZnGa_2O_4$ effectively blocks the flow of electrons through In_2O_3 and ZnO nanocrystals. Moreover, based on low-pressure chemical vapor deposition (LPCVD) technology, a series of ZnO NWs have been grown on SiO_2/Si and SiNWs/Si substrates. These composite nanomaterials have quasi-lasting photoconductivity and enhanced ohmic *I*–*V* characteristics [83]. Moon et al. [84] obtained ZnO–NiO mixed oxide nanostructures by electrospinning. After measuring their electrical conductivity, they found that the value increased with increasing temperature, which can be attributed to the thermal emission of the carrier from the grain boundary energy state. Park et al. [85] showed that, compared with pure ZnO, the prepared ZnO/N-doped carbon nanotube nanocomposite has higher electron mobility. In addition, the work of Khurana et al. [86] revealed that the electron transfer kinetics of ZnO graphene composites is faster than those of pure ZnO NWs.

The application of ZnO NW materials in energy storage is generally used as the window layer of solar cells or the photoanode of dye-sensitized solar cells. ZnO NWs can be combined with certain polymers, graphene, or quantum dots to form hybrid photovoltaic devices. In 2005, Lévy-Clément et al. [87] took the lead in using CdSe-sensitized p-CuSCN layer and n-ZnO NWs to form a heterojunction solar cell and achieved a conversion efficiency of 2.3%. Liu et al. [88] used ZnO NW/Sb_2S_3/P3HT structure to fabricate an organic/inorganic hybrid solar cell and achieved a conversion efficiency of 2.9%. Then, Park et al. [89] grew vertically arranged ZnO NWs on graphene surfaces and constructed a hybrid solar cell with PEDOT:PEG as the interface layer and PbS quantum dots as the hole transport layer, obtaining a conversion efficiency of 4.2% under AM1.5 light conditions. In addition, Chen et al. [90] produced a dye-sensitized solar cell with vertical array of ZnO NWs and studied the effect of NW length on the photovoltaic performance of solar cells. The results indicate that the length of the ZnO NW increases with increasing conversion efficiency.

1.1.3.3 Single Nanowire Electrochemical Energy Storage Device

Mai et al. [9] used the unique anisotropy of NWs to design and assemble an all-solid-state electrochemical energy storage device with single NWs that can be used for in situ diagnosis and micro-nanosystem support power supply. This device can detect the electrical conductivity and law of structural changes of the NWs during cycle test without adding conductive additives or binders. Using a single vanadium oxide NW as a working electrode shows that after a shallow discharge with a current of 100 pA, the conductance of the NW drops by 2 orders of magnitude. After shallow charging, the conductance returns to the original level. This result indicates that the structural change of the NWs caused by the insertion and extraction of lithium ions is reversible in the case of shallow charge and discharge. After 400 seconds of deep discharge at 100 pA, the conductivity of the NWs irreversibly decrease by 5 orders of magnitude, showing that the structure undergoes permanent changes. The direct connection between the NW electrical transport and charging and discharging states has been established through research on the in situ characterization platform based on the all-solid-state single NW device. This also inherently reveals the scientific law of the capacity attenuation of NW energy storage devices. In other words, the capacity decay of NW electrode material and rapid decline of conductivity are directly related to structural deterioration. Huang et al. [10] designed and constructed a battery system with single-crystal SnO_2 NWs as anode and lithium cobalt oxide as cathode. Using TEM technology to observe the morphological changes of the NWs during charging and discharging, it is found that when lithium ions are inserted into the NWs, it will cause violent volume expansion, elongation, and curling deformation. It is worth noting that no matter how severe the deformation is, the nanowire always maintains structural integrity during the cycle process, which further verifies the superior performance of the nanomaterial as an electrode.

Based on the characterization of single NW materials, the increasingly limited nature of their structure and morphology has been discovered. It needs to be modified to improve its electrochemical performance. Commonly used methods include pre-embedding technology to modify one-dimensional NWs, surface modification hierarchical nanowires, internal structure optimization NWs, and synergy optimization NWs. Pre-embedding technology is an effective method to increase the carrier concentration and improve the conductivity of the NW electrode without destroying the morphology [91]. Hierarchical structure NW materials include dendritic structure NW materials and coaxial NW materials. When dendritic NW materials are used as electrodes, they can provide more ion transmission paths, higher electronic conductivity, more active sites, and effectively inhibit the self-aggregation of nanomaterials [92]. The preparation of coaxial NW materials is relatively easy to realize. This structure is usually used to improve the structural stability of the material, enhance the electrical conductivity, and buffer the stress and strain of the internal material [93]. In addition, the specific surface area of the NW, the contact area between electrolyte and electrode material, and reactive sites can be improved by effectively combining the advantages of one-dimensional nanostructures and mesoporous structures to construct mesoporous NWs, providing free space for stress release in the reaction [94]. The synergistic effect of surfactants can change the orientation of the nanowires and reduce their interface resistance [95].

Single NW devices have good applications in different energy conversion devices. A single NW device used in a solar cell can increase the conversion efficiency of solar energy by up to 40%. The thermal conductivity of one-dimensional NW materials is completely different from that of bulk materials [96], as thermoelectric devices can effectively improve thermal energy utilization. In addition, a single nanodevice can be used as an efficient electrocatalyst in electrocatalytic hydrogen production devices. It can effectively enhance the charge injection of the electrode material and improve the catalytic activity of the catalyst by reducing the contact resistance between the electrode and the material [97].

It is necessary to understand some basic theories of NW energy storage materials when the composition of the material is determined, such as thermodynamic and kinetic characteristics and important electrochemical performance parameters. These will be introduced in detail in Chapter 2. After that, we will consider the method used to synthesize the required NW materials, whether hierarchical NWs or porous NWs, hydrothermal or electrospinning, which will be discussed in Chapter 3. In order to study the electrochemical mechanism of the synthesized NWs, in situ characterization methods are needed to provide convincing evidence. Chapter 4 lists some in situ characterization methods, including detailed descriptions of single NW devices, in situ electron microscopy, and in situ spectroscopy. Chapters 5–9 of this book show readers the application of NW electrode materials in a variety of electrochemical systems, leading readers to the forefront of nanowire energy storage materials and devices.

References

1 Schenning, A.P. and Meijer, E.W. (2005). Supramolecular electronics; nanowires from self-assembled π-conjugated systems. *Chemical Communications* 26: 3245–3258.

2 Yang, P., Yan, H., Mao, S. et al. (2002). Controlled growth of ZnO nanowires and their optical properties. *Advanced Functional Materials* 12 (5): 323–331.

3 Wagner, A.R. and Ellis, S.W. (1964). Vapor-liquid-solid mechanism of single crystal growth. *Applied Physics Letters* 4 (5): 89–90.

4 Givargizov, E.I. (1975). Fundamental aspects of VLS growth. In: *Vapour Growth and Epitaxy*, 20–30. Elsevier.

5 Morales, A.M. and Lieber, C.M. (1998). A laser ablation method for the synthesis of crystalline semiconductor nanowires. *Science* 279 (5348): 208–211.

6 Zhang, Y.F., Tang, Y.H., Wang, N. et al. (1998). Silicon nanowires prepared by laser ablation at high temperature. *Applied Physics Letters* 72 (15): 1835–1837.

7 Huang, M.H., Mao, S., Feick, H. et al. (2001). Room-temperature ultraviolet nanowire nanolasers. *Science* 292 (5523): 1897–1899.

8 Tian, B., Zheng, X., Kempa, T.J. et al. (2007). Coaxial silicon nanowires as solar cells and nanoelectronic power sources. *Nature* 449 (7164): 885–889.

9 Mai, L., Dong, Y., Xu, L., and Han, C. (2010). Single nanowire electrochemical devices. *Nano Letters* 10 (10): 4273–4278.
10 Huang, J.Y., Zhong, L., Wang, C.M. et al. (2010). In situ observation of the electrochemical lithiation of a single SnO_2 nanowire electrode. *Science* 330 (6010): 1515–1520.
11 Mourik, V., Zuo, K., Frolov, S.M. et al. (2012). Signatures of Majorana fermions in hybrid superconductor-semiconductor nanowire devices. *Science* 336 (6084): 1003–1007.
12 Gong, S., Schwalb, W., Wang, Y. et al. (2014). A wearable and highly sensitive pressure sensor with ultrathin gold nanowires. *Nature Communications* 5 (1): 1–8.
13 Yang, Z., Albrow-Owen, T., Cui, H. et al. (2019). Single-nanowire spectrometers. *Science* 365 (6457): 1017–1020.
14 Yang, Y., Wang, X., Wang, C. et al. (2020). Ferroelectric enhanced performance of a GeSn/Ge dual-nanowire photodetector. *Nano Letters* 20 (5): 3872–3879.
15 Numanoğlu, H.M., Akgöz, B., and Civalek, Ö. (2018). On dynamic analysis of nanorods. *International Journal of Engineering Science* 130: 33–50.
16 Kim, Y.A., Hayashi, T., Endo, M., and Dresselhaus, M.S. (2013). Carbon nanofibers. In: *Springer Handbook of Nanomaterials* (ed. Vajtal), 233–261. Berlin, Heidelberg: Springer.
17 Hwang, J.Y., Lee, S.H., Sim, K.S., and Kim, J.W. (2002). Synthesis and hydrogen storage of carbon nanofibers. *Synthetic Metals* 126 (1): 81–85.
18 De Jong, K.P. and Geus, J.W. (2000). Carbon nanofibers: catalytic synthesis and applications. *Catalysis Reviews* 42 (4): 481–510.
19 Chinthaginjala, J.K., Seshan, K., and Lefferts, L. (2007). Preparation and application of carbon-nanofiber based microstructured materials as catalyst supports. *Industrial & Engineering Chemistry Research* 46 (12): 3968–3978.
20 Ji, L. and Zhang, X. (2009). Fabrication of porous carbon nanofibers and their application as anode materials for rechargeable lithium-ion batteries. *Nanotechnology* 20 (15): 155705.
21 Ciselli, P. (2007). The potential of carbon nanotubes in polymer composites. Doctor Philosophy. Eindhoven University of Technology, Germany. pp: 53–58.
22 Iijima, S. (1991). Helical microtubules of graphitic carbon. *Nature* 354 (6348): 56–58.
23 Thess, A., Lee, R., Nikolaev, P. et al. (1996). Crystalline ropes of metallic carbon nanotubes. *Science* 273 (5274): 483–487.
24 Ma, W., Song, L., Yang, R. et al. (2007). Directly synthesized strong, highly conducting, transparent single-walled carbon nanotube films. *Nano Letters* 7 (8): 2307–2311.
25 Ma, R.Z., Liang, J., Wei, B.Q. et al. (1999). Study of electrochemical capacitors utilizing carbon nanotube electrodes. *Journal of Power Sources* 84 (1): 126–129.
26 Lv, R., Cui, T., Jun, M.S. et al. (2011). Open-ended, N-doped carbon nanotube-graphene hybrid nanostructures as high-performance catalyst support. *Advanced Functional Materials* 21 (5): 999–1006.

27 Sharma, P. and Ahuja, P. (2008). Recent advances in carbon nanotube-based electronics. *Materials Research Bulletin* 43 (10): 2517–2526.

28 Pan, Z.W. and Wang, Z.L. (2001). Nanobelts of semiconducting oxides. *Science* 291 (5510): 1947–1949.

29 Wang, X., Song, J., Liu, J., and Wang, Z.L. (2007). Direct-current nanogenerator driven by ultrasonic waves. *Science* 316 (5821): 102–105.

30 Wang, Z.L., Gao, R.P., Pan, Z.W., and Dai, Z.R. (2001). Nano-scale mechanics of nanotubes, nanowires, and nanobelts. *Advanced Engineering Materials* 3 (9): 657–661.

31 Suenaga, K., Colliex, C., Demoncy, N. et al. (1997). Synthesis of nanoparticles and nanotubes with well-separated layers of boron nitride and carbon. *Science* 278 (5338): 653–655.

32 Zhang, Y., Suenaga, K., Colliex, C., and Iijima, S. (1998). Coaxial nanocable: silicon carbide and silicon oxide sheathed with boron nitride and carbon. *Science* 281 (5379): 973–975.

33 Meng, G.W., Zhang, L.D., Mo, C.M. et al. (1998). Preparation of β-SiC nanorods with and without amorphous SiO_2 wrapping layers. *Journal of Materials Research* 13 (9): 2533–2538.

34 Xu, C., Kim, M., Chung, S. et al. (2004). The formation of $SiGaN/SiO_xN_y$ nanocables and SiO_xN_y-based nanostructures using GaN as a resource of Ga. *Chemical Physics Letters* 398 (1–3): 264–269.

35 Luo, T., Chen, L., Bao, K. et al. (2006). Solvothermal preparation of amorphous carbon nanotubes and Fe/C coaxial nanocables from sulfur, ferrocene, and benzene. *Carbon* 44 (13): 2844–2848.

36 Qiu, T., Xie, H., Zhang, J. et al. (2011). The synthesis of Ag/polypyrrole coaxial nanocables via ion adsorption method using different oxidants. *Journal of Nanoparticle Research* 13 (3): 1175–1182.

37 Hu, L., Brewster, M.M., Xu, X. et al. (2013). Heteroepitaxial growth of GaP/ZnS nanocable with superior optoelectronic response. *Nano Letters* 13 (5): 1941–1947.

38 Wang, H., Zhou, H., Lu, J. et al. (2018). Electrodeposition of $CdSe/TiO_2$ coaxial nanocables for enhanced photocatalytic performance and H_2 evolution in visible light. *Journal of the Electrochemical Society* 165 (3): D160.

39 Wang, W., Lu, L., Cai, W., and Chen, Z.R. (2013). Synthesis and characterization of coaxial silver/silica/polypyrrole nanocables. *Journal of Applied Polymer Science* 129 (5): 2377–2382.

40 Li, X., Li, H., Song, G. et al. (2018). Preparation and magnetic properties of Nd/FM (FM = Fe, Co, Ni)/PA66 three-layer coaxial nanocables. *Nanoscale Research Letters* 13 (1): 1–5.

41 Llinas, R.R., Walton, K.D., Nakao, M. et al. (2005). Neuro-vascular central nervous recording/stimulating system: using nanotechnology probes. *Journal of Nanoparticle Research* 7 (2–3): 111–127.

42 Mishra, S., Song, K., Koza, J.A., and Nath, M. (2013). Synthesis of superconducting nanocables of FeSe encapsulated in carbonaceous shell. *ACS Nano* 7 (2): 1145–1154.

43 Tao, C.S., Jiang, J., and Tao, M. (2011). Natural resource limitations to terawatt-scale solar cells. *Solar Energy Materials and Solar Cells* 95 (12): 3176–3180.

44 Benato, A. and Stoppato, A. (2018). Pumped thermal electricity storage: a technology overview. *Thermal Science and Engineering Progress* 6: 301–315.

45 Chen, L., Zheng, T., Mei, S. et al. (2016). Review and prospect of compressed air energy storage system. *Journal of Modern Power Systems and Clean Energy* 4 (4): 529–541.

46 Amiryar, M. and Pullen, K. (2017). A review of flywheel energy storage system technologies and their applications. *Applied Sciences* 7 (3): 286.

47 Tixador, P. (2008). Superconducting Magnetic Energy Storage: status and Perspective. *IEEE/CSC&ESAS European Superconductivity News Forum (No. 3).*

48 Haas, O. and Cairns, E.J. (1999). Electrochemical energy storage. *Annual Reports Section "C" (Physical Chemistry)* 95: 163–198.

49 LITHIUM – U.S. Geological Survey, Mineral Commodity Summaries, January 2018. Retrieved from https://minerals.usgs.gov/minerals/pubs/commodity/lithiummcs-2018-lithi.pdf

50 Raza, S.S., Janajreh, I., and Ghenai, C. (2014). Sustainability index approach as a selection criteria for energy storage system of an intermittent renewable energy source. *Applied Energy* 136: 909–920.

51 Gong, H., Wang, M.Q., and Wang, H. (2013). New energy vehicles in China: policies, demonstration, and progress. *Mitigation and Adaptation Strategies for Global Change* 18 (2): 207–228.

52 Rodrigues, L.E.O.C. and Mansur, M.B. (2010). Hydrometallurgical separation of rare earth elements, cobalt and nickel from spent nickel–metal–hydride batteries. *Journal of Power Sources* 195 (11): 3735–3741.

53 Salkuti, S.R. (2020). Comparative analysis of electrochemical energy storage technologies for smart grid. *TELKOMNIKA* 18 (4): 2118–2124.

54 Rao, M.C. (2013). $LiMn_2O_4$ cathodes for solid state lithium-ion batteries-energy storage and conversion. *Journal of Optoelectronics and Biomedical Materials* 5 (1): 9–16.

55 Scipioni, R., Jørgensen, P.S., Ngo, D.T. et al. (2016). Electron microscopy investigations of changes in morphology and conductivity of $LiFePO_4$/C electrodes. *Journal of Power Sources* 307: 259–269.

56 Cui, L., Zhou, L., Kang, Y.M., and An, Q. (2020). Recent advances on rational design and synthesis of two-dimensional materials for multivalent ion batteries. *ChemSusChem* .

57 Halper, M.S. and Ellenbogen, J.C. (2006). *Supercapacitors: A Brief Overview*, 1–34. McLean, VA: The MITRE Corporation.

58 Liu, T.C., Pell, W.G., and Conway, B.E. (1999). Stages in the development of thick cobalt oxide films exhibiting reversible redox behavior and pseudocapacitance. *Electrochimica Acta* 44 (17): 2829–2842.

59 Poonam, Sharma, K., Arora, A., and Tripathi, S.K. (2019). Review of supercapacitors: materials and devices. *Journal of Energy Storage* 21: 801–825.

60 Ruffo, R., Hong, S.S., Chan, C.K. et al. (2009). Impedance analysis of silicon nanowire lithium ion battery anodes. *The Journal of Physical Chemistry C* 113 (26): 11390–11398.

61 Tang, J., Huo, Z., Brittman, S. et al. (2011). Solution-processed core–shell nanowires for efficient photovoltaic cells. *Nature Nanotechnology* 6 (9): 568–572.

62 Wu, M.C. and Lee, C.S. (2009). Field emission of vertically aligned V_2O_5 nanowires on an ITO surface prepared with gaseous transport. *Journal of Solid State Chemistry* 182 (8): 2285–2289.

63 Hu, P., Dong, S., Gui, K. et al. (2015). Ultra-long SiC nanowires synthesized by a simple method. *RSC Advances* 5 (81): 66403–66408.

64 Hochbaum, A.I., Gargas, D., Hwang, Y.J., and Yang, P. (2009). Single crystalline mesoporous silicon nanowires. *Nano Letters* 9 (10): 3550–3554.

65 Song, Y.J., Kim, J.Y., and Park, K.W. (2009). Synthesis of Pd dendritic nanowires by electrochemical deposition. *Crystal Growth and Design* 9 (1): 505–507.

66 Poizot, P.L.S.G., Laruelle, S., Grugeon, S. et al. (2000). Nano-sized transition-metal oxides as negative-electrode materials for lithium-ion batteries. *Nature* 407 (6803): 496–499.

67 Zhou, G., Xu, L., Hu, G. et al. (2019). Nanowires for electrochemical energy storage. *Chemical Reviews* 119 (20): 11042–11109.

68 Nelson, P.A. (1978). High-performance batteries for stationary energy storage and electric-vehicle propulsion. *Progress report (ANL-78-45)*. Argonne National Lab. (ANL), Argonne, IL (United States).

69 Sharma, R.A. and Seefurth, R.N. (1976). Thermodynamic properties of the lithium-silicon system. *Journal of the Electrochemical Society* 123 (12): 1763.

70 Weydanz, W.J., Wohlfahrt-Mehrens, M., and Huggins, R.A. (1999). A room temperature study of the binary lithium–silicon and the ternary lithium–chromium–silicon system for use in rechargeable lithium batteries. *Journal of Power Sources* 81: 237–242.

71 Gao, B., Sinha, S., Fleming, L., and Zhou, O. (2001). Alloy formation in nanostructured silicon. *Advanced Materials* 13 (11): 816–819.

72 Limthongkul, P., Jang, Y.I., Dudney, N.J., and Chiang, Y.M. (2003). Electrochemically-driven solid-state amorphization in lithium-silicon alloys and implications for lithium storage. *Acta Materialia* 51 (4): 1103–1113.

73 Li, J. and Dahn, J.R. (2007). An in situ X-ray diffraction study of the reaction of Li with crystalline Si. *Journal of the Electrochemical Society* 154 (3): A156.

74 Obrovac, M.N. and Christensen, L. (2004). Structural changes in silicon anodes during lithium insertion/extraction. *Electrochemical and Solid-State Letters* 7 (5): A93.

75 Yue, L., Wang, S., Zhao, X., and Zhang, L. (2012). Nano-silicon composites using poly(3,4-ethylenedioxythiophene): poly(styrenesulfonate) as elastic polymer matrix and carbon source for lithium-ion battery anode. *Journal of Materials Chemistry* 22 (3): 1094–1099.

76 Wang, S., Han, J., and Yin, S. (2019). The effect of silver-plating time on silicon nanowires arrays fabricated by wet chemical etching method. *Optics and Photonics Journal* 9 (8): 1–10.

77 Wang, D., Sheriff, B.A., McAlpine, M., and Heath, J.R. (2008). Development of ultra-high density silicon nanowire arrays for electronics applications. *Nano Research* 1 (1): 9–21.

78 Designing Nanowires for Energy Storage. (2009). Meeting Abstracts (Electrochemical Society).

79 Stiegler, J.M., Tena-Zaera, R., Idigoras, O. et al. (2012). Correlative infrared–electron nanoscopy reveals the local structure–conductivity relationship in zinc oxide nanowires. *Nature Communications* 3 (1): 1–7.

80 Özgür, Ü., Alivov, Y.I., Liu, C. et al. (2005). A comprehensive review of ZnO materials and devices. *Journal of Applied Physics* 98 (4): 11.

81 Bae, S.Y., Seo, H.W., and Park, J. (2004). Vertically aligned sulfur-doped ZnO nanowires synthesized via chemical vapor deposition. *The Journal of Physical Chemistry B* 108 (17): 5206–5210.

82 Choi, S.H., Jang, B.H., Park, J.S. et al. (2014). Low voltage operating field effect transistors with composite In_2O_3–ZnO–$ZnGa_2O_4$ nanofiber network as active channel layer. *ACS Nano* 8 (3): 2318–2327.

83 Yu, Y., Zha, G.W., Shang, X.J. et al. (2017). Self-assembled semiconductor quantum dots decorating the facets of GaAs nanowire for single-photon emission. *National Science Review* 4 (2): 196–209.

84 Moon, J., Park, J.A., Lee, S.J. et al. (2009). Structure and electrical properties of electrospun ZnO–NiO mixed oxide nanofibers. *Current Applied Physics* 9 (3): S213–S216.

85 Park, J.S., Lee, J.M., Hwang, S.K. et al. (2012). A ZnO/N-doped carbon nanotube nanocomposite charge transport layer for high performance optoelectronics. *Journal of Materials Chemistry* 22 (25): 12695–12700.

86 Khurana, G., Sahoo, S., Barik, S.K., and Katiyar, R.S. (2013). Improved photovoltaic performance of dye sensitized solar cell using ZnO–graphene nano-composites. *Journal of Alloys and Compounds* 578: 257–260.

87 Lévy-Clément, C., Tena-Zaera, R., Ryan, M.A. et al. (2005). CdSe-sensitized p-CuSCN/nanowire n-ZnO heterojunctions. *Advanced Materials* 17 (12): 1512–1515.

88 Liu, C.P., Wang, H.E., Ng, T.W. et al. (2012). Hybrid photovoltaic cells based on ZnO/Sb_2S_3/P3HT heterojunctions. *Physica Status Solidi B* 249 (3): 627–633.

89 Park, H., Chang, S., Jean, J. et al. (2013). Graphene cathode-based ZnO nanowire hybrid solar cells. *Nano Letters* 13 (1): 233–239.

90 Chen, W., Qiu, Y., Zhong, Y. et al. (2010). High-efficiency dye-sensitized solar cells based on the composite photoanodes of SnO_2 nanoparticles/ZnO nanotetrapods. *The Journal of Physical Chemistry A* 114 (9): 3127–3138.

91 Xu, X., Luo, Y.Z., Mai, L.Q. et al. (2012). Topotactically synthesized ultralong LiV_3O_8 nanowire cathode materials for high-rate and long-life rechargeable lithium batteries. *NPG Asia Materials* 4 (6): e20–e20.

92 Mai, L.Q., Yang, F., Zhao, Y.L. et al. (2011). Hierarchical $MnMoO_4$/$CoMoO_4$ heterostructured nanowires with enhanced supercapacitor performance. *Nature Communications* 2 (1): 1–5.

93 Mai, L., Xu, X., Han, C. et al. (2011). Rational synthesis of silver vanadium oxides/polyaniline triaxial nanowires with enhanced electrochemical property. *Nano Letters* 11 (11): 4992–4996.

94 Zhao, Y., Xu, L., Mai, L. et al. (2012). Hierarchical mesoporous perovskite $La_{0.5}Sr_{0.5}CoO_{2.91}$ nanowires with ultrahigh capacity for Li-air batteries. *Proceedings of the National Academy of Sciences* 109 (48): 19569–19574.

95 Wang, X., Peng, K.Q., Hu, Y. et al. (2014). Silicon/hematite core/shell nanowire array decorated with gold nanoparticles for unbiased solar water oxidation. *Nano Letters* 14 (1): 18–23.

96 Pop, E., Mann, D., Wang, Q. et al. (2006). Thermal conductance of an individual single-wall carbon nanotube above room temperature. *Nano Letters* 6 (1): 96–100.

97 Lukowski, M.A., Daniel, A.S., Meng, F. et al. (2013). Enhanced hydrogen evolution catalysis from chemically exfoliated metallic MoS_2 nanosheets. *Journal of the American Chemical Society* 135 (28): 10274–10277.

2

Fundamentals of Nanowire Energy Storage

2.1 Physical and Chemical Properties of Nanowires

Nanomaterials with different dimensions often display different physical and chemical characteristics such as the electronic state distribution. This section focuses on the physicochemical properties of one-dimensional nanowire materials.

2.1.1 Electronic Structure

Physicochemical properties of a material are greatly influenced by its electronic structure, and at the same time, its electronic structure can be affected by its size. For nanowire, its axial direction is the macroscopic scale defined by solid-state physics, while the other directions are the nanoscale. The lattice periodic potential field is truncated, so it can be considered that there is an infinite potential barrier at the boundary, i.e. inside the nanowire is an infinite square potential well. The potential energy function is:

$$U(\boldsymbol{r}) = \begin{cases} 0 & \boldsymbol{r} \text{ inside the well} \\ \infty & \boldsymbol{r} \text{ outside the well} \end{cases} \tag{2.1}$$

since the barrier is infinite, there is no electron outside the well, i.e. $\phi(\boldsymbol{r}) = 0$. The electron motion in the potential well satisfies the stationary Schrodinger equation:

$$-\frac{\hbar^2}{2m^*}\nabla^2\phi(\boldsymbol{r}) = E\phi(\boldsymbol{r}) \tag{2.2}$$

where m^* is the effective mass of the electron, and it is assumed to be isotropic.

Let the nanowire be a rectangle with a cross-section of nanoscale, the length and width of the rectangle are represented by a and b (along the x and y directions, respectively), and the length of the nanowire is the macroscopic scale L (along the z-direction). In this case, the nanowire can be regarded as a rectangular cross-section

Nanowire Energy Storage Devices: Synthesis, Characterization, and Applications, First Edition.
Edited by Liqiang Mai.

infinite deep potential well, and the solution of Eq. (2.2) obtained under the standing wave boundary condition is:

$$\phi(x,y,z) = e^{ik_z z}\sqrt{\frac{4}{ab}}\sin\frac{l\pi x}{a}\sin\frac{l\pi y}{b}\ (l, m = 1, 2, 3, \ldots) \tag{2.3}$$

The corresponding intrinsic energy is:

$$E_{l,m,k_z} = \frac{\hbar^2\pi^2}{2m^*}\left(\frac{l^2}{a^2}+\frac{m^2}{b^2}\right)+\frac{\hbar^2 k_z^2}{2m^*} \tag{2.4}$$

where

$$E_{l,m} = \frac{\hbar^2\pi^2}{2m^*}\left(\frac{l^2}{a^2}+\frac{m^2}{b^2}\right),\quad E_{k_z} = \frac{\hbar^2 k_z^2}{2m^*}$$

Equation 2.4 can be further denoted as:

$$E_{l,m,k_z} = E_{l,m} + E_{k_z} \tag{2.5}$$

E_{k_z} represents the energy of the quasi-free movement of the electron in the z direction, and $E_{l,m}$ represents the quantized energy level generated after the electron is constrained in the (x, y) direction. At each sub-energy level, the number of allowed electronic states is:

$$Z(E) = \frac{2L}{2\pi}2k = \frac{2Lk}{\pi}\frac{\sqrt{2m^*E_z}}{\hbar} \tag{2.6}$$

Then the density of states at each energy level is:

$$D_{1\text{D}} = \frac{1}{\pi}\left(\frac{2m^*}{\hbar^2}\right)^{\frac{1}{2}}\frac{L}{\pi\hbar}\sqrt{\frac{2m^*}{E_z}} \tag{2.7}$$

i.e. in one dimension, the energy relationship between the density of states and electrons in that direction is proportional to the reciprocal of $\sqrt{E_z}$.

For the entire nanowire, the total density of states is:

$$D_{1\text{D}}^{\text{tot}} = \frac{1}{\pi}\left(\frac{2m^*}{\hbar^2}\right)^{\frac{1}{2}}\sum_{l,m}(E-E_{l,m})^{-\frac{1}{2}} \tag{2.8}$$

It means that there is a peak density of states corresponding to each sub-energy level, as shown in Figure 2.1, which is quite different from the bulk material, and the quasi-continuous band disappears. It should be emphasized that the geometric shape and the confinement potential of the actual nanowire are related to the formation conditions of the low-dimensional structure, rather than the simple infinitely deep flat bottom potential well. Therefore, Poisson equation and Schrodinger equation should be solved simultaneously to perform self-consistent calculation when solving its electronic structure. Compared with bulk materials, the electronic structure of nanowires is very different, which is also the origin of many physical and chemical characteristics of nanowires.

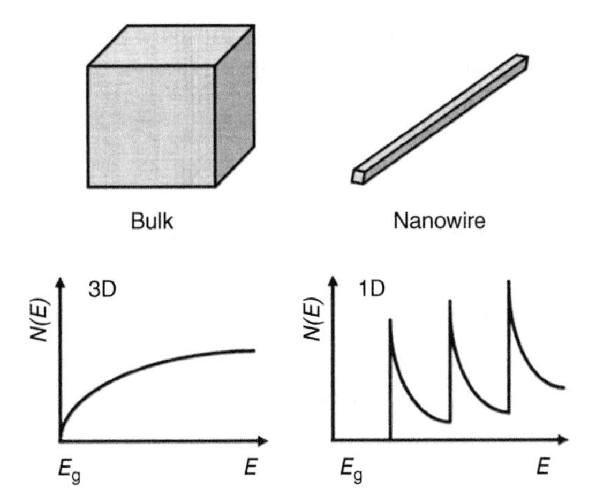

Figure 2.1 Electron energy density of bulk materials and nanowires.

2.1.2 Thermal Properties

Due to the surface effect, small size effect, quantum size effect, and other special effects, nanowires have many differences from bulk materials in their thermal properties. In the following, the differences between nanowires and bulk materials are discussed in terms of two basic thermal properties, including melting point and thermal conduction.

2.1.2.1 Melting Point

The melting point of a solid is directly proportional to its binding energy, i.e. the sum of its bond energies. When the material contains only one atom, the binding energy of the bulk material can be expressed as:

$$E_0 = \frac{1}{2} n\beta_c \varepsilon \tag{2.9}$$

where E_0 is the binding energy of the bulk material, n is the total number of atoms in the material, β_c is the coordination number of atoms in the crystal, and ε is the energy of each bond. For nanowires, the surface atoms account for a large proportion, and the difference between surface atoms and internal atoms cannot be ignored. The binding energy is as follows:

$$E_{1D} = \left(\frac{1}{2} n_i \beta_c + \frac{1}{2} n_s \beta_s\right) \varepsilon = \frac{1}{2} \left(n\beta_c - n_s(\beta_c - \beta_s)\right) \varepsilon \tag{2.10}$$

where E_{1D} is the binding energy of the nanowire material, n_i and n_s are the total number of atoms inside and on the surface of the nanowire, and β_s is the coordination number of surface atoms. Because the melting point is proportional to the binding energy, therefore:

$$\frac{T_{1D}}{T_0} = \frac{E_{1D}}{E_0} = \frac{n\beta_c - n_s(\beta_c - \beta_s)}{n\beta_c} = 1 - \frac{n_s}{n} \frac{(\beta_c - \beta_s)}{\beta_c} \tag{2.11}$$

where T_{1D} and T_0 are the melting points of nanowire material and bulk material, respectively. Let $q = \beta_s/\beta_c$, i.e. q is the ratio of coordination numbers between surface and internal atoms, then:

$$\frac{T_{1D}}{T_0} = 1 - (1-q)\frac{n_s}{n} \tag{2.12}$$

Set the cross-section radius of the nanowires as r, length as L ($r \ll L$), and atomic radius as r_a. Take the outer monolayer atom as the surface, i.e. $q = 1/2$, then:

$$\frac{T_{1D}}{T_0} = 1 - \frac{n_s}{2n} \approx 1 - \frac{V_s}{2V} \approx 1 - \frac{\pi\left(r^2 - (r-r_a)^2\right)L}{2\pi r^2 L} = 1 - \frac{r_a}{r} + \frac{r_a^2}{2r^2} \tag{2.13}$$

Inside the nanowire, r and r_a differ by more than one order of magnitude, and the quadratic term can be ignored. Thus, there are:

$$\frac{T_{1D}}{T_0} \approx 1 - \frac{r_a}{r} \tag{2.14}$$

$$\Delta T = T_0 - T_{1D} \propto \frac{1}{r} \tag{2.15}$$

The smaller the radius of the nanowires, the lower the melting point. The actual situation is far more complicated than the assumption, but the relation (2.15) still holds. The decrease in the melting point is due to the surface atoms. The smaller the radius of the nanowire, the larger the proportion of surface atoms, and the greater the reduction in melting point.

In addition, the experimental results reported by many researchers are also consistent with the relationship (2.15). For example, the melting point of Ge bulk material is 1203 K, while the Ge nanowires with a radius of 55 nm begin to melt at about 923 K, and the melting starts from the two ends with a higher proportion of surface atoms [1]. In like manner, the melting point of the Sn block material is 505 K, and when the radius of the Sn nanowires decreases from 30 to 7 nm, the melting point decreases from 502 to 486 K, and the relationship between the melting point and the material radius also satisfies Eq. (2.15) [2].

2.1.2.2 Thermal Conduction

Because of the structural difference, the thermal conduction mechanism of different substances is not the same, but the thermal conduction of all substances is the result of the collision and transmission of the carrier of material heat. The carriers that conduct heat in solids include electrons, phonons (lattice waves), photons (high-frequency electromagnetic radiation), and magnetic excitations.

The thermal conductivity k of all solid materials can be expressed mathematically as follows:

$$k = \frac{1}{3}\sum_i C_{vi} v_i \Lambda_i \tag{2.16}$$

where C_v is the heat capacity of thermal conduction carrier in unit volume, v is the average velocity of thermal conduction carrier in unit volume, Λ is the average free path of thermal conduction carrier in unit volume, and i represents different thermal conduction carriers.

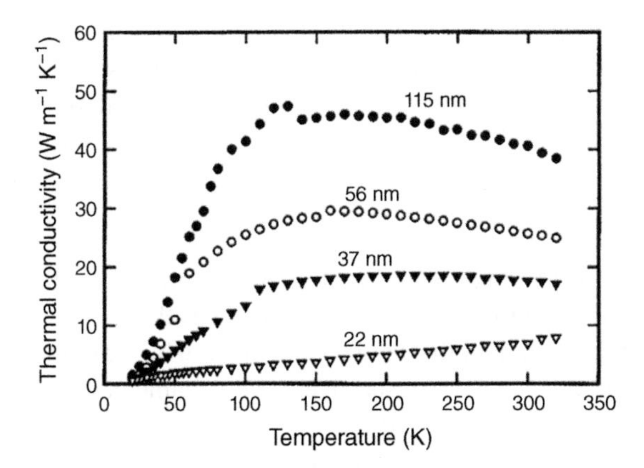

Figure 2.2 Thermal conductivity of Si nanowire with different diameters at different temperatures. Source: Li et al. [3]/with permission of American Institute of Physics.

In conductive materials such as metals and alloys, electrons are the main heat carriers. In nonconductors, the conduction of heat energy is realized by lattice vibration, i.e. the phonon conduction mechanism is the main mechanism. Whether electrons or phonons are used as heat conduction carriers, the reduction of the nanowires' diameter will enhance boundary scattering and reduce the mean free path of the thermal conduction carriers, thus leading to the decrease in thermal conductivity.

Figure 2.2 shows the thermal conductivity of Si nanowire with the same orientation and different diameters as a function of temperature. The thermal conductivity of Si nanowire decreases with decreasing diameter, which indicates that the enhanced boundary scattering with decreasing size has a significant impact on phonon motion in Si nanowire [3]. The heat conduction properties of nanowire electrode materials have an important impact on the thermal control and safety performance of batteries.

2.1.3 Mechanical Properties

Young's modulus is a physical quantity describing the ability of solid materials to resist deformation and is an important mechanical property of materials. At the macroscopic scale, Young's modulus is only related to the type of material and has nothing to do with the size of the external force or the shape of the material. For example, Young's modulus of steel wire is independent of the thickness of the steel wire. However, at the nanoscale, due to surface effects, Young's modulus varies with the size of the nanowires.

Young's modulus is greatly affected by atomic spacing. Compared with the inside of the nanowire, the atoms on the surface of the nanowire are loosely arranged, i.e. there is a larger atomic spacing, so the inside of the nanowire and the surface have different Young's modulus. The smaller the diameter of the nanowire, the larger the proportion of surface atoms and the greater the influence of surface atoms on Young's modulus.

Based on the difference between the surface and the interior of the nanowires, the nanowires can be regarded as core–shell structure for analysis. The atomic spacing on the surface of the nanowire varies with depth, but to simplify the calculation, the surface is approximated as a shell with uniform atomic spacing. Ignoring shear deformation:

$$EI = E_c I_c + E_s I_s \tag{2.17}$$

where, E, E_c, and E_s are the effective Young's modulus of the whole nanowire, the inside of the nanowire, and the surface layer of the nanowire, respectively; I, I_c, and I_s are the moments of inertia of the section of the whole nanowire, the inside of the nanowire, and the surface layer of the nanowire, respectively. If the diameter of the nanowire is d and the surface layer thickness is d_s, then:

$$E = E_c\left[1 + 8\left(\frac{E_s}{E_c} - 1\right)\left(\frac{d_s}{d} - 3\frac{d_s^2}{d^2} + 4\frac{d_s^3}{d^3} - 2\frac{d_s^4}{d^4}\right)\right] \tag{2.18}$$

Equation (2.18) reflects the relationship between Young's modulus of nanowire and its diameter. When the diameter of nanowire decreases, Young's modulus increases. As shown in Figure 2.3, Young's modulus of ZnO nanowires with different diameters measured by experiment can be well consistent with (2.18), which proves the correctness of this relation [4]. Mechanical properties such as Young's modulus are important parameters to be considered when nanowire electrode materials are applied to flexible energy storage devices.

2.1.4 Adsorption and Surface Activity

For nanowire, due to the large specific surface area (SSA), the number of surface atoms accounts for a large proportion of the total number of atoms, and the presence of abundant suspended and unsaturated bonds caused by unsaturated coordination of surface atoms, therefore, nanowire has stronger absorbability and higher chemical activity compared with bulk materials.

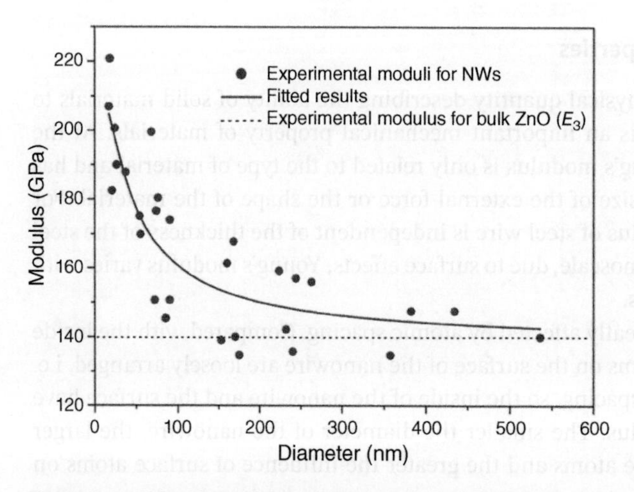

Figure 2.3 Young's modulus of ZnO nanowires with different diameters. Source: Chen et al. [4]/with permission of American Physical Society.

2.1.4.1 Adsorption

Adsorption includes two types: physical adsorption and chemical adsorption. Physical adsorption refers to the bonding of adsorbent and solid surface atoms by weak physical forces such as van der Waals force, while chemical adsorption is strongly bonded by chemical bonds. Due to the large SSA and the lack of coordination of the surface atoms, the nanowires have stronger absorbability compared with the same bulk materials. At the same time, the adsorption properties of nanowires are affected by the material properties, adsorbent properties, solution properties, and other factors.

When nanowires are applied to energy storage devices, they usually work in electrolyte solution, so the adsorption phenomenon of nanowires in electrolyte solution is mainly discussed here. Electrolytes exist in the form of ions in solution, and the adsorption of ionized ions is mainly through electrostatic attraction, and the adsorption capacity is determined by Coulomb force. There are many suspended and unsaturated bonds on the surface of the nanowires, so the surface of the nanowires is charged. Therefore, in electrolyte solution, nanowire often attracts ions with opposite charges to the surface through Coulomb force to maintain the electrical neutrality of the surface, so the adsorption of nanowire in electrolyte solution is mostly physical adsorption. This physical adsorption tends to be hierarchical. Generally speaking, the layer close to the surface of the nanowire is a strong adsorption layer, whose role is to balance the charge on the surface of the nanowire, called the adsorption layer. The electrolyte ions farther away from the nanowires form a weaker adsorption layer called the diffusion layer. The above two layers constitute a double electric layer, and a potential descent gradient from the surface of the nanowire is generated in the entire adsorption layer. Taking the nanowire surface as the origin, the potential at any position d away from the nanowire surface in the solution can be expressed as follows according to Debye–Huckel theory:

$$\Psi = \Psi_0 e^{-\kappa d} \tag{2.19}$$

$$\kappa = \left(\frac{2 e_0{}^2 N_A c Z^2}{\varepsilon k_B T} \right)^{\frac{1}{2}} \tag{2.20}$$

when $d \rightarrow \infty$, $\Psi = 0$, Ψ_0 is the surface potential of the nanowire, κ is the spread of the double electric layer, e_0 is the amount of electron charge, N_A is Avogadro's constant, c is the molar concentration of the strong electrolyte in solution, Z is the valence, ε is the dielectric constant, k_B is Boltzmann's constant, and T is the absolute temperature. $1/\kappa$ is the thickness of the double electric layer. It can be seen from Eq. (2.20) that $1/\kappa$ is inversely proportional to Z and $\sqrt{c}$, i.e. the double electric layer is very thin under the concentration of high ion and high strength electrolyte.

2.1.4.2 Surface Activity

The SSA of a nanowire is related to its size. For a nanowire with radius r and length L, its volume SSA is:

$$\frac{S}{V} = \frac{2\pi r L}{\pi r^2 L} = \frac{2}{r} \tag{2.21}$$

That is, the smaller the radius of the nanowires, the larger the SSA, and the unsaturation of the surface atoms leads to a large number of suspended and unsaturated bonds, thus leading to high surface activity. Therefore, ultrafine nanowires exhibit high catalytic activity and are widely concerned in the field of catalysis.

2.2 Thermodynamics and Kinetics of Nanowires Electrode Materials

Chemical reactions are influenced by both thermodynamics and kinetics. Thermodynamics determines whether a chemical reaction can proceed, and kinetics determines the rate of a chemical reaction. The unique physicochemical properties of nanowires compared to bulk materials were discussed in Section 2.1, and this section will explore the thermodynamic and kinetic characteristics of nanowire electrodes compared to bulk materials.

2.2.1 Thermodynamics

Gibbs free energy is a very important state function in thermodynamics. Compared to the bulk, the nanowire composed of the same number of atoms has a higher Gibbs free energy. With the bulk as the starting state and the nanowire as the final state, the increment of the Gibbs free energy of the system comes from the increase of the Gibbs free energy at the surface, and the expression is as follows:

$$\Delta G_{\mathrm{NWs}} = G_{\mathrm{NWs}} - G_{\mathrm{Bulk}} = \gamma \Delta A \tag{2.22}$$

where ΔG_{NWs} is Gibbs free energy increment of the system; G_{NWs} is Gibbs free energy of the nanowire; G_{Bulk} is Gibbs free energy of the block; γ is the specific surface energy; ΔA is the SSA increment. According to Eq. (2.21), the smaller the radius of nanowires, the larger the SSA, and the larger the incremental SSA, which means the higher Gibbs free energy.

Compared with bulk materials, the high surface Gibbs free energy of nanowire materials leads to changes in their physicochemical properties, such as lower melting point and increased surface activity. The high surface activity of nanowires has attracted much attention in the energy storage field. Take lithium-ion batteries (LIBs) as an example, nanowire electrodes with high SSA can provide more surface activity sites, which is conducive to the improvement of capacity. How to balance the high surface activity and the agglomeration of nanowire electrode is one of the key research directions in this field.

2.2.2 Kinetics

The power density of energy storage devices is closely related to the kinetics in the electrode materials, and optimizing the electron conduction and ion diffusion kinetics of the electrode materials is the focus of current research, such as LIBs.

For solid-state diffusion of Li in electrode materials, the mean diffusion (or storage) time is calculated according to the following formula:

$$t = \frac{L^2}{2D} \tag{2.23}$$

where t is the average diffusion time, D is the diffusion coefficient, and L is the diffusion distance.

Therefore, it is clear from Eq. (2.23) that there are two ways to optimize the kinetics of electrode materials, i.e. enhancing the diffusion coefficient D and shortening the diffusion distance L, and the latter is more effective. For nanowire electrode materials, if the ions diffuse radially along the nanowire, the diffusion process can be completed just through the nanoscale diffusion path, which drastically reduces the diffusion distance. For example, reducing L from 10 μm to 100 nm for materials with $D = 10^{-10}$ cm^2 s^{-1} (typical diffusion coefficient of electrode materials), the t reduces from 5000 to 0.5 seconds. Not only that, but the nanowire structure can also significantly reduce the path length for electron conduction inside the electrode material. Therefore, compared to bulk materials, nanowire electrodes have significantly enhanced kinetics and can achieve an increased power density of energy storage devices.

In addition, the enhanced kinetics can enable the utilization of certain thermodynamically permissible but kinetically limited energy storage reactions, i.e. certain "inactive" electrode materials become "active" electrode materials after nanosizing. For example, for rutile TiO_2, Li^+ ions embedding is thermodynamically permissible, but the diffusion coefficient of Li^+ ions along its *ab* surface is extremely low ($D_{ab} \sim 10^{-15}$ cm^2 s^{-1}). For rutile TiO_2 with a size of 10 μm, the distance L_{ab} at which lithium ions need to diffuse along the *ab*-plane is 5 μm, leading to a diffusion time t_{ab} of about four years, which greatly limits the lithium storage reaction. Therefore, micron-sized rutile TiO_2 is considered to be "inactive." However, the rutile TiO_2 nanorods with a diameter of 10 nm exhibit significantly higher lithium storage activity due to a dramatic reduction in t_{ab} to about two minutes after reducing the L_{ab} to 5 nm [5].

Not only that, the ion diffusion kinetics of nanowire electrode materials can be further enhanced by crystal structure modification and nanowire structure design. The significant advantages of nanowire electrode materials in ion transport kinetics make them occupy an important position in the field of electrochemical energy storage devices.

2.3 Basic Performance Parameters of Nanowires Electrochemical Energy Storage Devices

The performance parameters of the nanowire electrochemical energy storage devices are the focus of attention in practical applications. Taking battery as an example, the basic performance parameters of nanowire electrochemical energy storage devices are introduced below.

2.3.1 Electromotive Force

The electromotive force of a battery is determined by the potential difference between the positive and negative electrodes:

$$E = \varphi_{+} - \varphi_{-} \tag{2.24}$$

where E is the electromotive force, φ_{+} is the electrode potential of the positive electrode, φ_{-} is the electrode potential of the negative electrode. The electrode potential of positive and negative electrodes has the following relationship with the Gibbs free energy change of electrode reaction:

$$\Delta G_{+} = -F\varphi_{+} \tag{2.25}$$

$$\Delta G_{-} = -F\varphi_{-} \tag{2.26}$$

where ΔG_{+} and ΔG_{-} are the Gibbs free energy changes of positive and negative reactions, respectively. F is Faraday's constant ($F = 96\,485\ \text{C mol}^{-1}$). Gibbs free energy change ΔG for the total reaction of the battery is:

$$\Delta G = -FE \tag{2.27}$$

2.3.2 Operating Voltage

Operating voltage is also known as discharge voltage or load voltage. Due to the effect of battery internal resistance, the operating voltage of the battery is not equal to the electromotive force and has the following relationship:

$$U = E - IR_{\text{i}} \tag{2.28}$$

U is the operating voltage; E is the electromotive force; I is the working current; R_{i} is the internal resistance, including ohmic resistance and polarization resistance. In practical applications, the operating voltage of a battery is affected by many factors, including discharge current, discharge cut-off voltage, and ambient temperature.

2.3.3 Capacity and Specific Capacity

Capacity refers to the amount of electricity released by a battery under certain discharge conditions. At constant current discharge, the capacity is:

$$C = It \tag{2.29}$$

where C is battery capacity; I is the discharge current; t is the discharge time.

Specific capacity refers to the capacity of electrode or battery per unit mass, volume, or area, which can be divided into mass-specific capacity, volume-specific capacity, and area-specific capacity. There are:

$$C_m = \frac{C}{m} \tag{2.30}$$

$$C_V = \frac{C}{V} \tag{2.31}$$

$$C_S = \frac{C}{S} \tag{2.32}$$

where C_m is mass-specific capacity; C_V is volume-specific capacity; C_S is area-specific capacity; C is the capacity of electrode or battery; m is the mass; V is the volume; S is the area. In the absence of special instructions, the electrode material capacity mentioned later is the mass-specific capacity.

The theoretical mass-specific capacity of the electrode material is:

$$C_0 = \frac{nF}{3.6M} \tag{2.33}$$

where C_0 is the theoretical capacity; n is the electron transfer number of reaction; F is Faraday's constant; M is the relative molecular mass of the active substance. Taking $LiCoO_2$ cathode material as an example, its theoretical reaction is $LiCoO_2 \leftrightarrow CoO_2 + Li^+ + e^-$, so its theoretical specific capacity is:

$$\frac{1 \times 96\,485}{3.6 \times 97.87} = 273.8\ \text{mAh g}^{-1}$$

2.3.4 Energy and Specific Energy

The energy of a battery refers to the electric energy output under certain discharge conditions, which is determined by the capacity and working voltage of the battery:

$$W = \int U(C)dC = C \cdot U_a \tag{2.34}$$

where W is battery energy; $U(C)$ is the relationship between operating voltage and discharge capacity; C is the battery capacity. U_a is the average operating voltage of the battery.

Specific energy, also known as energy density, is divided into mass energy density and volume energy density, respectively, referring to the amount of energy released per unit mass and per unit volume of battery, that is:

$$W_m = \frac{W}{m} \tag{2.35}$$

$$W_V = \frac{W}{V} \tag{2.36}$$

where W_m is mass energy density; W_V is volumetric energy density; W is battery energy; m is battery quality; V is the battery volume.

For micro/nano batteries, two-dimensional flexible batteries, and other energy storage devices, the area energy density is the energy released per unit area of the battery, which is also one of its important performance parameters. The calculation formula is as follows:

$$W_S = \frac{W}{S} \tag{2.37}$$

where W_S is the area energy density; W is battery energy; S is the battery area.

2.3.5 Current Density and Charge–Discharge Rate

For batteries, current density is divided into area current density and mass current density, which respectively refer to the ratio of charge and discharge current to the area or mass of the electrode or battery, namely:

$$I_s = \frac{I}{S} \tag{2.38}$$

$$I_m = \frac{I}{m} \tag{2.39}$$

where I_s is the area current density; I_m is the mass current density; I is the current through the electrode or battery; S is the area of battery or electrode; m is the electrode or battery mass.

Charge and discharge ratio is a measure of the speed of charge and discharge. C/n is the charge and discharge rate, which refers to the completion of a rated capacity charge or discharge in n hours. For example, a discharge completed in two hours is called $C/2$ (or $0.5C$) charge/discharge rate.

2.3.6 Power and Specific Power

Power refers to the energy output per unit time of a battery under certain discharge conditions. Specific power refers to the power output per unit mass or unit volume of a battery. The actual power of the battery is related to the discharge current and the operating voltage under the discharge current:

$$P = UI \tag{2.40}$$

where P is the actual power of the battery; I is the discharge current; U is the operating voltage under the discharge current. Substituting Eq. (2.28) into Eq. (2.40), it can be obtained:

$$P = IE - I^2R_i \tag{2.41}$$

2.3.7 Coulombic Efficiency

Coulombic efficiency refers to the percentage of battery discharge capacity and charge capacity under certain discharge conditions, also known as charge and discharge efficiency.

2.3.8 Cycle Life

Cycle life generally refers to the cycling number experienced by the battery before the battery capacity decays to a specific value (generally 80% or 60% of the initial capacity) under certain discharge conditions.

2.4 Interfacial Properties of Nanowires Electrode Materials

2.4.1 Interface Between Nanowire Electrode Materials and Electrolytes

Nanowire electrode materials have become a hot spot in energy storage field, due to their advantages of large SSA, short ion diffusion distance, and ability to adapt to volume expansion. The interface between electrode and electrolyte has a great impact on the performance, so the research on the interface between nanowires and electrolyte deserves people's attention.

In general, in the system of organic electrolytes, the decomposition of the electrolyte often occurs, and a passivation layer is formed on the electrode surface, also known as the solid electrolyte interphase (SEI). Take LIBs as an example: during the first charge–discharge process, SEI will form on surface of electrode materials when the Li^+ ions embedding potential (~0.1–0.4 V) is lower than the SEI generation potential (~0.8 V). This interface layer acts like a solid electrolyte, insulating electrons but allowing Li^+ ions to pass through. In the process of battery cycle, the volume expansion of electrode material will lead to the rupture of SEI on the surface, and contact between the exposed new electrode material surface and electrolyte will make SEI continue to grow, as shown in Figure 2.4a. This gradual thickening process of SEI not only consumes electrolyte but also increases the diffusion distance of Li^+ ions, so the stability of SEI is extremely important to the improvement of battery cycle performance [6].

Nanowire materials have a high SSA, which will aggravate the generation of SEI, thus having a significant impact on their electrochemical performance. Some researchers have used the method of surface modification to improve the overgrowth of SEI. Cui et al. reported a nanotube electrode structure with active silicon as the inner layer and silicon oxide as the outer layer, as shown in Figure 2.4c. In the process of Li^+ ions embedding, ions diffuse through the silicon–oxygen-restricted layer and react with the active silicon in the inner layer. The active silicon in the

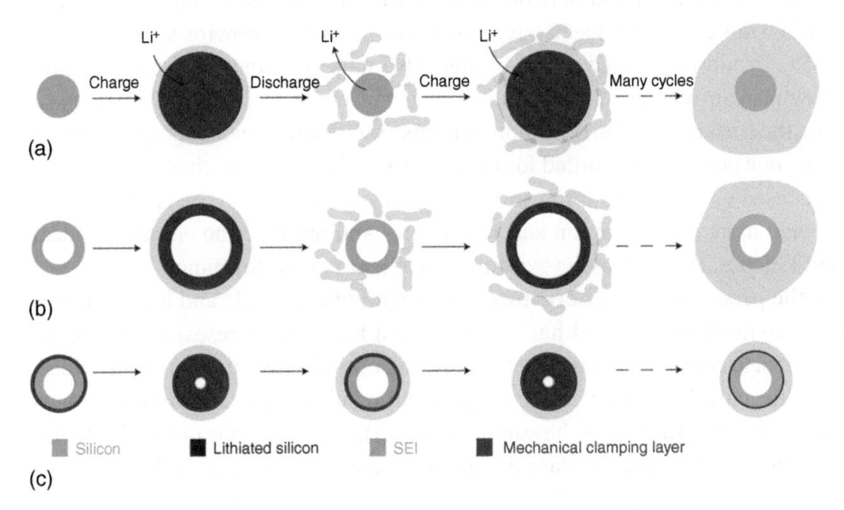

Figure 2.4 Schematic diagram of SEI growth changes in nanowire, nanotube, and mechanical protective layer. (a) Volume expansion occurs during lithiation of silicon nanowire in the charging process, and volume contraction leads to SEI breakage during discharge, and new SEI continues to grow in the new charging process; (b) similar processes occur in silicon nanotubes, and the solid–liquid interface remains unstable; (c) design mechanical confinement layer coated silicon nanotubes to stabilize SEI. Source: Wu et al. [6]/with permission of Springer Nature.

inner layer expands to the inner hollow area, and the active part of the inner layer will shrink back after the lithium ion is released. Compared with Figure 2.4b, the structure of these confined nanotubes significantly alleviates the problem of SEI overgrowth, and the capacity retention rate reaches 88% after 6000 cycles of testing. In addition, some other modification methods such as pre-coated artificial SEI and atomic layer deposition can be beneficial in minimizing the loss of irreversible capacity.

On the other hand, in view of the reasons for the formation of SEI, the researchers selectively designed the electrode materials. When the reaction potential of the electrode material is within the stable potential window of the electrolyte, the establishment of SEI will be greatly avoided, so that the advantages of the nanowire electrode material can be better utilized. Taking $Li_4Ti_5O_{12}$ (LTO) as an example, the voltage plateau of LTO (1.5 V versus Li^+/Li) is higher than the voltage of graphite (0.05 V versus Li^+/Li), which greatly avoids the formation of SEI. The LTO nanotubes obtained a reversible capacity of ~156 mAh g^{-1} at 0.1*C* [7]. Compared with LTO, the one-dimensional structure of $Li_{0.91}TiO_2$ also has better performance. Peter G. Bruce et al. reported that the $Li_{0.91}TiO_2$ nanowires have high capacity (305 mAh g^{-1}) and good cycling performance [8].

Outside the scope of the SEI discussion, the advantages of nanowire electrode materials are also reflected in the enhancement of surface pseudocapacitive response. Increasing the SSA of the material is of great help in improving the battery capacity, which in turn contributes to the improvement of the rate performance. Mai et al. used iron–manganese-based oxide porous nanowires to construct an interlaced three-dimensional network framework [9]. In the test from 1000 mA g^{-1} back to 100 mA g^{-1}, the capacity retention rate of this nanowire structure was about 95%. In addition, based on the unique advantages of nanowires, modification of nanowires can further enhance the surface capacitive reaction. In the work of Lei et al., the surface reaction of carbon nanofibers with a nitrogen doping content of 13.8% (atomic percent) accounted for as high as 90% [10]. Among them, the surface defects caused by nitrogen doping and the more exposed edges of the graphite layer can provide more potassium ion adsorption sites and improve the electrochemical reaction kinetics, thus achieving extremely excellent rate performance.

From the perspective of performance, the pursuit of long-cycle and high-rate performance to meet the demand has always been a hot spot for researchers. It is an effective idea to construct nanowire structure and optimize its solid–liquid interface. However, the current understanding of the reaction mechanism of the nanowire surface layer and the influence of the surface microstructure on SEI is not yet in-depth, and novel and convenient interface design ideas need to be further explored.

2.4.2 Heterogeneous Interfaces in Nanowire Electrode Materials

Through designing heterointerface, a multiphase composite structure with high grain boundary density can be created, which can promote the transport of electrons and ions and is also beneficial to improve the rate and long-cycle performance.

Therefore, the heterointerface design of nanowire electrode materials is of great significance for optimizing their electrochemical performance.

The core–shell composite nanowire structure provides a simple and efficient way to realize long-life, high-performance LIBs. Beng Kang Tay et al. developed a new high-performance carbon nanotube–nickel–silicon (CNT–Ni–Si) one-dimensional core–shell nanostructure. By introducing a Ni layer, the CNT core and Si shell can be effectively improved. The interface between them plays a crucial role in stress engineering [11]. Based on rational interface design, the CNT–Ni–Si core–shell nanowires exhibit good capacity retention and rate performance. Besides, changing the interface or introducing an interfacial layer is a significant method to enhance the bonding strength of nanowire electrode materials with planar substrates.

Many researches have shown that LIBs anode materials with nanowire structures have better performance and lifespan than micron-scale bulk materials. This is attributed to the unique geometry of the nanowires as well as their strong deformation-strain adaptability during cycling. Huang et al. found that the shape change of SnO_2 nanowires during charging and discharging is not only due to lithiation but is a comprehensive result of electrochemical-mechanical interaction, in which stress-induced plasticity plays an important role effect [12]. The study found that the nanowires did not break despite the large strain and conformational changes after charge–discharge cycles. This further demonstrates the better mechanical stability of nanowires relative to micron-scale electrode materials.

For the nanowire array composite structure grown on the substrate, the open space between adjacent nanowires makes it easier for the electrolyte to diffuse into the inner region of the electrode, thereby reducing the internal resistance and greatly improving the rate capability. In addition, each nanowire is in contact with the bottom substrate, which ensures that each nanowire participates in the electrochemical reaction. At the same time, the use of nanowire arrays can save the tedious process of mixing active materials with auxiliary materials such as carbon black and polymers. Wu et al. used a mild template-free method to fabricate Co_3O_4 mesoporous nanowire arrays in large areas [13]. The test results show that the mesoporous nanowire array has good performance of cycling and high-rate capability, and the nanowire array has only slight bending deformation after dozens of cycles. Due to their ease of fabrication and good performance, nanowire arrays have promising applications in secondary batteries.

In addition, the heterointerface of nanowires also plays a significant role in electrocatalysis in metal-air batteries. Herein, we synthesized porous NiO/CoN nanowire arrays with abundant oxygen vacancies and strongly coupled NiO/CoN heterointerfaces by in situ nitridation. The nanowire array exhibited significantly improved catalytic activity and stability when used as a bifunctional electrocatalyst for oxygen evolution reaction and oxygen reduction reaction [14]. The strong coupling nanoheterointerface between NiO/CoN units and the abundant oxygen vacancies at the interface are the reasons for the good bifunctional electrocatalytic performance of NiO/CoN nanowire arrays. Xi et al. reported a NiS_2/CoS_2 porous heteronanowire

[15]. The nanowires have abundant NiS_2/CoS_2 heterointerfaces and provide abundant active sites, thus exhibiting excellent catalytic activity for oxygen evolution reaction.

In conclusion, the nanowire heterointerface plays a significant role in the application of nanowire electrode materials. The nanowire heterointerface can reduce the transport time of Li^+ ions and electrons, resulting in structural stabilization and improved lithium storage kinetics. Furthermore, the rich heterointerface of the nanowires increases ion storage sites. In the nanowire array, the nanowires are well dispersed and fixed on the substrate, which avoids using polymer binders and conductive additives, greatly reduces the proportion of "inactive" components in the electrode material, and can enhance structural stability and improve charge transport dynamics of materials. Nanowire heterointerfaces are rich in oxygen vacancies and pore structures, and the nanowire interface design for promoting catalysis can be extended to other nanowire materials, which provides a promising way to fabricate multifunctional materials for efficient renewable energy applications.

2.5 Optimization Mechanism of Electrochemical Properties of Nanowires Electrode Materials

To boost the electrochemical performance of nanowire electrode materials, researchers have explored many effective optimization strategies and proposed corresponding optimization mechanisms. Among them, the mechanism of electron/ion bicontinuous transport and the self-buffering mechanism have shown significant effects on enhancing the magnification performance and cycle stability of nanowire electrode materials. This section will take these two mechanisms as the representative, introducing the electrochemical performance optimization mechanism of one-dimensional nanowire electrode materials.

2.5.1 Mechanism of Electron/Ion Bicontinuous Transport

One of the key objectives of electrochemical energy storage systems is to simultaneously achieve high energy density and high power density. However, the current electrode materials are generally slow in electrical transport kinetics in the electrochemical process, which makes it difficult to keep high energy density at high power density. The electric transport behavior of electrode materials is limited to the side with slow electron/ion transport. The optimization of electrode materials usually starts from the ion or electron side, but the improvement effect is limited and even has a negative impact on the non-optimized side. Therefore, how to optimize the electron/ion transport dynamics at the same time is of great significance for the realization of electrochemical energy storage devices with high power and energy densities.

Based on single nanowire electrochemical devices constructed by Mai et al., the study found that the tightly coated conductive layer would impede the diffusion rate of ions in the electrolyte to the nanowire electrode material, which could not realize the coordinated optimization of electron/ion transport kinetics [15, 16]. As a result, the group from the construction of electron/ion transport pathways optimized the structure of conductive layer and the composite form of nanowire electrode materials constructed various nanowire electrode materials with electron/ion bicontinuous transport property, such as V_3O_7/graphene scrolls coaxial semihollow nanowires and manganese oxide/porous carbon coaxial semi-hollow nanorods (one-dimensional "egg yolk-shell" structure) [17, 18]. The charge transport characteristics of coaxial semihollow nanowire and tightly coated nanowire (Figure 2.5) can be described by the following formula: where the ionic conductivity can be expressed by formulas (2.42) (coaxial bihollow nanowire) and (2.43) (tightly coated nanowire), respectively. The electronic conductivity of both is controlled by the surface conductive layer; therefore, the electronic conductivity is the same. It is uniformly expressed by formula (2.44).

For example, Li^+:

$$G_i = 2\pi \left(\frac{\sigma_{il}}{\ln \frac{r_{NW}+d_l}{r_{NW}}} + \frac{\sigma_{iNW}}{\ln \frac{r_{NW}}{r'_{NW}}} \right) \tag{2.42}$$

$$G'_i = 2\pi \left(\frac{\sigma_{iG}}{\ln \frac{r_{NW}+d_G}{r_{NW}}} + \frac{\sigma_{iNW}}{\ln \frac{r_{NW}}{r'_{NW}}} \right) \tag{2.43}$$

$$G_e = \frac{\pi}{L} \left[\sigma_{eG} d_G (d_G + 2r_{NW} + 2d_l) + \sigma_{eNW} {r_{NW}}^2 \right] \tag{2.44}$$

where G_i and G'_i are the total ionic conductivity of coaxial semihollow nanowire and tightly coated nanowire, respectively; G_e is the total electronic conductivity; σ_{eG} is the electronic conductivity of conductive layer, σ_{eNW} is the electronic conductivity of nanowire, σ_{iG} is the ionic conductivity of graphene; σ_{iNW} is the ionic conductivity of nanowire; and σ_{il} is the ionic conductivity of liquid electrolyte. L is the length of the nanowire; r_{NW} is the radius of the nanowire; r'_{NW} is the radius of the ion-embedded part of the nanowire; d_G is the thickness of graphene, and d_l is the thickness of the cavity filled with liquid electrolyte. Since $\sigma_{il} \gg \sigma_{iG}$, $G_i \gg G'_i$, namely coaxial semihollow nanowires, not only have high electron conductivity but also high ion conductivity, which realizes the bicontinuous ion/electron transport effect. Thanks to the bicontinuous electron/ion transport effect, these nanowire electrode materials still maintain high energy density at high power density.

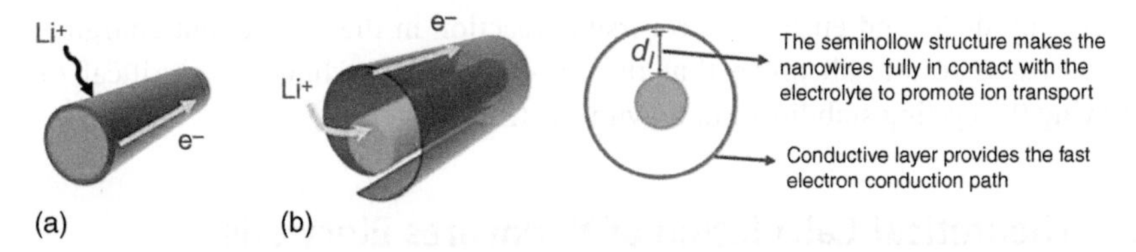

Figure 2.5 Design of coaxial semihollow structure. (a) Schematic diagram of tightly coated nanowires and their ion/electron transport; (b) schematic diagram of coaxial semi-hollow nanowires and their ion/electron transport.

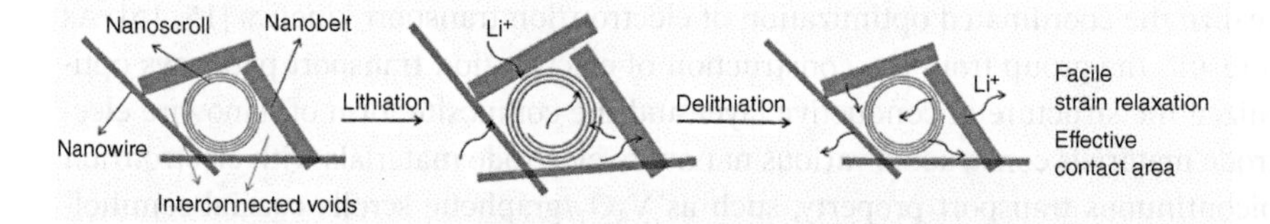

Figure 2.6 Schematic diagram of the "self-buffering" mechanism of the nanoribbon/nanoscroll composite structure. Source: Mai et al. [21]/with permission of Wiley-VCH.

In addition, starting from the material structure, the pre-embedding method is adopted to embed alkali metal and alkaline earth metal ions between the layers of the electrode material in advance, which can introduce impurity energy levels, provide traveling electrons, increase the layer spacing, lower the ions' diffusion resistance, and also realize the bicontinuous transmission of electrons/ions [19].

2.5.2 Self-Buffering Mechanism

Cycle stability, which is directly related to the structural stability of electrode materials in the charge–discharge process, is a key performance indicator of electrochemical energy storage systems. One of the main causes of structural damage and capacity decay is the structural stress brought on by the electrode material's volume change during the charge–discharge operation. In order to improve the cycling stability of electrode material, it is crucial to consider the buffer expansion/contraction stress.

Mai et al. found various self-buffering optimization mechanisms. In graded nanowires with nanorod orientation laps, where the lap space can accommodate lattice expansion and contraction [20]. Moreover, the slip at the interface can significantly relieve the stress concentration in the nanoelectrode material, which is an effective self-buffering structure. Compared with ordinary nanowire electrode materials, the cycling stability of graded nanowire electrode materials has been greatly improved. In the nanoribbon/nanoscroll composite structure, the nanoscrolls and the interconnected voids between hybrid nanostructures as the buffered section could be able to adjust the internal expansion stress distribution of the material. Contributing to the stability of the structure, which in turn improves the cyclic stability of the electrode material (Figure 2.6) [21]. After the alkaline earth metal ions are pre-embedded in the conversion reaction type nanowires, the "self-maintenance" effect can be produced to prevent the agglomeration of the discharge products and ensure the reversible reaction in the subsequent charging process. These results all prove that the self-buffering mechanism is critical to improving the cycling stability of nanowire electrode materials.

2.6 Theoretical Calculation of Nanowires Electrode Materials

The rise of nanotechnology has enabled a range of nanomaterials to be used in various fields. The urgent need for clean energy has created an urgent need for the development of efficient and inexpensive energy storage devices. One-dimensional nanowire electrode materials are an important component in the design and construction of energy storage devices due to their unique electron transport properties. First-principles calculations can theoretically provide a means to study the electronic structure of nanowires, ion diffusion paths, anomalous cross-sections, and changes in electromagnetic properties caused by doping. Also, theoretical calculations give us a viable way of predicting the properties of some materials that cannot be synthesized under current experimental conditions and explaining experimental phenomena that cannot be observed under limited equipment conditions.

VASP is a software package for ab initio molecular dynamics calculations using pseudopotentials and plane wave basis sets. VASP obtains the electronic states and energies of the system by using efficient matrix diagonalization techniques to approximate the Schrödinger equation, and a hybrid scheme of Broyden and Pulay densities is used to accelerate the convergence of the self-consistent cycle during the iterative solution. The simulations are performed for atoms, molecules, clusters, nanowires, thin films, crystals, quasicrystals, uncharacteristic materials, surface systems, and solids with periodic boundaries. The structural parameters (bond lengths, bond angles, lattice constants, atomic positions, etc.) and configurations of materials can be calculated; the equation of state and mechanical properties of materials (bulk elastic modulus and elastic constants); the electronic structure of materials (energy levels, charge density distribution, energy bands and electron density of states, etc.); the optical, magnetic, and lattice dynamics properties of materials: and the simulation of surface systems (reconstructions, surface states).

Gaussian is a very widely used quantum chemistry software for doing semi-empirical and ab initio calculations to study molecular energies and structures, transition state energies and structures, chemical bonding, reaction energies, molecular orbitals, dipole and multipole moments, atomic charges and potentials, vibrational frequencies, infrared and Raman spectra, polarizabilities and hyper-polarizabilities, thermodynamic properties, and reaction paths. Computational simulations can be performed on systems in the gas phase and the solution phase, and in addition, Gaussian is a powerful tool for studying such things as substitution effects, reaction mechanisms, potential energy surfaces, and excited state energies.

Quantum ATK is a complete atomic simulation toolkit, and a first-principles electronic structure calculation program capable of simulating the electrical and quantum transport properties of nanostructured systems and nanodevices. Quantum ATK's modeling ranges from classical force fields for large, complex material systems to ab initio tools with high-accuracy results for small systems. On this basis, it can deal with the electron transport properties of nanodevices under external bias and the associated magnetic and spin transport problems in nanodevices.

QUICKSTEP is a density general function theory program using a mixture of Gaussian and plane wave basis sets for the study of complex large systems (e.g. liquids, crystals, proteins, and interfaces), which can simulate both static properties (e.g. calculation of spectral and total energy derivative properties) and dynamic properties (e.g. dispersion based on molecular dynamics).

QMD-FLAPW is suitable for studying metals, semiconductors, insulators, densely stacked structures, and open structures. The self-consistent shielding exchange LDA and model GW options provide engineering accuracy for predicting semiconductor

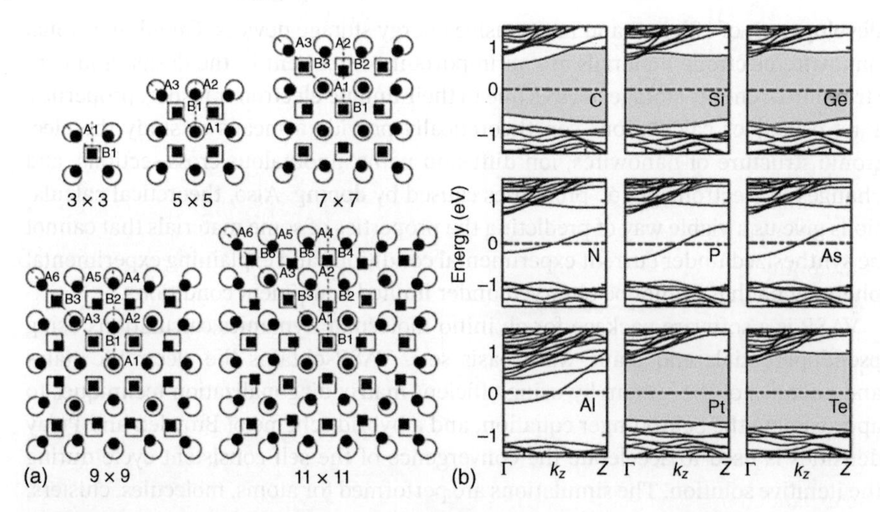

Figure 2.7 Electronic structure changes of nanowires with different interface sizes and elemental doping. (a) The positions of the atoms Ai and Bi on the representative A (circles) and B (squares) (001) layers before (open symbols) and after (closed symbols) relaxation for the 3 × 3, 5 × 5, 7 × 7, 9 × 9, and 11 × 11 Fe nanowires. Source: Wang et al. [22]/with permission of Elsevier. (b) Energy band structures of adatom (impurity) adsorbed Si nanowires. The dash-dotted line represents the Fermi level. Source: Durgun et al. [23]/with permission of American Physical Society.

and insulator bandgaps and are based entirely on ab initio algorithms for computing the optical spectra of the system.

Zhang et al. studied the relaxation structure and electronic properties of Fe nanowires with different cross-sections based on the first-nature principle (Figure 2.7a) and found that Fe nanowires with different cross-sectional sizes all maintain C_4 symmetry [22]. There is a tendency for the relaxation displacement in the interior to increase with increasing size. The large displacement of the edge atoms to the interior and the small displacement of the side atoms to the interior lead to the "rounded corners" phenomenon. Batra et al. theoretically investigated nanowires with various pentagonal structures [24]. The results show that nanowires with different types of elements, such as alkali metals, simple metals, transition metals, noble metals, and noble gas atoms, are structurally stable due to the interlocking pentagons that form a linear structure perpendicular to the plane of the pentagon and passing through its center in the structural model. The bond angles of this structure are close to those of the icosahedral structure. Among these structures, the one composed of silicon elements is more energetically favorable. These quasi-nanowire structures have higher bonding energies compared to other nanowire structures and are therefore easier to synthesize experimentally.

Silicon is an attractive anode material for LIBs due to its high theoretical specific capacity (4200 mAh g^{-1}). However, the chalking and the detachment of the electrode material from collector fluid are due to the volume expansion of the silicon anode by about 400% during the Li^+ ions inserting process, which in turn leads to capacity decay. Si nanowires are used as electrode materials due to their efficient charge transport, and their conductivity can be improved by doping the Si nanowires during their growth. The doping of hydrogen-saturated nanowires with impurity elements such as Al, Ga, C, Si, Ge, N, P, As, Te, and Pt was investigated based on first principles by Ciraci et al. [23]. For high concentrations of impurity element doping in the IIIA and VA groups, the energy bands of the doped Si nanowires were shown to be metallic. However, for IVA and VIA group element doping, the nanowires show semiconductors. Usually, low doping concentrations result in Si nanowires exhibiting n-type and p-type doping properties like bulk silicon. Therefore, the electronic structure properties of Si nanowires can be tuned by surface impurity element doping, as shown in Figure 2.7b.

Vanadium-based materials have been widely studied due to their high theoretical capacity, and low prices. $VOPO_4$ material with a layer spacing of 1.42 nm was studied as the cathode material for Mg-ion batteries by Mai et al. [25]. It was found that the expanded interlayer structure of the material provided a fast transport space for $MgCl^+$. The diffusion paths and energy barriers of ions in the interlayer were calculated by density flooding, as shown in Figure 2.8a,b. $MgCl^+$ prefers the P1 path with a diffusion potential barrier of only 0.42 eV. Meanwhile, the group designed a multidimensional nanocomposite structure ($V_2O_3 \subset$ CNTs $\subset$ rGO) of zero-dimensional V_2O_3 nanoparticles-1D carbon nanotubes-2D reduced graphene oxide [26]. Thanks

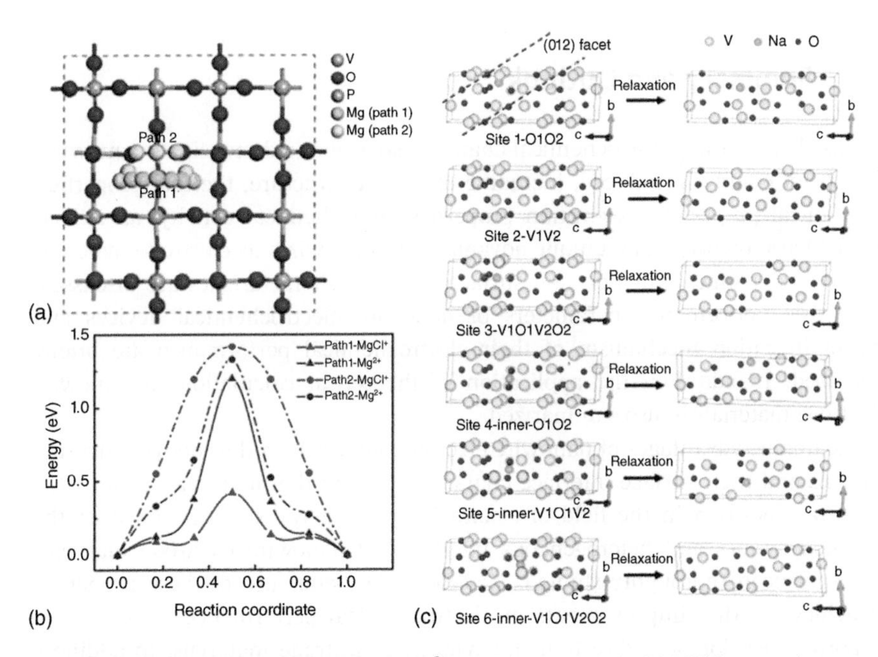

Figure 2.8 Diffusion barrier profiles of $Mg^{2+}/MgCl^+$ transport in PA-$VOPO_4$ nanosheets. (a) $Mg^{2+}/MgCl^+$ diffusion routine. (b) Diffusion energy barrier profiles of $Mg^{2+}/MgCl^+$ transport in PA-$VOPO_4$ nanosheets. Source: Zhou et al. [25]/with permission of Wiley-VCH. (c) Density flooding calculation of a model for sodium ions embedded in different sites of V_2O_3. Source: Tan et al. [26]/with permission of Wiley-VCH.

to the synergistic effect between its multiple dimensional nanostructures, this multidimensional nanostructure exhibits significantly improved cyclic stability and fast ion/electron transport kinetics. Calculation of the sodium ion embedding sites in V_2O_3 by density flooding theory reveals that sodium ions embedded in (012) crystalline planes have small binding energy or volume expansion. However, the sodium ions exhibit large volume expansion and severe structural disruption when embedded in the (012) interlayer, as shown in Figure 2.8c. The results indicate that sodium ions select their embedding sites according to the exposed crystallographic surfaces of V_2O_3 nanoparticles.

Nanowire electrode materials play an extremely important role in energy storage. Theoretical calculations give us a very efficient tool to give preliminary validation to some inferences that cannot be verified in experiments. These include the calculation of the structural relaxation of nanowire materials with different cross-sectional dimensions, the structural stability of nanowires with different cross-sectional shapes, the changes in the electronic structure due to different types and concentrations of impurity elements doping on the nanowire surface, the selection of different embedding sites for ions in the electrode material, and the calculation of different diffusion paths and diffusion potentials. Modeling simulations to calculate the electronic structure and other properties provide a solid foundation for the application of nanowire electrode materials.

2.7 Summary and Outlook

In this chapter, the physicochemical characteristics of one-dimensional nanowires have been mainly introduced, including electronic structure, thermal properties, mechanical properties, adsorption properties, and chemical activity, as well as related basic theories. The unique advantages of nanowires as electrode materials in thermodynamics, kinetics, and surface interface characteristics are discussed. The basic performance parameters of nanowire electrochemical devices and the optimization mechanism of their electrochemical performance are briefly described. In addition, the application of theoretical calculation in nanowire electrode materials is also summarized.

The unique advantages of nanowire electrode materials in thermodynamics and kinetics and their surface interface characteristics are the main reasons for their important position in the field of electrochemical energy storage. However, the thermodynamics and heterogeneous interfaces of nanowire electrode materials have not been thoroughly studied. How to make better use of these properties to achieve further improvements in electrochemical performance is one of the important development directions for nanowire electrode materials. In addition, theoretical calculation are important for understanding the physicochemical properties and electrochemical behavior of nanowire electrode materials, but in order to make the calculation results more valuable, it is necessary to establish a model closer to the real state of electrochemical energy storage devices.

References

1 Wu, Y. and Yang, P. (2001). Melting and welding semiconductor nanowires in nanotubes. *Advanced Materials* 13 (7): 520–523.

2 Shin, H.S., Yu, J., and Song, J.Y. (2007). Size-dependent thermal instability and melting behavior of Sn nanowires. *Applied Physics Letters* 91 (17): 173106.

3 Li, D., Wu, Y., Kim, P. et al. (2003). Thermal conductivity of individual silicon nanowires. *Applied Physics Letters* 83 (14): 2934–2936.

4 Chen, C., Shi, Y., Zhang, Y.S. et al. (2006). Size dependence of Young's modulus in ZnO nanowires. *Physical Review Letters* 96 (7): 075505.

5 Hu, Y.S., Kienle, L., Guo, Y.G. et al. (2006). High lithium electroactivity of nanometer-sized rutile TiO_2. *Advanced Materials* 18 (11): 1421–1426.

6 Wu, H., Chan, G., Choi, J.W. et al. (2012). Stable cycling of double-walled silicon nanotube battery anodes through solid–electrolyte interphase control. *Nature Nanotechnology* 7 (5): 310–315.

7 Lee, S.C., Lee, S.M., Lee, J.W. et al. (2009). Spinel $Li_4Ti_5O_{12}$ nanotubes for energy storage materials. *Journal of Physical Chemistry C* 113 (42): 18420–18423.

8 Armstrong, A.R., Armstrong, G., Canales, J. et al. (2005). Lithium-ion intercalation into TiO_2-B nanowires. *Advanced Materials* 17 (7): 862–865.

9 Wang, X., Xu, X., Niu, C. et al. (2017). Earth abundant Fe/Mn-based layered oxide interconnected nanowires for advanced K-ion full batteries. *Nano Letters* 17 (1): 544–550.

10 Xu, Y., Zhang, C., Zhou, M. et al. (2018). Highly nitrogen doped carbon nanofibers with superior rate capability and cyclability for potassium ion batteries. *Nature Communications* 9 (1): 1720.

11 Lu, C., Fan, Y., Li, H. et al. (2013). Core–shell CNT–Ni–Si nanowires as a high performance anode material for lithium ion batteries. *Carbon* 63: 54–60.

12 Huang, J.Y., Zhong, L., Wang, C.M. et al. (2010). In situ observation of the electrochemical lithiation of a single SnO_2 nanowire electrode. *Science* 330 (6010): 1515–1520.

13 Li, Y., Tan, B., and Wu, Y. (2008). Mesoporous Co_3O_4 nanowire arrays for lithium ion batteries with high capacity and rate capability. *Nano Letters* 8 (1): 265–270.

14 Yin, J., Li, Y., Lv, F. et al. (2017). NiO/CoN porous nanowires as efficient bifunctional catalysts for Zn–air batteries. *ACS Nano* 11 (2): 2275–2283.

15 Yin, J., Li, Y., Lv, F. et al. (2017). Oxygen vacancies dominated NiS_2/CoS_2 interface porous nanowires for portable Zn–air batteries driven water splitting devices. *Advanced Materials* 29 (47): 1704681.

16 Hu, P., Yan, M., Wang, X. et al. (2016). Single-nanowire electrochemical probe detection for internally optimized mechanism of porous graphene in electrochemical devices. *Nano Letters* 16 (3): 1523–1529.

17 Yan, M., Wang, F., Han, C. et al. (2013). Nanowire templated semihollow bicontinuous graphene scrolls: designed construction, mechanism, and enhanced energy storage performance. *Journal of the American Chemical Society* 135 (48): 18176–18182.

18 Cai, Z., Xu, L., Yan, M. et al. (2015). Manganese oxide/carbon yolk–shell nanorod anodes for high capacity lithium batteries. *Nano Letters* 15 (1): 738–744.

19 Mai, L.Q., Hu, B., Chen, W. et al. (2007). Lithiated MoO_3 nanobelts with greatly improved performance for lithium batteries. *Advanced Materials* 19 (21): 3712–3716.

20 Mai, L.-Q., Yang, F., Zhao, Y.-L. et al. (2011). Hierarchical $MnMoO_4/CoMoO_4$ heterostructured nanowires with enhanced supercapacitor performance. *Nature Communications* 2 (1): 381.

21 Mai, L., Wei, Q., An, Q. et al. (2013). Nanoscroll buffered hybrid nanostructural VO_2 (B) cathodes for high-rate and long-life lithium storage. *Advanced Materials* 25 (21): 2969–2973.

22 Wang, S.-F., Zhang, Y., Zhang, J.-M. et al. (2010). First-principles study on the relaxed structures and electronic properties of Fe nanowires. *Physica B* 405 (12): 2726–2732.

23 Durgun, E., Akman, N., Ataca, C. et al. (2007). Atomic and electronic structures of doped silicon nanowires: a first-principles study. *Physical Review B* 76 (24): 245323.

24 Sen, P., Gülseren, O., Yildirim, T. et al. (2002). Pentagonal nanowires: a first-principles study of the atomic and electronic structure. *Physical Review B* 65 (23): 235433.

25 Zhou, L., Liu, Q., Zhang, Z. et al. (2018). Interlayer-pacing-regulated $VOPO_4$ nanosheets with fast kinetics for high-capacity and durable rechargeable magnesium batteries. *Advanced Materials* 30 (32): 1801984.

26 Tan, S., Jiang, Y., Wei, Q. et al. (2018). Multidimensional synergistic nanoarchitecture exhibiting highly stable and ultrafast sodium-ion storage. *Advanced Materials* 30 (18): 1707122.

3

Design and Synthesis of Nanowires

The morphology of nanomaterials depends on the preparation process, and the preparation methods of nanomaterials have always been an important part of the field of nanoscience. Generally speaking, the synthetic methods of nanomaterials include physical methods and chemical methods. Compared with other morphologies of nanomaterials, it is unique to synthesize nanowires (NWs) in that more attention is paid to the design and regulation in a single direction at the atomic and molecular levels in order to form the dominant growth in one direction (while inhibiting the growth in other directions). The directional regulation of NWs can be divided into two types: one is to control the growth of the dominant crystal face according to the anisotropy of crystal materials; the other is to force the directional growth by artificially inducing. The external performances are displayed differently because the physical and chemical properties of the NWs prepared by the two types of regulation methods have a degree of difference. In this chapter, the common synthesis methods and growth mechanisms of six different types of NWs (conventional, porous, hierarchical, heterogeneous, hollow, and array) are briefly introduced based on the classification of NWs with different morphologies.

3.1 Conventional Nanowires

Conventional NWs are a kind of common one-dimensional (1D) NWs that are nanoscale in two dimensions in space. With this characteristic, the materials exhibit nanoeffects that are different from macroscopic ones. The preparation process of conventional NWs generally includes wet chemical, dry chemical, and physical methods.

3.1.1 Wet Chemical Methods

Wet chemical methods are those that contain liquid-phase chemical processes while synthesizing NWs. Some of the commonly used methods will be introduced in the following.

Nanowire Energy Storage Devices: Synthesis, Characterization, and Applications, First Edition.
Edited by Liqiang Mai.

3.1.1.1 Hydrothermal/Solvothermal Method

Due to its simple equipment, convenient operation, and good crystallization, the hydrothermal/solvothermal method is a widely used liquid-phase chemical method. Through the process of dissolution and recrystallization, the target product can be obtained in a closed container with water or organic solvent as the reaction medium under high temperature and pressure.

Many variables, such as the pH value, reaction temperature/pressure/time, solvent species, and so on, are the important factors in the morphologies of products. Therefore, many NWs are prepared by controlling the above variables during the synthesis. Using $TiCl_4$ as the titanium source and ammonia solution as the solvent, Liu et al. from Sun Yat-Sen University obtained single-crystal anatase TiO_2 NWs through a one-step hydrothermal method for the first time (Figure 3.1a–c) [1]. In addition, Liu and coworkers from the Iberian International Nanotechnology Laboratory first used selenium urea and foamed cobalt as raw materials to prepare Co_3Se_4 NWs with a diameter of 10–20 nm by a simple hydrothermal method (Figure 3.1d–f) [2]. Besides, Mai and coworkers from Wuhan University of

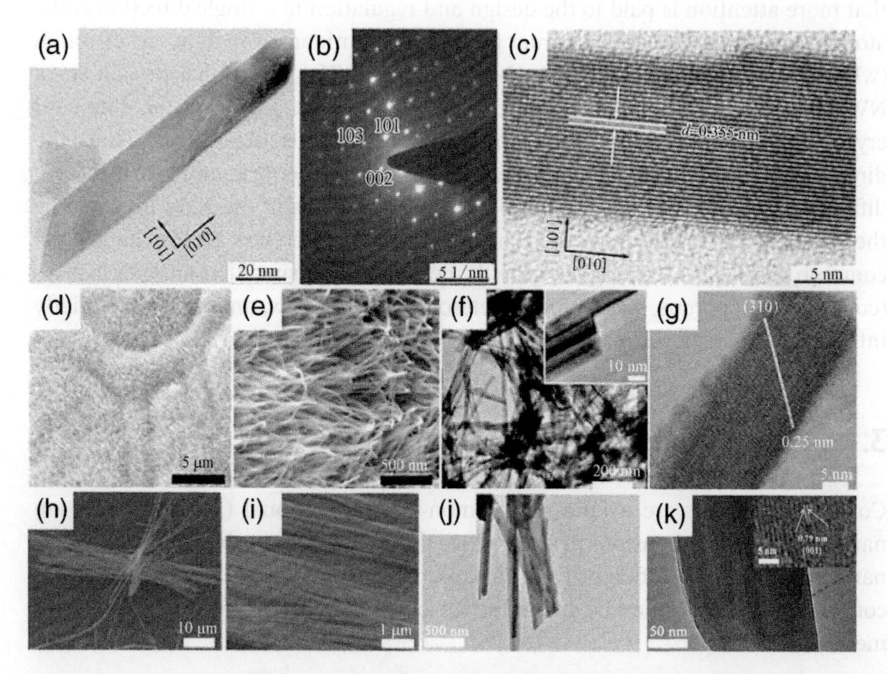

Figure 3.1 Morphology characterization of conventional NWs prepared by hydrothermal/solvothermal method. (a) Transmission electron microscope (TEM), (b) Selected area electron diffraction (SAED), and (c) High-resolution transmission electron microscope (HRTEM) images of TiO_2. (d, e) Scanning electron microscope (SEM), (f) TEM, and (g) HRTEM images of Co_3Se_4, respectively. (h, i) SEM images of $Na_2V_6O_{16}\cdot1.63H_2O$. (j, k) TEM images of $Na_2V_6O_{16}\cdot1.63H_2O$. Source: Refs. [1–3] / Royal Society of Chemistry / with permission of JOHN WILEY & SONS, INC / American Chemical Society.

Technology directly prepared $Na_2V_6O_{16}·1.63H_2O$ NWs as shown in Figure 3.1h–k by hydrothermal method [3].

The homogeneous conventional NWs can be prepared simply and quickly by the hydrothermal/solvothermal method. Although there are problems such as the choice of suitable solvent and whether the solvent is completely removed during the product collection process after the reaction, the hydrothermal/solvothermal method is still a common and reliable method for preparing conventional NWs.

3.1.1.2 Sol–Gel Method

The sol–gel method is to homogeneously mix compounds, including chemically active components, under liquid-phase conditions, followed by hydrolysis and polymerization reactions. As a result, a stable, transparent sol system will be formed in solution. In the later aging process, the sol colloidal particles are gradually polymerized to a gel, which is dried and calcined to obtain the target product. The application technology of this method is gradually improving and has been widely used.

Wang et al. from Xiangtan University prepared a sol–gel solution of citric acid and $Mn(CH_3COO)_2$ to get MnO_2 NWs with uniform length and diameter using porous anodic alumina (AAO) as a template (Figure 3.2a,b). The diameter and length of the NWs are determined by the size of the hole and the thickness of the template [4]. In addition, Zheng and coworkers from Nanjing University, using $CdCl_2$ and Na_2S

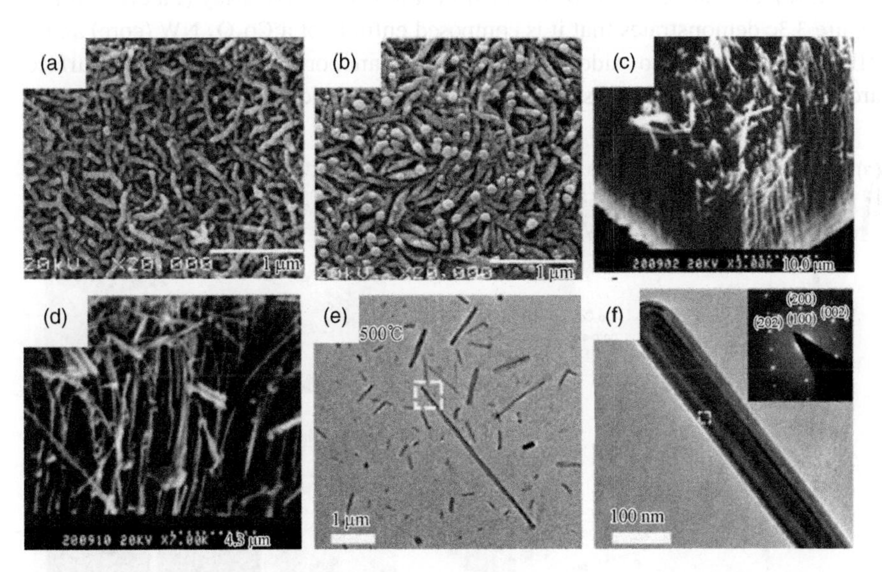

Figure 3.2 Morphology characterization of conventional NWs prepared by Sol–Gel method. (a, b) SEM images of MnO_2 NWs. (c, d) SEM images of CdS NWs. (e, f) TEM images of VO_2 NWs. Source: Refs. [4–6] / with permission of Elsevier / with permission of JOHN WILEY & SONS, INC / Royal Society of Chemistry.

as cadmium and sulfur sources, successfully made CdS NWs with a length of 60 μm by adjusting the reaction time and calcination system based on the sol–gel method (Figure 3.2c,d) [5]. In addition, Kim's group obtained precursor samples by sol–gel method and treated them with different calcination systems to adjust the tensile stress needed in the process of nucleation growth. The obtained VO_2 NWs is shown in Figure 3.2e,f [6].

The advantage of this method is easy to control the nanostructure of the prepared materials. Besides, it has the ability to obtain desirable materials with specific functions and structures, which makes it popular in energy storage and catalysis.

3.1.1.3 Coprecipitation Method

The coprecipitation method is a common method for high-purity NWs. The suitable precipitating agent is added to a mixed metal salt solution containing two or more metal ions to form a uniform precipitate through a precipitation reaction, and the precipitate is thermally decomposed to obtain a high-purity nanopowder material. Among them, the influencing factors of the coprecipitation reaction are pH value, temperature, the solubility of the reactants, and so on. Although the coprecipitation process is relatively complicated, it is also often used to prepare incompatible inorganic salt NWs due to its obvious advantages: low cost, simple process, short synthesis cycle, and easy control of preparation conditions.

Yu et al. used the hydrothermal coprecipitation method to homogeneously coat the ZIF-67 derivative Ni/Co oxide shell on the Co_3O_4 NW array (Figure 3.3) [7]. Figure 3.3c demonstrates that it is composed entirely of a Co_3O_4 NW (core) and a ZIF-67 derivative Ni/Co oxide (shell). The inner nanopores can provide more surface area while shortening the distance of ion diffusion, which improves performance.

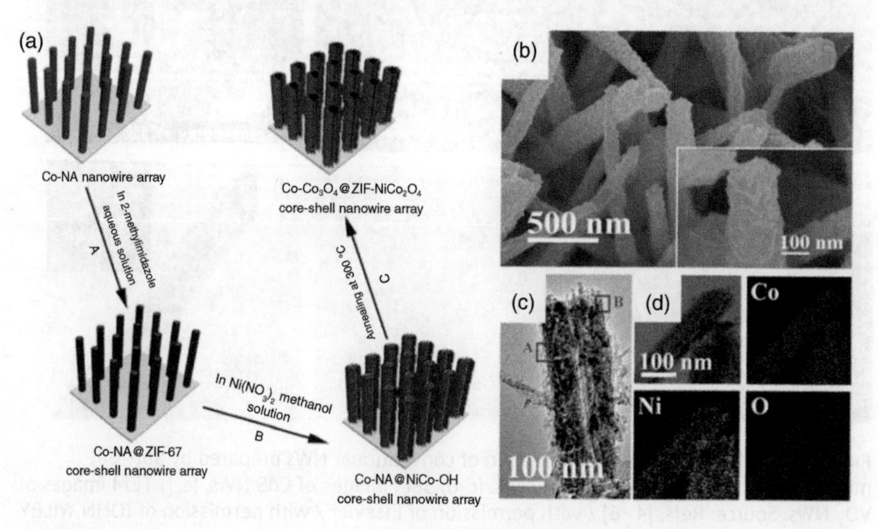

Figure 3.3 (a) Synthetic diagram of Co-Co_3O_4@ZIF-$NiCo_2O_4$ core-shell NW array electrode based on nickel foam. (b) SEM, (c) TEM, and (d) EDX mapping images of Co-Co_3O_4@ZIF-$NiCo_2O_4$ (inset: corresponding high-magnification SEM image). Source: Yu et al. [7] / Royal Society of Chemistry.

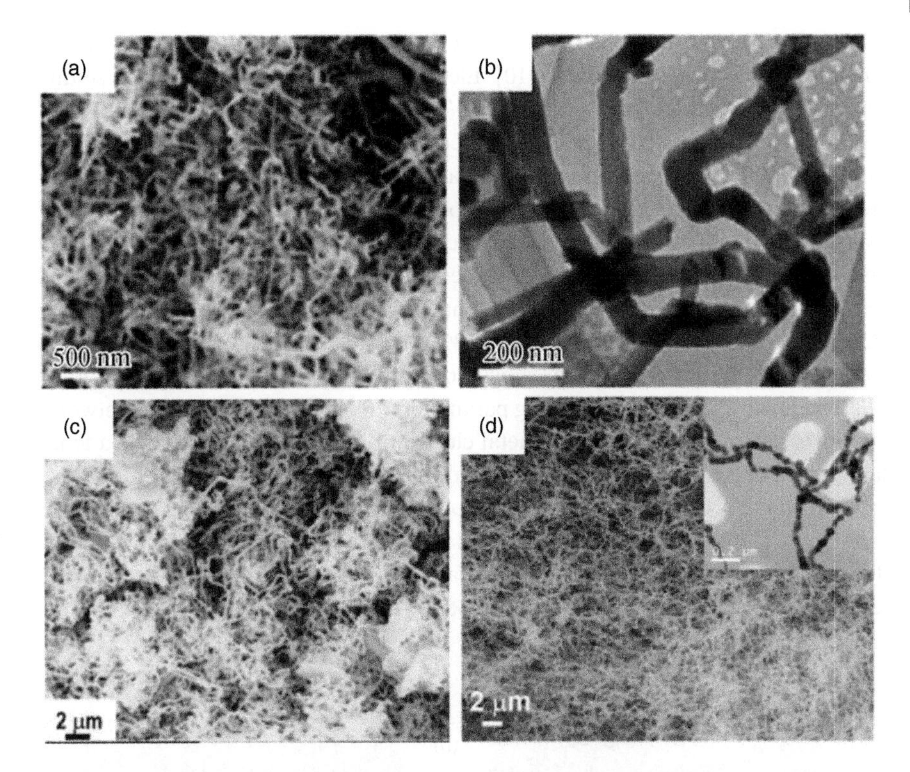

Figure 3.4 Morphology characterization of conventional NWs prepared by ultrasonic spray pyrolysis. (a) SEM and (b) TEM images of Zn NWs, respectively. (c) SEM image of Cd NWs; (d) SEM image of Co NWs (inset image: the corresponding TEM image). Source: [8] / with permission of JOHN WILEY & SONS, INC.

3.1.1.4 Ultrasonic Spray Pyrolysis Method

Ultrasonic spray pyrolysis is a process in which the precursor solution is atomized through the atomizer to form aerosols (solid or liquid droplets suspended in the gas), and the aerosols are carried by the airflow for thermal reaction. The characteristic of this method is that the submicron powders composed of nanometer primary particles can be obtained directly, and the particle distribution is relatively uniform.

Rao et al. used aqueous solutions of zinc acetate, chromium acetate, and cobalt acetate as precursors to prepare Zn, Cd, and Co metal single-crystal NWs by ultrasonic spray pyrolysis. By morphological characterization of the obtained NWs, it can be seen that the diameter of Zn single-crystal NWs is distributed in the range of 50–100 nm and several hundred microns in length (Figure 3.4a,b). The nanostructures of Cd and Co can be seen in Figure 3.4c,d, respectively [8].

3.1.1.5 Electrospinning Method

Electrospinningrefers to making a charged polymer melt or solution under the action of a high-voltage electrostatic field and the mouth sprays out a stable stream, which is accelerated, stretched, split, solidified, or solvent volatilized, and finally formed into a fibrous substance [9, 10]. Electrospinning can effectively realize the preparation of NWs and composite NWs. The diameter of the obtained fibers is generally at micron or nanolevel. The microscopic morphology of the product is not restricted by the crystal growth orientation of the material itself. To a certain extent, this method greatly enriches the types of NWs. At the same time, it has the advantages of being a simple device, low cost, diversified raw material selection, controllable process, etc. Therefore, the electrospinning method is also one of the most important ways to prepare conventional NWs.

Cui et al. from Stanford University used 8% polyvinyl butyral (PVB) and 9% anhydrous tin chloride in *n*-butanol as precursors and obtained PVB/$SnCl_2$ ultra-long NWs by adjusting the electrospinning parameters (Figure 3.5a,b). The NW network can provide a continuous path for metal electron transmission and provide a template for the deposition of metal [11]. In addition, Mai et al. from Wuhan University of Technology dissolved a certain amount of ammonium metavanadate, lithium acetate, PEO, and PVA in water and stirred it in a water bath at 80 °C for six hours. After that, LiV_3O_8 NWs with smooth surface and uniform size were obtained by electrospinning and calcination (Figure 3.5c,d) [12].

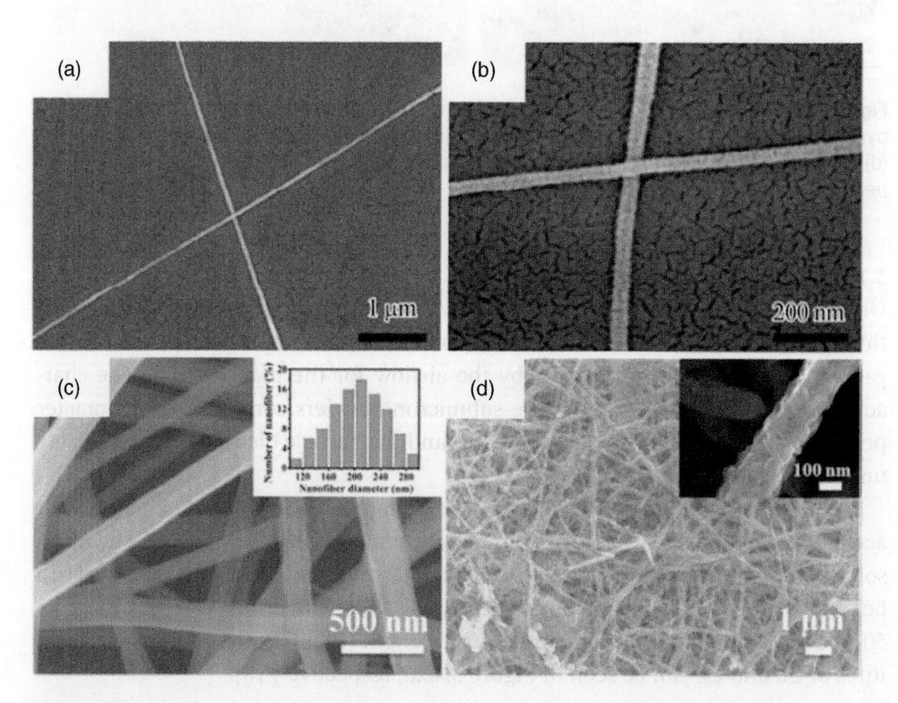

Figure 3.5 Morphology characterization of conventional nanowires prepared by electrospinning. Source: Refs. [11, 12] / American Chemical Society / Royal Society of Chemistry. (a) SEM and (b) TEM images of PVB/$SnCl_2$ nanowires. (c) SEM image of LiV_3O_8-PEO/PVA precursor (the inset image: the fiber diameter distribution diagram); (d) SEM image of LiV_3O_8 after calcination.

3.1.2 Dry Chemical Method

Dry chemical methods are those that do not use a liquid-phase chemical process while synthesizing NWs. Some dry chemical methods will be introduced in the following.

3.1.2.1 High-Temperature Solid-State Method

High-temperature solid-state method is to prepare target products through solid-state reaction at high temperature, which has the advantages of low cost, large yield, simple preparation technology, no agglomeration, good filling property, and so on. However, due to the unique driving force required for the synthesis of NWs, it is difficult to obtain NWs with uniform morphology by the general solid-state method. And usually, reasonable regulation of calcination system and feeding ratio is required.

Li et al. obtained $LiCoO_2$ NWs by keeping the Co_3O_4 NW precursor at 500 °C for three hours, thoroughly mixing it with $Li(OH){\cdot}H_2O$, and heating it to 750 °C for two hours. The SEM and TEM images show that the NWs were composed of many nanoparticles with diameters ranging from 150 to 200 nm (Figure 3.6a,b) [13]. Jia et al. from Zhengzhou University used silicon powder and phenolic resin as raw

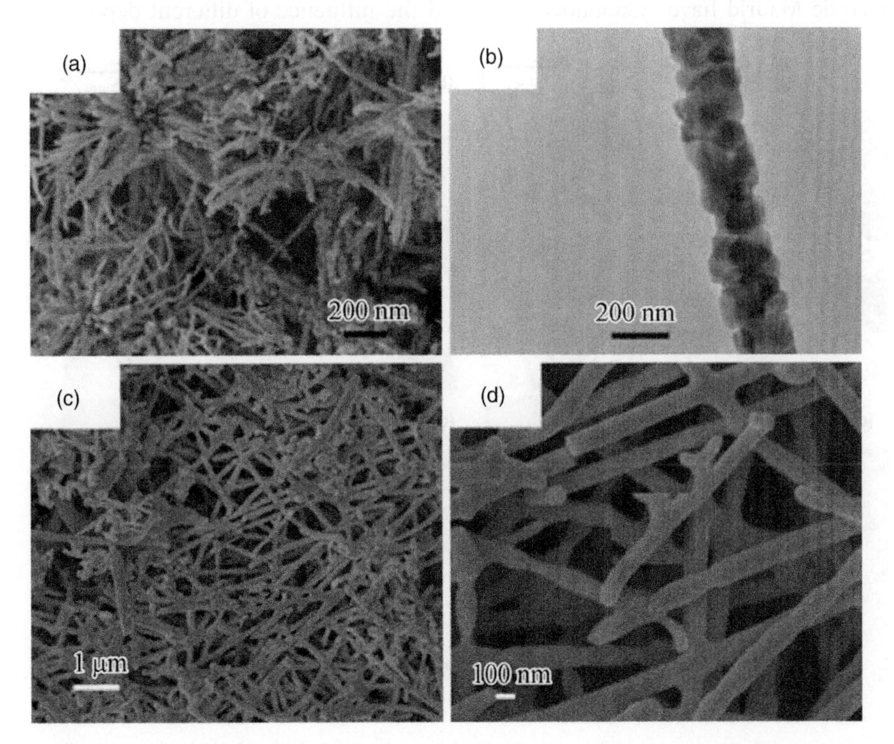

Figure 3.6 Morphology characterization of conventional $LiCoO_2$ NWs prepared by high-temperature solid-state method. (a) SEM and (b) TEM images of $LiCoO_2$ NWs, respectively. (c, d) SEM images of SiC NWs. Source: Refs. [13, 14] / Springer Nature / with permission of Elsevier.

materials and added NaCl and NaF to the molten salt. When the temperature of the molten salt rises to 1450 °C, SiC NWs with a length of about 10 μm and a diameter of 100–150 nm can be obtained (Figure 3.6c,d) [14].

3.1.2.2 Chemical Vapor Deposition Method

Chemical vapor deposition (CVD) is a method of chemically reacting gaseous substances and depositing them on a substrate in an atomic state to obtain a target product. When using CVD to prepare conventional NWs, firstly, chemical means must be used to form a certain active site on the surface of the substrate to induce preferential nucleation of the NWs, followed by regulating the growth conditions to induce the directional growth of the NWs. Among all the methods for preparing NWs, substrate CVD is a common method, which has the advantages of low cost and large scale.

Li and coworkers from Northwestern Polytechnical University obtained SiC NWs as shown in Figure 3.7a,b by using CH_3SiCl_3 and H_2 as precursors without adding a metal catalyst [15]. Suzuki et al. from the National Institute of Materials Science in Japan used gold as catalyst and adjusted the deposition temperature and deposition rate to obtain Si NWs with a diameter of 10–100 nm and a length of tens to hundreds of microns on silicon wafers (Figure 3.7c) [16]. In addition, Iriarte from Politecnico de Madrid have systematically studied the influence of different deposition

Figure 3.7 Preparation of conventional NWs by chemical vapor deposition. (a, b) SEM images of SiC NWs. (c) SEM images of Si NWs grown between gold quantum dots; (d) SEM image of NiS_2 NW. Source: Refs. [15–17] / with permission of Elsevier / The Japan Institute of Metals and Materials.

conditions on the growth of $NiSi_2$ NWs, which were obtained by controlling the reaction of SiH_4 with nucleation site of Ni (Figure 3.7d) [17].

3.1.3 Physical Method

Physical vapor deposition (PVD) is a kind of physical method in which the material source is vaporized through physical processes such as evaporation or sputtering under vacuum conditions, and the target material is deposited on the surface of the substrate. The PVD method has a higher deposition rate, is cleaner, and is more repeatable than the solution-based method. With the continuous development of technology, the PVD method also has certain applications in the synthesis of conventional NWs.

Shen et al. used a vacuum DC arc as the evaporation heat source to directly react metal aluminum and N_2 and deposit aluminum nitride (AlN) NWs with a diameter of about 45 nm and a length of about 5 μm on the molybdenum cathode (Figure 3.8a,b) [18]. For the first time of preparing GaN NWs, Xue et al. from

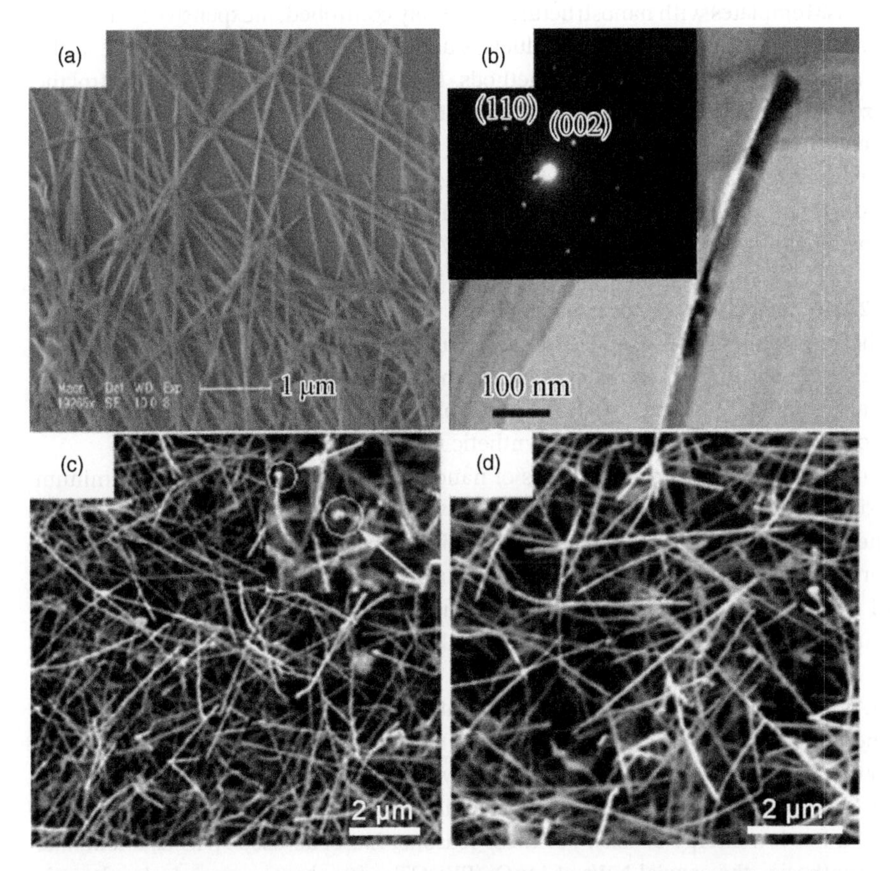

Figure 3.8 Morphology characterization of conventional NWs prepared by physical vapor deposition method. (a) SEM image of AlN NWs. (b) TEM image of AlN NWs (inset: SAED image). (c, d) SEM images of GaN NWs. Source: Refs. [18, 19] / EDP Sciences.

Shenyang University of Science and Technology used magnetron sputtering to deposit Ta and Ga_2O_3 films on Si substrates following with ammonifying process (Figure 3.8c,d) [19].

3.2 Porous Nanowires

Porous NWs, as one of the NWs with many pore structures, have high specific surface areas. Porous NWs can make the electrolyte and the active material contact more adequately when applied to electrochemical energy storage. Common methods for synthesizing porous NWs include the template method and the self-assembly method.

3.2.1 Template Method

The template method is a highly effective method for preparing 1D porous nanomaterials. Templates with nanostructures are easily controlled, inexpensive and readily available. The material for the products is deposited into the pores or surface of the template by physical or chemical methods. The method could effectively control the size, morphology, and structure of the product materials through spatial confinement or orientation induction effects. The material and structure of the template can be designed by the corresponding synthesis method to meet actual needs. At the same time, it can realize the integration of nanomaterial assembly and synthesis, which helps to solve the dispersion stability problem of nanomaterials [20–23].

3.2.1.1 Template by Nanoconfinement

The nanoconfinement template method is used to prepare 1D nanostructured materials by depositing target materials in the nanopores of the template and based on the spatial confinement effect of the template while realizing control of the size, morphology, and structure of the synthetic material [24].

One of the most important kinds of nano-confinement is the anodic aluminum oxide (AAO) template. As early as 1932, it was recognized by the porous layer and the barrier layer. Wherein the porous layer is composed of nanopores arranged uniformly, and the dense oxide layer is a layer of an insulating barrier. With the rise of research on self-assembled nanostructure systems, it should also be noted that nanomaterials with highly ordered nanoscale array pores have attracted researchers' attention.

As an electrochemical energy storage material, MnO_2 has the advantages of high energy density, lower cost, and environmental friendliness, but its conductivity is low. It is noteworthy that Polyethylenedioxythiophene (PEDOT) has the advantages of high electrical conductivity, good stability, and excellent mechanical flexibility, but it has low energy density. For these reasons, Liu et al. used AAO as a template to synthesize the coaxial NWs of MnO_2/PEDOT using the one-step electrophoretic co-deposition method [24]. Figure 3.9 displays the synthesis's schematic diagram. The SEM and TEM images of MnO_2/PEDOT coaxial NWs are shown in Figure 3.10.

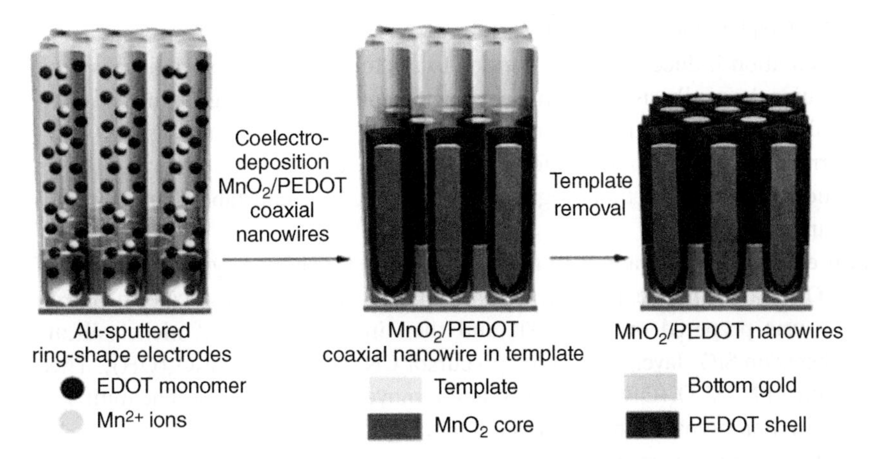

Figure 3.9 The coaxial nanowires of MnO_2/PEDOT by one-step synthesis. Source: Ran Liu [24] / American Chemical Society.

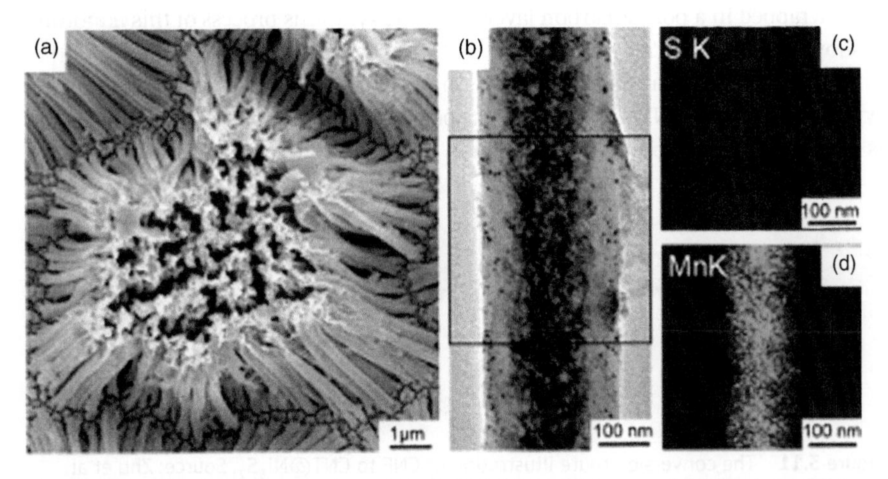

Figure 3.10 The coaxial nanowires of MnO_2/PEDOT. (a) SEM image (at 0.75 V). (b) TEM image of a single nanowire (at 0.75 V). (c, d) EDS mappings of S and Mn elements. Source: Ran Liu [24] / American Chemical Society.

It should also be noted that the PEDOT shell, with its high conductivity, porosity, and flexibility, is conducive to electron transport and ion diffusion, whose benefits can reduce the structural stress caused by charging and discharging processes, which makes it less collapse, shrinkage, and fracture in the structure.

In addition, it can be well controlled by adjusting the diameter of the aperture and the film of the sphere. Li et al. from University of New Orleans, who used porous membranes and small spheres as templates, first stacked two sections of porous membranes with different pore sizes (such as AOO) in parallel. Then the polymer (such as polystyrene) sphere is filtered onto the upper membrane. After removing the underlying porous film, the above polymer is used as the electrodeposition electrode. In the electrodeposition process, porous NWs are obtained [25].

3.2.1.2 Template by Orientation Induction

The orientation-induced template method uses a template with a specific morphology to guide the synthesis of nanomaterials with similar morphology to the template.

It is also worth noting that carbon nanotubes (CNT) and carbon nanofibers (CNF) are currently the most popular effective orientation templates for preparing porous 1D nanomaterials due to their original 1D nanostructure, adjustable diameter, large yield, and easy removal.

Zhu et al. from Nanyang Technological University show a typical example of using CNT as template [26]. By hydrolyzing tetraethyl orthosilicate, a silicon coating was prepared on the CNT, and then in situ growth of nickel silicate nanosheets on SiO_2 layer formed the precursor CNT@SiO_2@$Ni_3Si_2O_5(OH)_4$. Then they convert $Ni_3Si_2O_5(OH)_4$ into Ni_3S_2 and remove the SiO_2 coating. The multistep transformation preparation route is shown in Figure 3.11.

Besides, inorganic orientation templates can also be used in composites of porous NWs. Zhou et al. from Nanyang Technological University put forward that single-crystal NWs of MnO_x are used as templates to design and synthesize an SnO_2 layer wrapped in a porous carbon layer [27]. The synthesis process of this nanotube is shown in Figure 3.12.

Firstly, SnO_2 nanocrystals are grown on MnO_x NWs to form the core and shell types of MnO_x/SnO_2 NWs with a higher aspect ratio. Subsequently, a dopamine layer is coated on the surface of the MnO_x/SnO_2 nanomaterials, and after carbonization,

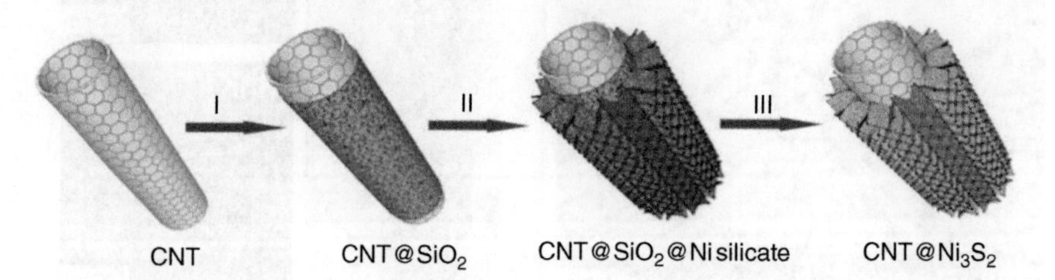

Figure 3.11 The conversion route illustration of CNF to CNT@Ni_3S_2. Source: Zhu et al. [26] / John Wiley & Sons.

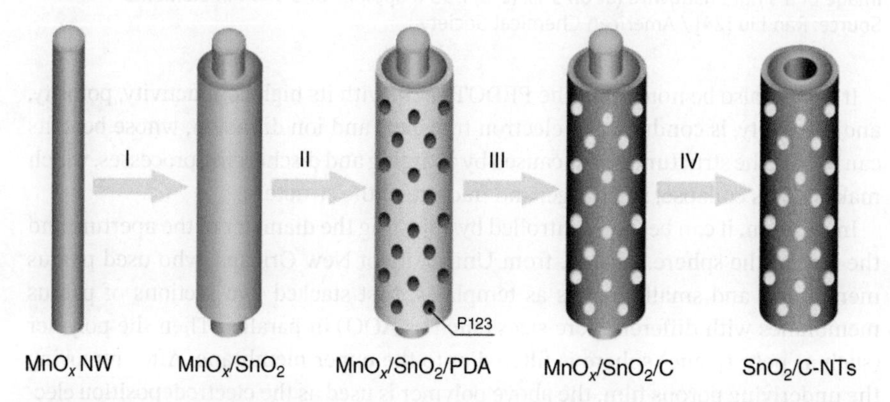

Figure 3.12 The preparation route illustration for SnO_2/CNTs. Source: Zhou et al. [27] / The Royal Society of Chemistry.

the dense carbon coating is formed, which enhances the electronic conductivity and stabilizes the morphology of the NWs. Finally, oxalic acid is used to selectively etch away the MnO_x NWs to obtain porous carbon-coated SnO_2 nanotubes.

3.2.2 Self-Assembly Method

The self-assembly method is a kind of multifunctional method, which uses the affinity group, for example, the electrostatic force, coordination bond, hydrogen bond, and other functions to assemble the basic structural units into an ordered structure. The synthesis method is easy to implement and has a wide range of universality; it has been widely used in the preparation of porous NWs.

$Co(OH)_2$ is a low-cost material substitute for the latest RuO_2, which is attributable to its excellent electronic conductivity, wide availability, and environmental friendliness. Yuan et al. from the Anhui University of Technology suggested that the $Co(OH)_2$ nanoparticles are self-assembled to synthesize porous NWs with a novel lysine-assisted hydrothermal process [28].

Researchers constructed a novel hybrid metal oxide core and shell NWs array, in which both the core and shell materials are good pseudocapacitance metal oxides just like Co_3O_4 and MnO_2 [29]. The porous NWs with hierarchical heterostructure composed of Co_3O_4 NWs and MnO_2 ultrathin nanosheets were synthesized by the self-assembly method. The general electrode design process is shown in Figure 3.13a. The SEM images of the Co_3O_4/MnO_2 nanowire array are shown in Figure 3.13b,c.

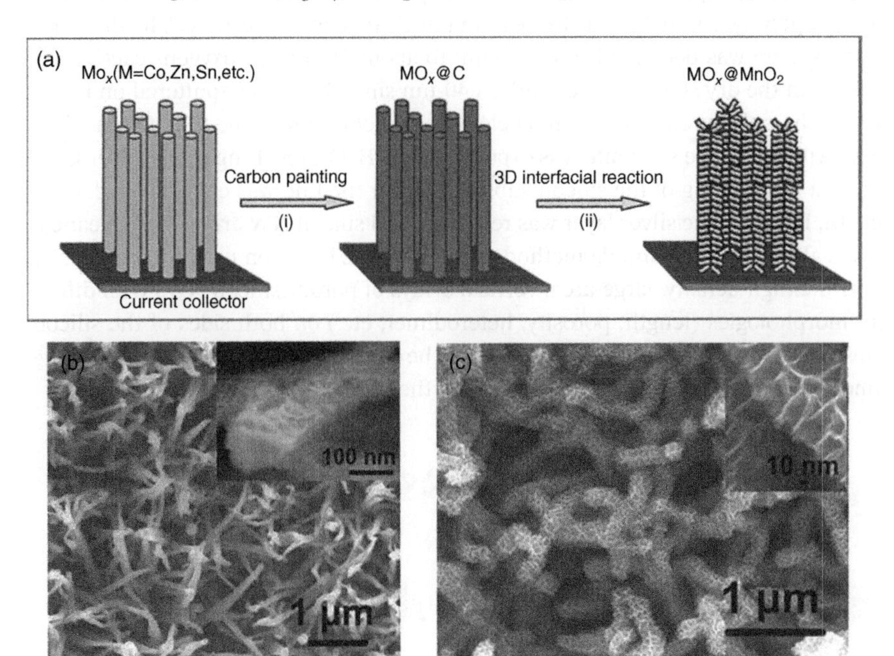

Figure 3.13 (a) The schematic diagram of the electrode design process. (b) SEM images of the Co_3O_4/MnO_2 nanowire arrays with one hour growth. (c) SEM images of the Co_3O_4/MnO_2 nanowire arrays with 5-hours growth. Source: Liu et al. [29] / with permission of John Wiley & Sons, Inc.

Moreover, Yu et al. from Nanyang Technological University synthesized the core-shell heterostructured porous NWs composed of $NiCo_2O_4$ and MnO_2 by self-assembly method [30].

3.2.3 Chemical Etching Method

Among the various fabrication techniques for synthesizing porous NWs, the metal-assisted chemical etching (MACE) method is simple in process with easily controllable parameters (such as diameter, length, crystal orientation, cross-sectional shape, doping concentration, doping type) and good repeatability. As a result, it can synthesize a highly ordered SiNWs array with a high aspect ratio, which is expected to realize industrialized mass production [31–34].

Herein, Bang et al. used commercially available low-cost silicon powder to generate porous NWs and ultrafine-sized cores to form multidimensional silicon through a combination of metal deposition and metal-assisted chemical etching [35]. This multidimensional silicon can be applied to anode materials for LIBs. In the bulk silicon particles, the length of the formed porous silicon NW is 5–8 μm and the pore diameter is 10 nm.

Veerbeek et al. from the University of Twente noted that the porous NWs are prepared by nanolithography technology and MACE synergistically. The synthesis process of this NW array is shown in Figure 3.14 [36].

Since most samples are made using nanosphere lithography technology, a dense array of polystyrene spheres is formed on the clean silicon substrate. The diameter of the sphere was decreased from 447 nm to about 280 nm by oxygen-reactive ion etching in the dry state. Subsequently, a 40-nm silver layer was sputtered on top of it, and the polystyrene balls were peeled off in dichloromethane to obtain a silver layer with gaps. The substrate was exposed to HF/H_2O_2, resulting in the silver layer under a wet etching of the silicon substrate. After the line was etched to a desired length, HNO_3 in the silver layer was removed, and silicon NW arrays were cleaned.

It is also true that a simple method such as MACE has been used to successfully fabricate high-density, large-area, vertical arrays of porous silicon NWs with different morphologies (length, porosity, heterodimer, etc.) on both sides of the silicon substrate [37]. The porosity of the NWs could be controlled by prolonging the etching time, which can increase the surface area of the porous NWs.

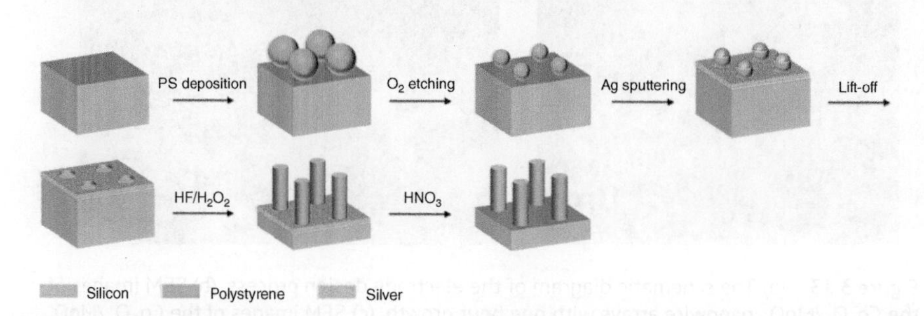

Figure 3.14 The preparation process for Si nanowires. Source: Veerbeek et al. [36] / The Royal Society of Chemistry.

3.3 Hierarchical Nanowires

The formation of nanostructures with controlled size and morphology has been the focus of intensive research in recent years [38, 39]. Such nanostructures are important in the development of nanoscale devices and the exploitation of various properties of nanomaterials [40]. The controlled growth of hierarchical structures has received particular attention recently because it can improve the properties of materials and devices. Although the fabrication of complex structures observed in natural materials or biominerals remains a significant challenge, substantial progress has been made to achieve hierarchical crystals. For example, the synthesis of branched tetrapod-shaped structures of controlled size and crystal structure has been reported [12]. Branched Cu_2O crystals exhibiting multiform geometries have also been fabricated by a kinetically controlled growth method [41, 42]. Despite progress in hierarchical fabrication, there are still relatively few works about the precise control of product size and branching degree. Synthesizing complex structures, such as multilevel branched structures, will open the door to more complex 3D materials and devices and provide more opportunities to exploit the unique properties of nanomaterials.

Different nanoelements combine to form hierarchical NWs. The NW structure is in the microscale category of nanometer scale in the radial direction and the macroscale category of micron or even centimeter scale in the axial direction. Through the introduction of different hierarchical units, the obtained hierarchical structure not only has the controllable size of NWs, radial electron transport, and other characteristics, but also has the unique properties of hierarchical nano-units. Compared with a single component, it shows higher specific surface area, more surface-active sites, and better electrolyte wettability, which make it show great potential in the field of energy storage.

Mai et al. created ultralong hierarchical vanadium oxide NWs with diameters of 100–200 nm and lengths of several millimeters using a low-cost precursor (NH_4VO_3, PVA), electrospinning, and annealing. Around 50 nm in diameter and 100 nm in length, vanadium oxide nanorods made up the outer layer of the hierarchical NWs [43] (Figure 3.15).

3.3.1 Self-Assembly Method

A popular technique for creating hierarchical NWs is the "self-assembly method" which combines the "directed attachment" and "self-assembly" crystal growth mechanisms. The "self-assembly" process refers to the formation of the product morphology structure in the process of nanocrystalline growth, which is spontaneous under certain conditions and without artificial control; "Directional attachment" mechanism is usually the self-assembly behavior of the adjacent elementary particles in the process of material growth, and they possess the same self-growth trigger mechanism, which makes it easy to get rid of the defects. In the same environment, each particle on the interface forms a similar aggregation state. Therefore, the final growth orientation of each unit crystal is the same,

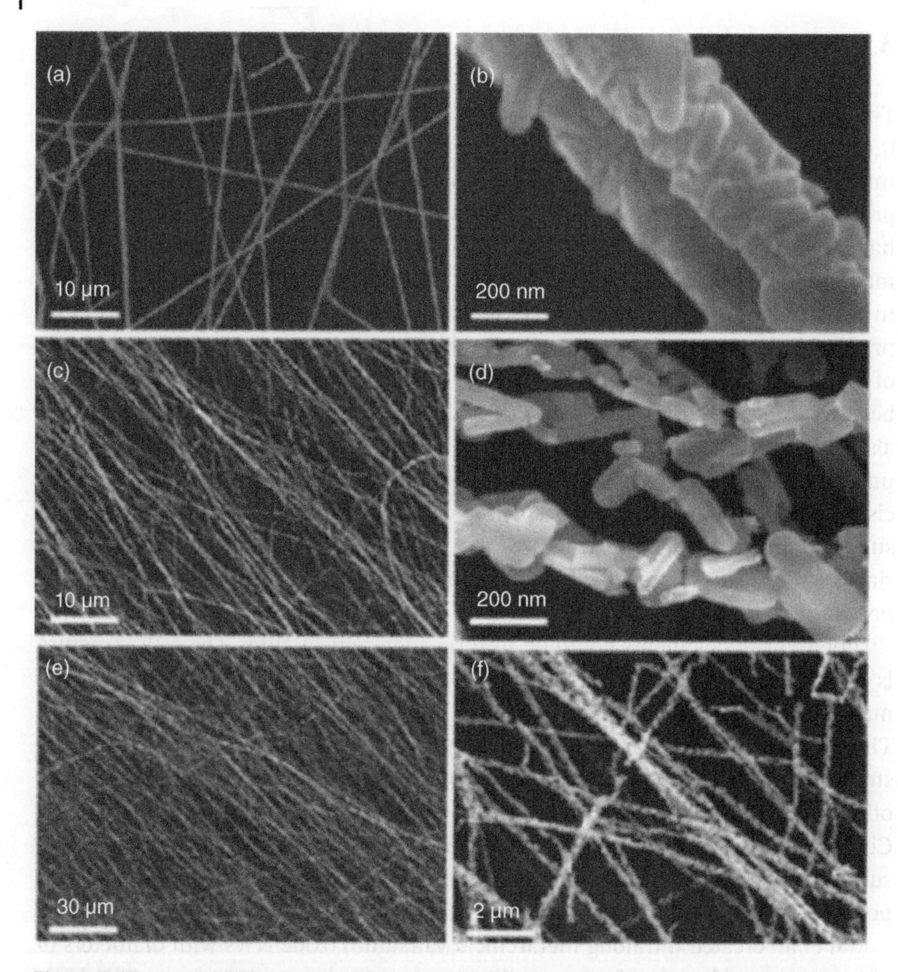

Figure 3.15 (a, b) FESEM images of electrospinning NH_4VO_3/PVA composite nanowires. (c–f) FESEM images of the ultralong hierarchical vanadium oxide nanowires after annealing. Source: Mai et al. [43] / American Chemical Society.

forming a morphology structure with the same growth direction, thus achieving the effect of "orientation overlap." This method has the advantages of simple process, high universality, low cost, and does not need a template or external force. The self-assembled structure has higher order than independent components. However, its disadvantages are also prominent, such as the strict requirements on experimental conditions and material ratios, which make it difficult to control in the assembly process. The influencing factors of this method are material ratio, solvent type, component structure, etc.

Using $MnMoO_4$ NWs as the substrate, $CoMoO_4$ self-assembly growth in aqueous solution without surfactant or stabilizer was directed by Mai et al. When the Co source was added, a supersaturated solution containing a large amount of $CoMoO_4$ crystals was formed. Due to nanoparticles' high surface energy and thermodynamic instability, they can be adsorbed on the surface of $MnMoO_4$, resulting in the decrease in surface energy. The matrix $MnMoO_4$ has similar lattice parameters, which can control the self-assembly and directional attachment crystallization of $CoMoO_4$ nanoparticles and finally form $MnMoO_4/CoMoO_4$ hierarchical NWs. The preparation process and morphology characterization of the $MnMoO_4/CoMoO_4$ hierarchical structure are shown in Figure 3.16a,b [44].

In addition to inorganic substrates, there are also some polymer substrates, Schmidt et al. proposed a novel strategy to fabricate hierarchical MoS_2/PANI NWs in a mild condition (≤200 °C), as shown in Figure 3.17. Starting from the 1D

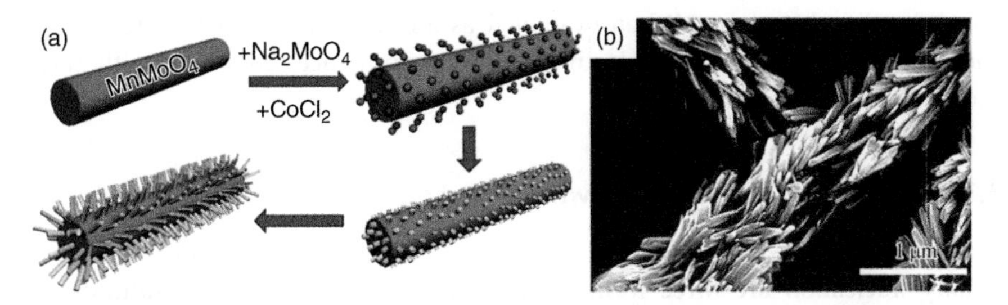

Figure 3.16 Synthesis process and morphology characterization of hierarchical nanowires synthesized by self-assembly orientation lapping method. (a) Synthesis diagram of $MnMoO_4/CoMoO_4$ hierarchical nanowires and (b) corresponding SEM image. Source: Yang et al. [44] / Springer Nature.

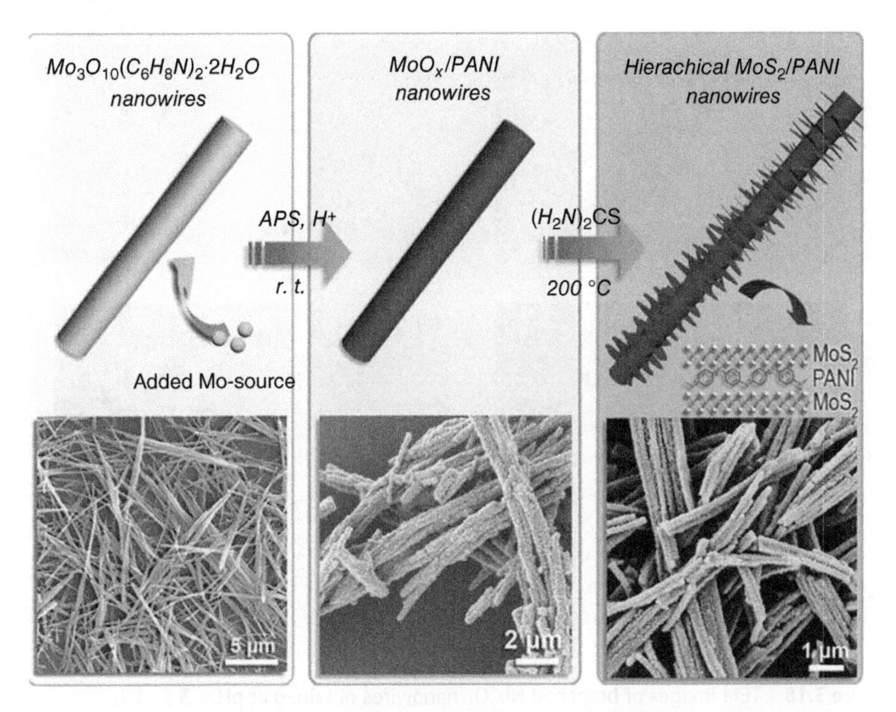

Figure 3.17 Scheme for the fabrication of hierarchical MoS_2/PANI nanowires through facile polymerization and hydrothermal treatment of $Mo_3O_{10}(C_6H_8N)_2{\cdot}H_2O$ precursor. Source: Yang et al. [45] / with permission of John Wiley & Sons, Inc.

precursor of $Mo_3O_{10}(C_6H_8N)_2{\cdot}H_2O$ (anilinium trimolybdate [ATM]), MoO_x/PANI nanowires were synthesized after polymerization, which were then converted to MoS_2/PANI via hydrothermal process with thiourea. The as-obtained products evenly integrate MoS_2 ultrathin nanosheets with PANI into the primary 1D architecture, resulting in novel hierarchical and polymer-hybrid NWs. Meanwhile, the contents of MoS_2 and PANI in NWs could be easily tuned by adding a varied amount of additional $Mo_7O_{24}(NH_4)_6{\cdot}4H_2O$ (ammonium heptamolybdate [AHM]), during polymerization. Such NWs of MoS_2/PANI were expected to exhibit enhanced conductivity, and fulfill the advantages of both nanocomposites and hierarchical architectures. When evaluated as an anode material for LIBs, they showed remarkably improved electrochemical performances compared with bare MoS_2 [45].

3.3.2 Secondary Nucleation Growth Method

Primary homogeneous nucleation, primary heterogeneous nucleation, and secondary nucleation are three forms of nucleation. Under high supersaturation,

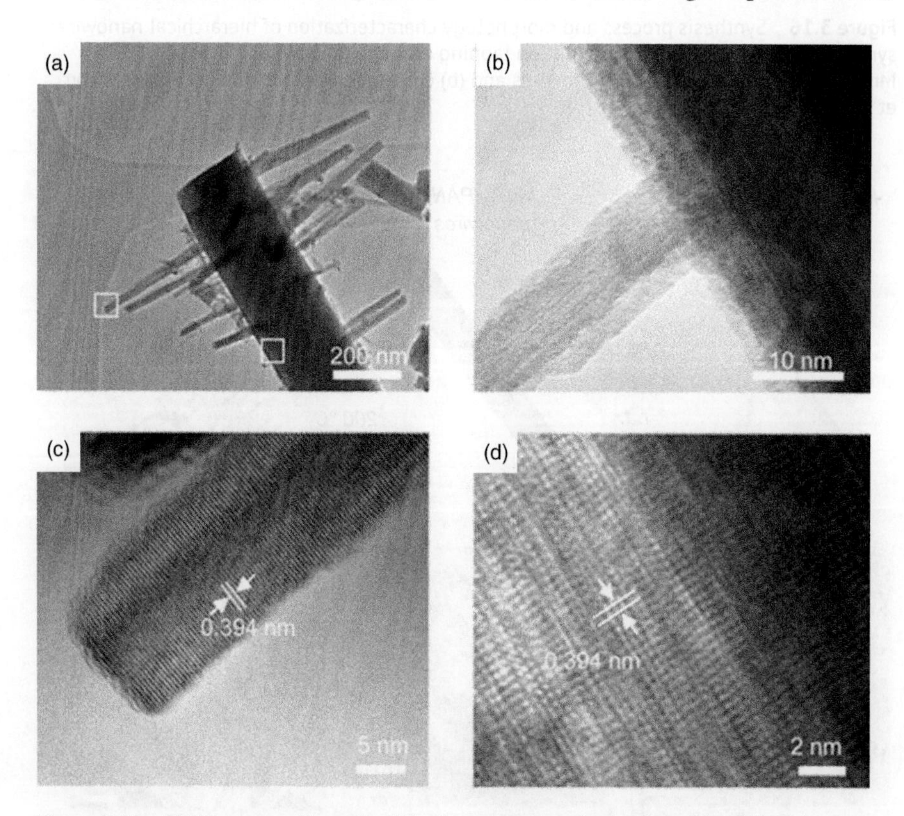

Figure 3.18 TEM images of branched Nb_2O_5 nanowires obtained at pH = 3.3–3.4. (a) Low-magnification SEM image of the typical branched nanowire. (b) SEM image of the adjunction part; (c, d) HRTEM images recorded from the white frames in (a). Source: Liu et al. [46] / IOP Publishing.

the process of nucleation is called primary homogeneous nucleation. Primary heterogeneous nucleation is the process of nucleation in solution induced by foreign matter (such as dust in atmosphere). Nucleation in solutions containing solute crystals is called secondary nucleation. Secondary nucleation is also a heterogeneous nucleation process, which occurs when crystals collide with each other or with other solid particles (wall, stirrer, etc.).

Xue et al. reported the synthesis of dendritic Nb_2O_5 NW arrays by treating niobium foil in an aqueous solution of hydrofluoric acid using secondary nucleation growth method under hydrothermal conditions (Figure 3.18). By controlling the dynamic parameters of the system, the Nb_2O_5 structure with the required branching degree could be obtained. This may have unique application potential in the development of photon, electron, and sensing nanodevices [46].

3.4 Heterogeneous Nanowires

Heterogeneous NWs are NW materials composed of different materials in contact with each other. It not only has the advantages of different components, but also integrates the special physical and chemical properties at the heterogeneous interface, which produces an enhanced effect or a synergistic effect that is better than a single component. The synthesis of heterogeneous NWs includes heterogeneous nucleation, secondary modification, and other methods.

3.4.1 Heterogeneous Nucleation

The preparation of heterogeneous NWs by the secondary hydrothermal method generally uses the NW materials synthesized in the previous step as crystal nucleus, and the heterogeneous nucleation growth occurs during the hydrothermal reaction, that is, the new assembly is carried out while maintaining the morphology of the NWs. The growth of sub-crystals will finally result in heterogeneous NWs composed of different components.

Zheng et al. synthesized the unique $Co_3O_4/\alpha\text{-}Fe_2O_3$ NWs by a simple two-step hydrothermal reaction. Firstly, they synthesized Co_3O_4 NWs on titanium foil substrate by hydrothermal reaction. Then, they used Co_3O_4 NWs as crystal nucleus, added substances containing iron sources and precipitants, and finally, $Co_3O_4/\alpha\text{-}Fe_2O_3$ NWs were obtained by secondary hydrothermal reaction (Figure 3.19a,b) [47]. Fan et al. synthesized SnO_2 NWs by CVD, and $FeOOH/SnO_2$ NWs were synthesized by hydrothermal heterogeneous nucleation and crystallization. And after being calcined in air, $\alpha\text{-}Fe_2O_3/SnO_2$ NWs were obtained (Figure 3.19c,d). They also found that the length of $\alpha\text{-}Fe_2O_3$ branches could be adjusted with specific reactant concentrations and reaction times [48]. Fan and coworkers synthesized SnO_2/ZnO NWs by the same method (Figure 3.19e,f) and added trisodium citrate to control the morphology of ZnO [49]. Shen from Huazhong University of Science and Technology synthesized $TiO_2@VS_2$ NWs by two-step hydrothermal method. TEM characterization shows that VS_2 particles grew uniformly on the surface of TiO_2 NWs (Figure 3.19g,h) [50].

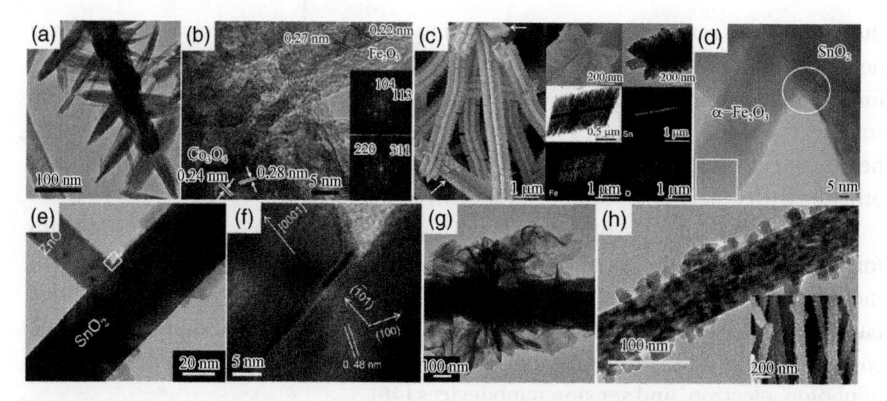

Figure 3.19 (a) TEM image of Co_3O_4/α-Fe_2O_3 heterogeneous nanowires. (b) HRTEM image of Co_3O_4/α-Fe_2O_3. (c) SEM and (d) TEM images of α-Fe_2O_3/SnO_2 nanowires. (e) TEM and (f) HRTEM image of SnO_2/ZnO heterogeneous nanowires. (g) TEM image of SnO_2/ZnO nanowires synthesized by adding trisodium citrate. (h) TEM image of TiO_2-B@VS_2 nanowires, with the inset showing its corresponding SEM image. Source: Refs. [47–50] / Springer Nature / with permission of John Wiley & Sons, Inc / American Chemical Society / with permission of Elsevier.

Figure 3.20 SEM images of (a) Ag nanowire, (b, c) necklace-shaped Ag nanowires/Ag_3PO_4 cube heterogeneous nanowire, (d–f) Ag/Ag_3PO_4 heterogeneous nanowires obtained under different concentrations of $[Ag(NH_3)_2]^+$ complex. Source: Bi et al. [51] / Royal Society of Chemistry.

In order to improve the charge separation efficiency of Ag_3PO_4 semiconductors, Lv and coworkers coupled Ag_3PO_4 submicrocubes and high conductivity Ag NWs into a new necklace-like coaxial heterogeneous NW [51]. They first produced single-crystal Ag NWs with an average diameter of 50 nm through the polyol process (Figure 3.20a), and used them as the initial template for the selective growth and assembly of Ag_3PO_4 crystals. When the Ag NWs reacted with $[Ag(NH_3)_2]^+$ and Na_2HPO_4 in aqueous solution at room temperature, homogeneous and ordered

Ag_3PO_4 submicron cube could be generated on the Ag NWs by a heterogeneous nucleation process. As shown in Figure 3.20b,c, a series of adjacent Ag_3PO_4 submicrocubes were drilled through by one Ag NW, and they formed a 1D necklace-like heterostructure. In addition, by changing the concentration of $[Ag(NH_3)_2]^+$ complex, the position and quantity of Ag_3PO_4 cube on the Ag NW could be further rationally controlled. The experimental results are shown in Figure 3.20d–f. When the concentration of the $[Ag(NH_3)_2]^+$ complex substance is 0.2 mol L^{-1}, the Ag NWs are completely covered by Ag_3PO_4 cubes. As the amount of $[Ag(NH_3)_2]^+$ substance decreased, the number of Ag_3PO_4 cubes formed on Ag NWs also decreased. When the concentration dropped to 0.05 mol L^{-1}, only a small amount of Ag_3PO_4 cubes formed on the Ag NWs. This novel necklace-shaped Ag nanowire/Ag_3PO_4 cube heterogeneous NWs showed higher activity than pure Ag_3PO_4 cube and Ag NW.

3.4.2 Secondary Modification

Using polyacrylonitrile (PAN) as spinning medium and carbon source, Zhang et al. prepared the precursor of Si/PAN composite NWs (Figure 3.21a) by electrospinning, which were carbonized to obtain Si/C NWs (Figure 3.21b). Then, Al_2O_3 was coated on Si/C NWs by the atomic layer deposition (ALD). Al_2O_3-coated Si/C heterostructure NWs were prepared and studied by controlling the deposition times of atomic layer. Investigators looked into how the thickness of the Al_2O_3 coating affected the electrochemical characteristics of the Si/C anode. Figure 3.21c–h are SEM images of Si/C NWs with different coating thickness of Al_2O_3. It is discovered that the mechanical integrity of the electrode material structure can be enhanced and that it is possible to effectively prevent the side reaction between the electrode and the electrolyte [52].

Zheng et al. successfully deposited Co_3O_4, TiO_2, ZnO, and NiO nanoparticles on CuO NWs using the sol-flame method and synthesized a "nanoparticle@nanowire" hybrid heterogeneous NW structure material (Figure 3.22b–e). They firstly immersed CuO NWs in the metal-ion sol to obtain the modified CuO NWs with a

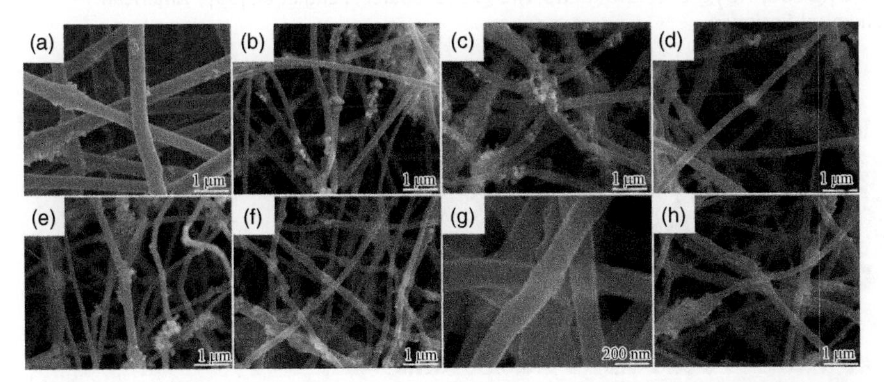

Figure 3.21 SEM images of (a) Si/PAN composite nanowire precursors, (b) carbonized Si/C nanowires. (c–g) 7, 14, 21, 28, and 35 times deposited Si/C nanowires, respectively. (h) SEM image of Al_2O_3-coated Si/C-based heterogeneous nanowires. Source: Li et al. [52] / Royal Society of Chemistry.

thin shell of metal salt. Then, the thin shell layer was converted into nanoparticle chains through rapid annealing, and finally the "nanoparticles@nanowires" hybrid heterogeneous NWs were obtained. The preparation process is shown in Figure 3.22a. The uniqueness of this synthetic method is that the high temperature and the short annealing time can make the solvent evaporate quickly and obtain a high nucleation rate, as well as inhibit grain growth and agglomeration to finally form a tight adhered nanoparticle chain around the NWs. This method has also been extended to synthesize "ternary metal oxide/noble metal nanoparticles@nanowires" and other hybrid heterogeneous NWs [53].

The necklace-like MgO/TiO_2 heterostructures were synthesized by a simple and economical precursor-calcination method. As shown in Figure 3.23a, researchers obtained the necklace-like precursor using a handy solution method. After being calcinated at a high temperature, the products still maintained a necklace-like structure

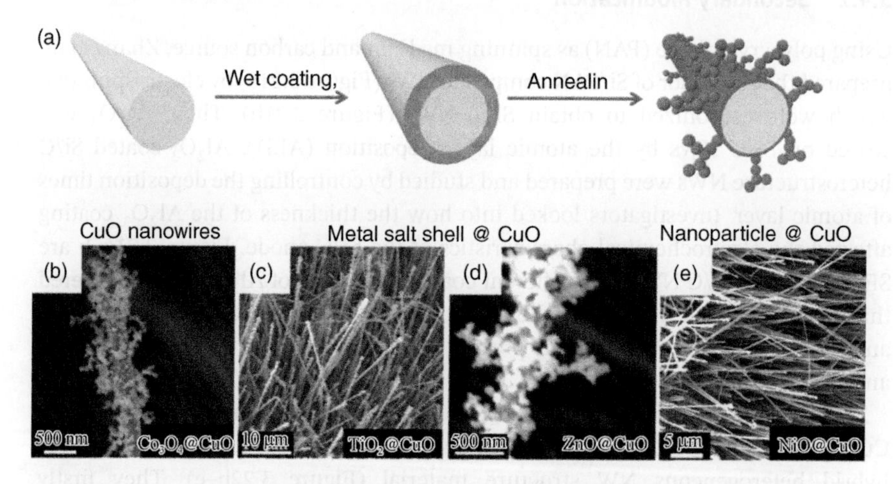

Figure 3.22 (a) Schematic diagram of the synthesis process. (b–e) SEM images of Co_3O_4@CuO, TiO_2@CuO, ZnO@CuO, NiO@CuO. Source: Feng et al. [53] / American Chemical Society.

Figure 3.23 (a) SEM image of necklace-like MgO/TiO_2 nanowire precursors. (b) SEM image of calcined necklace-like MgO/TiO_2 nanowire. (c) TEM image of calcined necklace-like MgO/TiO_2 nanowire. Source: Jia et al. [54].

(Figure 3.23b). The TEM image shows that in this unique structure, TiO_2 spheres with diameters of 100–200 nm were located on the side of MgO NWs or penetrated by NWs (Figure 3.23c). At the same time, the effect of tetrabutyl titanate (TBT) on the product morphology was studied by changing the amount of TBT in the reaction system [54].

3.5 Hollow Nanowires

Hollow-structured NWs are often referred to as nanotubes with 1D hollow structures [55]. Compared with conventional NWs, hollow NWs are widely applied in energy storage for their low mass density, high porosity, and large specific surface area [56]. At present, researchers have synthesized hollow NWs through wet chemical methods, template methods, gradient electrospinning, and so on. The methods are briefly introduced below.

3.5.1 Wet Chemical Method

The hydrothermal/solvothermal method, based on liquid-phase reactions, is one of the typical wet chemical methods [57]. By adjusting the reaction temperature and the ratio of reactants, products with low, intermediate valence or other special properties can be synthesized. Hollow structures can also be formed through similar reduction reactions.

As shown in Figure 3.24, using Co/Ni as catalyst, Qian and coworkers produced multiwall hollow NWs by reducing hexachlorobenzene with potassium metaborate at 350 °C [58]. Mai and coworkers mixed aniline, vanadium solution, and reduced graphene oxide (rGO) in a certain proportion. The semi-hollow discontinuous graphene scrolled V_3O_7 NWs (VGS) with a length of more than 30 μm were finally obtained by regulating the hydrothermal conditions (Figure 3.25) [59]. Self-supported hollow Co_3O_4 NWs were made using a straightforward hydrothermal synthesis process, according to Xia et al. They could grow vertically onto the substrates on vast surfaces to create aligned NW arrays [60].

Apart from the hydrothermal/solvothermal method, there are other ways to produce hollow NWs. Mai et al. proposed a new method by combining the self-assembly method with the rheological reaction method to synthesize vanadium oxide nanotubes [61]. In this work, V_2O_5 and cetylamine were added to a mixed solvent (ethanol and water), completely grinding and mixing. And then vanadium oxide nanotubes were obtained through the hydrothermal reaction of rheological products at 180 °C (Figure 3.26).

3.5.2 Template Method

Template synthesis is one of the most popular ways, for its simplicity and variability, to control the chemical composition and morphology of 1D nanostructures in both lateral and longitudinal dimensions [62]. In general, the template approach may be

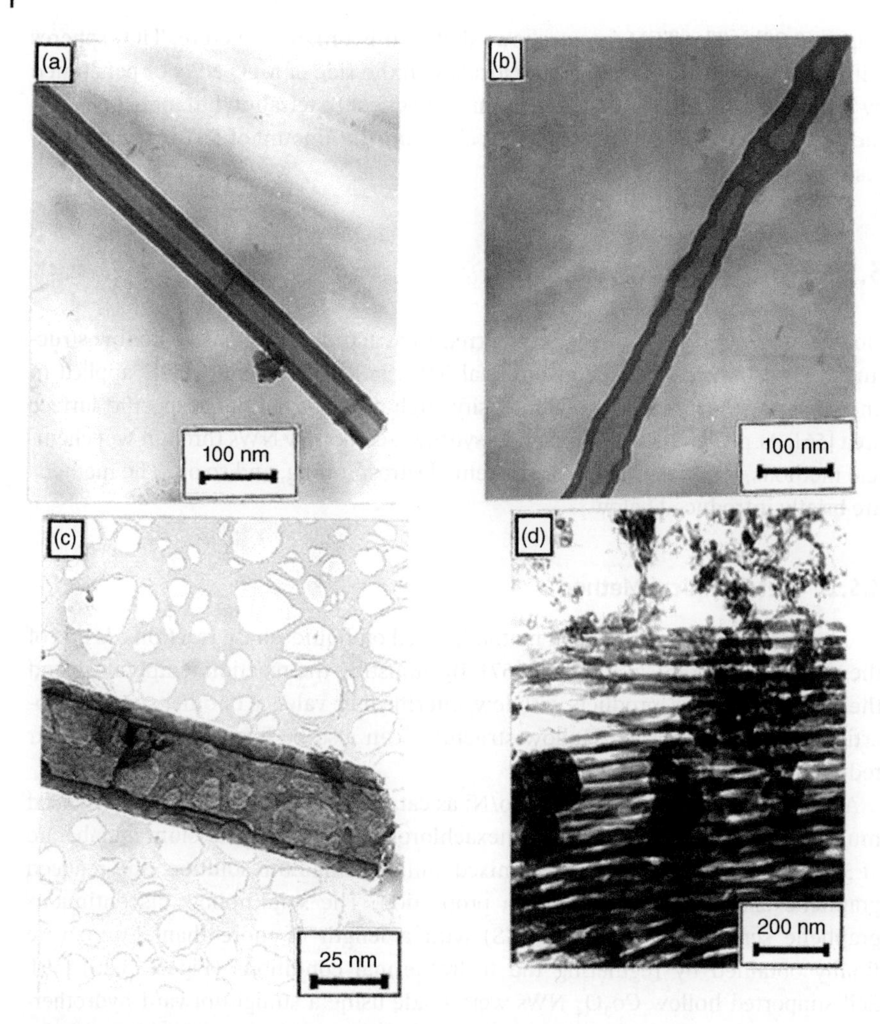

Figure 3.24 TEM images of multiwall carbon nanotubes. Source: Jiang et al. [58] / American Chemical Society.

broken down into three steps: the creation of the template, the stimulation of target material development, and the removal of the template. In the synthesis of hollow NWs, templates can be any material with nanostructure. The template methods can be divided into physical ones which involve only physical reactions and chemical ones, which involve both physical and chemical reactions, according to chemical properties [62, 63]. The common methods of removing templates can also be divided into physical methods such as direct dissolution and chemical methods including high-temperature solid-phase sintering, chemical etching, ion exchange, and electrochemical replacement.

As shown in Figure 3.27, Yu and coworkers used ultrathin Te NWs as templates to synthesize uniform and ultrathin Pt hollow NWs by electrochemical

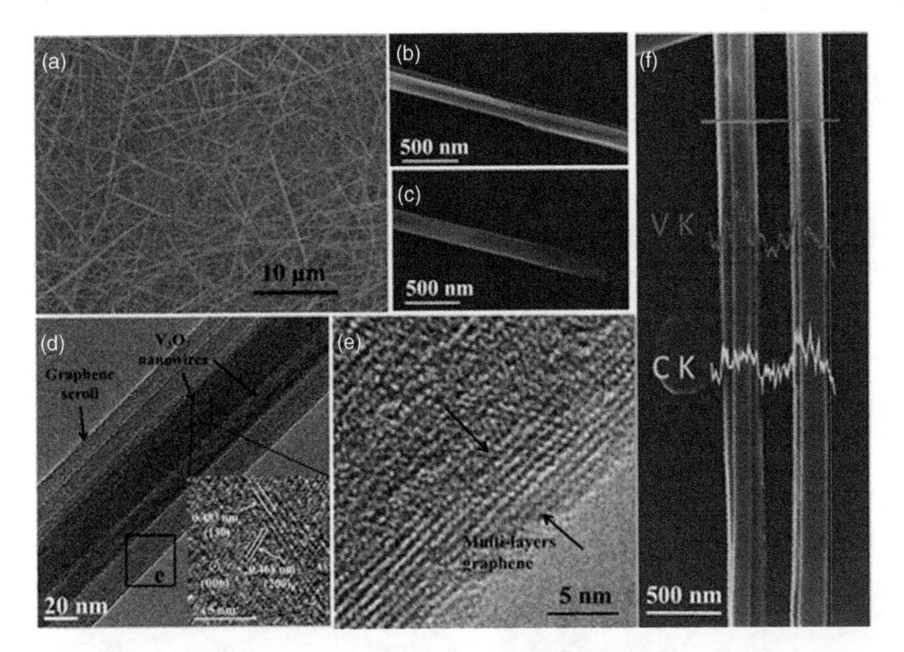

Figure 3.25 Morphological characterization of VGS. (a) SEM image of synthesized VGS at low magnification, (b, c) SEM image of synthesized VGS at high magnification. TEM images of (d) VGS; (e) the edge of VGS. (f) SEM image and corresponding EDS line scan of VGS. Source: Yan et al. [59] / American Chemical Society.

Figure 3.26 SEM image of vanadium oxide nanotubes. Source: Liqiang et al. [61] / Wuhan University of Technology.

replacement [64]. Besides, Lee et al. reported a method to prepare hollow ZnO NWs. In their experiments, ZnO layer was deposited on cyclopore polycarbonate membranes with different pore sizes via ALD [65]. The templates were removed by combustion at 450 °C over four hours. The structure of hollow ZnO NWs is shown in Figure 3.28. Besides, Caruso et al. used polymer fibers, which were prepared by electrospinning, as templates to prepare the hollow TiO_2 NWs. During the fabrication process, TiO_2 was coated on the polymer fibers by sol–gel coating, and hollow TiO_2 NWs were obtained after heat treatment [66]. This work provided a

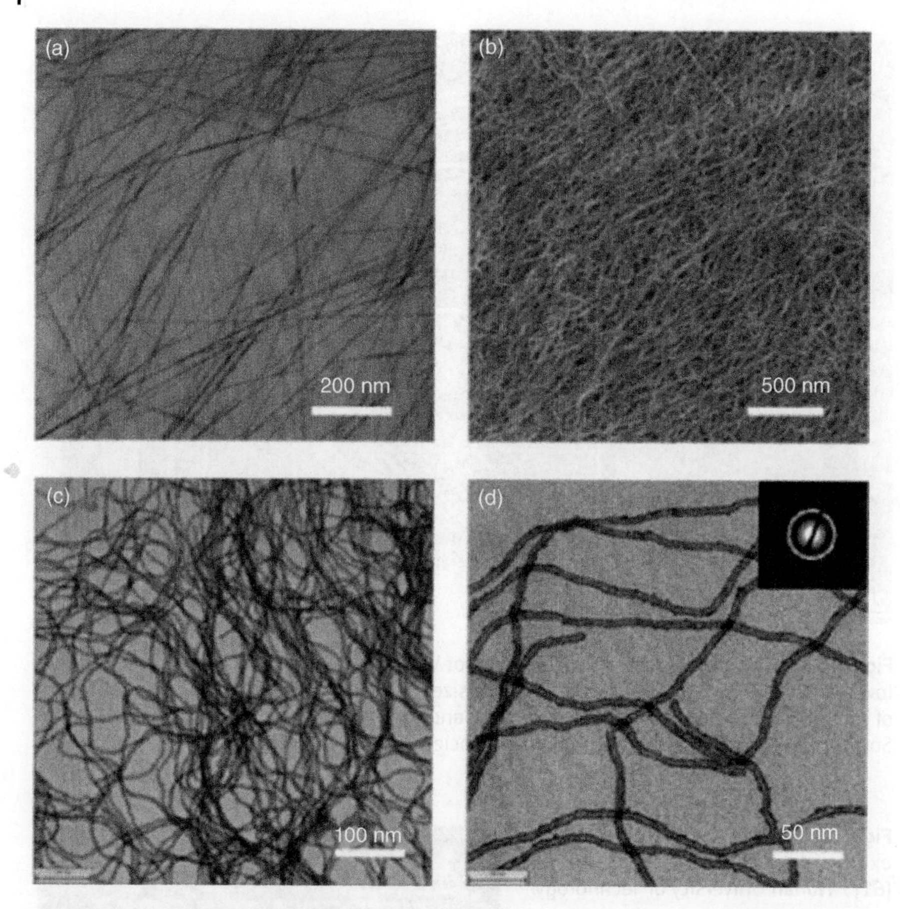

Figure 3.27 (a) TEM image of Te nanowires. (b) SEM image, (c, d) TEM images of Pt hollow nanowires. Source: Liang et al. [64] / American Chemical Society.

good reference for the regulation and preparation of secondary inorganic materials with uniform wall thickness and structure.

3.5.3 Gradient Electrospinning

Electrospinning is a widely used and convenient method to synthesize hollow NWs. Most gradient electrospinning methods include the following three processes: preparing polymer precursor solutions, electrospinning, and sintering. Besides, many advanced electrospinning technologies have been developed by adjusting these processes [67].

The molecular weight of the polymer and temperature of heat treatment are factors that influence the properties and structures of products in gradient electrospinning. Niu et al. synthesized various kinds of pea-like nanotubes and mesoporous

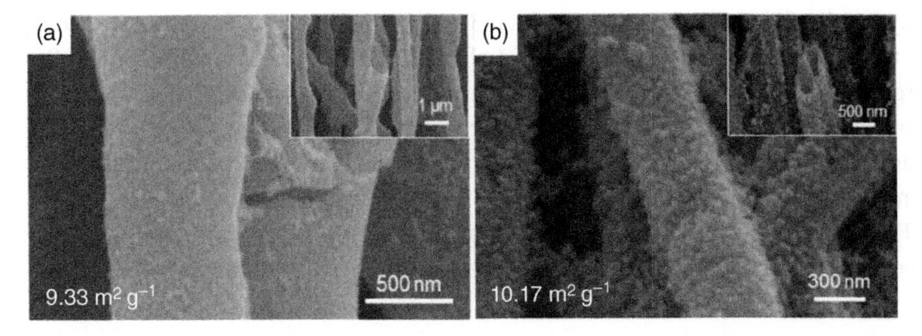

Figure 3.28 SEM images of hollow ZnO NWs with different surface areas: (a) 9.33 m^2 g^{-1}. (b) 10.17 m^2 g^{-1}. Source: Lee et al. [65] / Springer Nature.

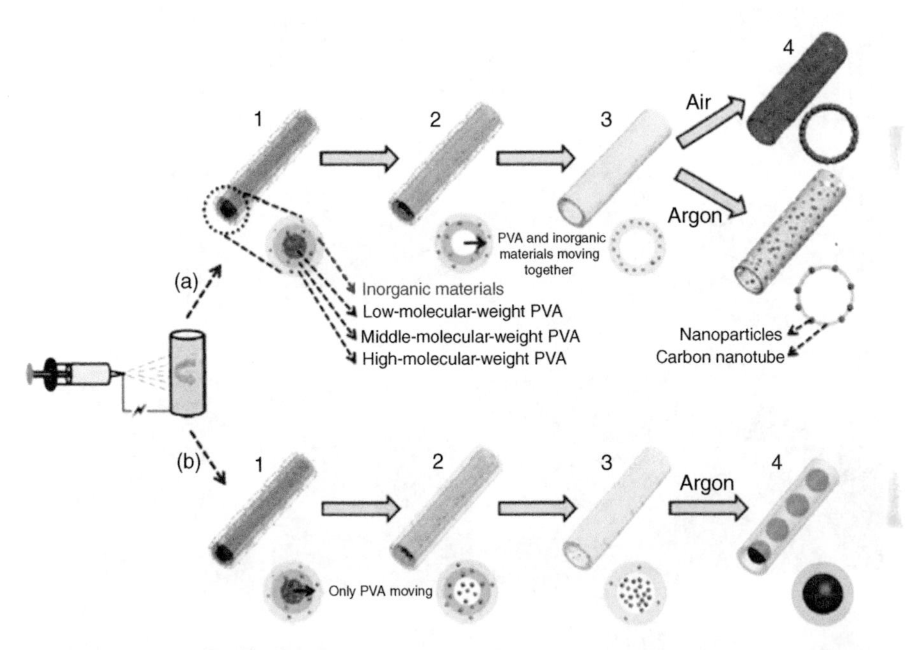

Figure 3.29 Schematics of the gradient electrospinning and controlled pyrolysis methods. Source: Niu et al. [68] / CC BY 4.0 / Public domain.

nanotubes by combining gradient electrospinning method with controlled pyrolysis method (Figure 3.29) [68]. They used PVA aqueous solution with three molecular weights as precursors and controlled the temperature of heat treatment to fabricate different nanotubes, including pea-like nanotubes consisting of outer carbon nanotubes and inner nanoparticles and hollow nanotubes consisting of inner carbon nanotubes and outer nanoparticles. During the electrospinning process, driven by electrostatic force, the PVA of three molecular weights was distributed in a coaxial gradient in a single NW when reaching the receiving plate [69, 70]. This unique polymer distribution in NWs provided a new way to prepare complex nanostructures by using differences in the physical and chemical properties of polymers with different molecular weights. This method makes it possible to fabricate a variety of inorganic nanotube materials and structures. In addition, this method allowed the batch preparation of various nanotubes serving as electrode materials by using conventional single needle head.

Figure 3.30 shows the multiple inorganic nanotubes prepared by gradient electrospinning, including simple hollow nanotubes of single-metal oxides (CuO,

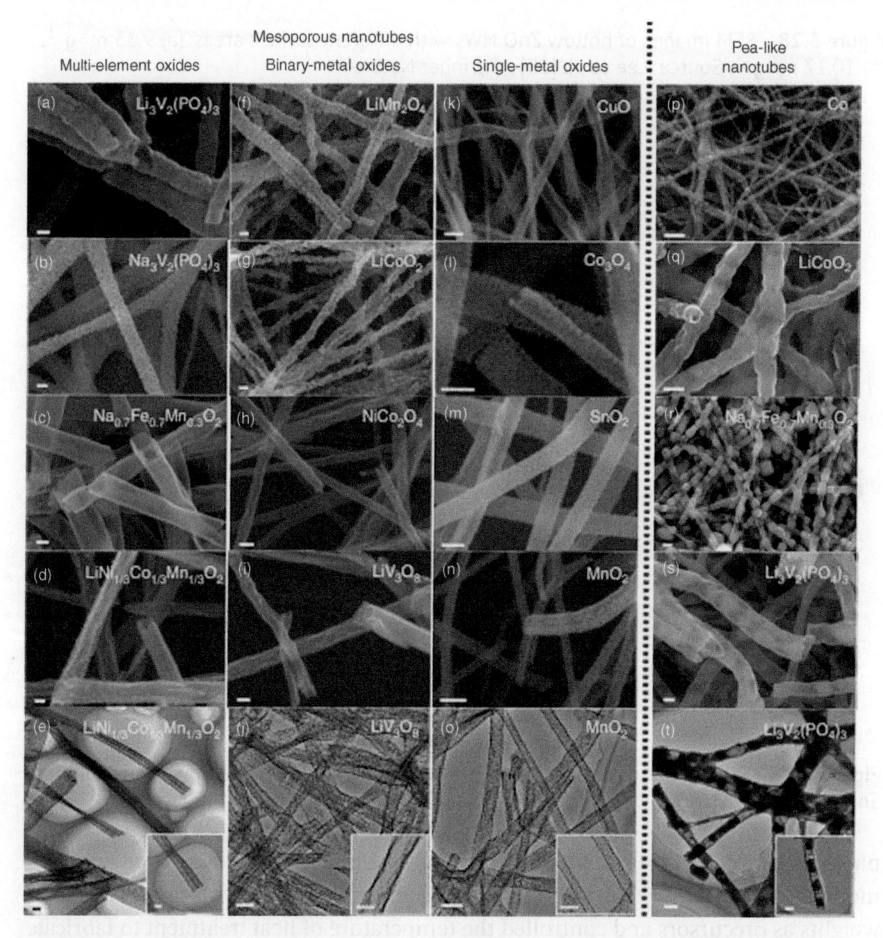

Figure 3.30 Expansion of the gradient electrospinning and controlled pyrolysis method. (a–o) SEM and TEM images of multi-element oxides ($Li_3V_2(PO_4)_3$, $Na_3V_2(PO_4)_3$, $Na_{0.7}Fe_{0.7}Mn_{0.3}O_2$ and $LiNi_{1/3}Co_{1/3}Mn_{1/3}O_2$), binary-metal oxides ($LiMn_2O_4$, $LiCoO_2$, $NiCo_2O_4$ and LiV_3O_8) and single-metal oxides (CuO, Co_3O_4, SnO_2 and MnO_2) mesoporous nanotubes, respectively, scale bars, 100 nm. (p–t) SEM and TEM images of pea-like nanotubes (Co, $LiCoO_2$, $Li_3V_2(PO_4)_3$ and $Na_{0.7}Fe_{0.7}Mn_{0.3}O_2$) from different species with scale bars at 200 nm. The scale bars for the inset TEM images (e, j, o, t) are 100 nm. Source: Niu et al. [68] / Springer Nature.

Co_3O_4, SnO_2, and MnO_2), bimetallic phosphates ($Li_3V_2(PO_4)_3$, $Na_3V_2(PO_4)_3$), bimetallic oxides ($LiMn_2O_4$, $LiCoO_2$, $NiCo_2O_4$, and LiV_3O_8), polymetallic oxides ($Na_{0.7}Fe_{0.7}Mn_{0.3}O_2$ and $LiNi_{1/3}Co_{1/3}Mn_{1/3}O_2$), and pea-like nanotubes which are internally modified by 0D particles. As could be seen, the mesoporous structure is uniformly distributed on the tube wall and serves as both a conduit for material transport during the chemical reaction and a stress-relieving buffer.

Based on these simple nanotubes, Mai and coworkers developed a universal preparation process for hierarchical nanotubes [70]. By regulating the calcination process, they adjusted the structure of gradient electrospinning polymer fibers by taking full advantage of the differences in thermal stability of three PVAs with different molecular weights at 280 °C. Wire-in-tube nanotubes, tube-in-tube nanotubes, and multi-walled nanotube structures were obtained. The results show that modifying the sintering process is useful for the secondary structure of NWs. The molecular type and molecular weight of polymers are the main factors in the formation of mesoporous structures and the limitation of nanocrystalline size.

3.6 Nanowire Arrays

Many NWs have been made using a variety of techniques, including electrospinning, the wet chemical method, template-assisted growth, sol–gel, laser-assisted catalytic growth, CVD/PVD, and thermal evaporation. After modification, the performance of NWs can be significantly improved, and their application areas will be broader by controlling the structure of NW arrays accurately.

NW arrays have significant advantages in energy storage: (i) An increase in the specific surface area of the electrode and electrolyte's contact leads to better electrochemical performance. (ii) Continuous channels formed by NWs promote electron transport and charge transfer, so that the conductivity and reaction rate of the electrode can be improved.

3.6.1 Template Method

Li and coworkers first synthesized Ni NW arrays using the AAO template [71]. As shown in Figure 3.31a, the average diameter of the prepared Ni NWs was 200 nm, and the average spacing between each NW was 130 nm. To carry more active materials on the Ni NWs, researchers used electrolytic polishing to increase the spacing distance between NWs. The SEM image is shown in Figure 3.31b. The specific parameters are as follows: before electropolishing, the average diameter of the NW was 100 nm; after that, the spacing increased to 230 nm. Subsequently, a simple and low-cost electroforming method was used to manufacture NW arrays with a 3D Ni–Si core-shell structure. Figure 3.31c shows the SEM image of Ni NW array deposited by Si. Each NW was wrapped in a Si shell with a thickness of 52 ± 10 nm. Only the diffraction signal peak of Ni was found in the XRD pattern, which indicates that the shell Si is in

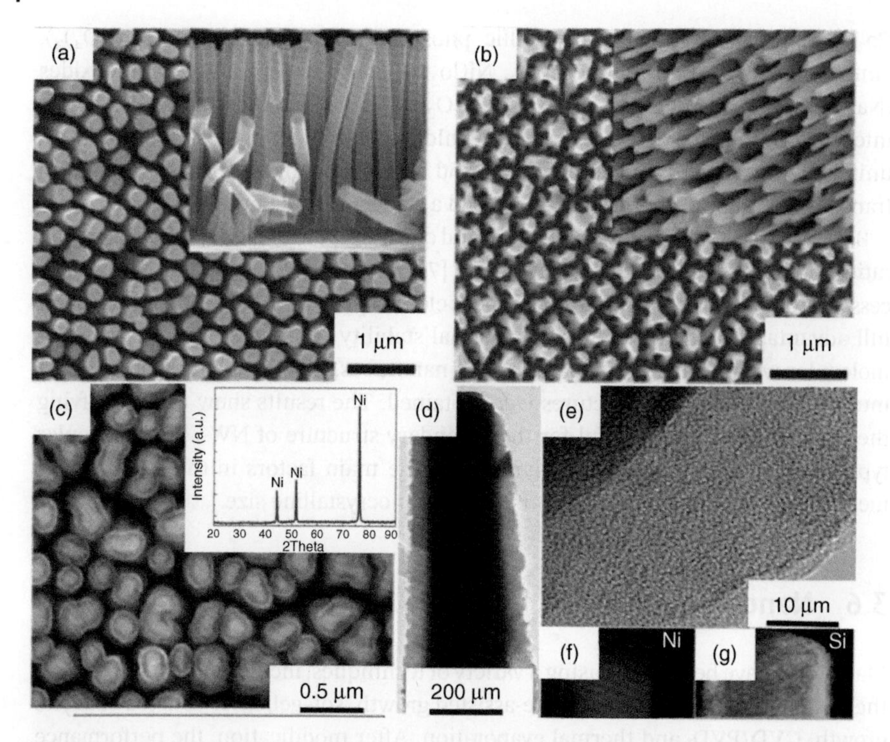

Figure 3.31 The morphology and characterization of the Ni nanowire array and the Ni–Si composite core-shell nanowire layout structure. SEM images of the Ni nanowire array (a) after growth and (b) after electropolishing. (c) SEM image of Si electrodeposited on the Ni nanowire array, the inset image: corresponding XPS spectrum. (d, e) TEM images of the typical Ni–Si nanowires in the array. (f, g) EDS mapping of the sample. Source: Liu et al. [71] / Royal Society of Chemistry.

an amorphous state. The TEM results in Figure 3.31d,e show that a single Ni–Si NW presents a core-shell structure with a contrast of light shell and dark core. It can be seen from the element diagram of the EDS that Ni and Si are located in the core and shell regions, respectively (Figure 3.31f,g). It was confirmed to be the amorphous state in high-resolution TEM image in the area of the shell. Compared with thin films of the same quality, the internal cavity of this composite core-shell NW array could adapt to the volume changes caused by ion deintercalation, relieve the internal stress generated in the material, and exhibit an improvement in electrochemical performance.

P. Forrer et al. [72] chose anodically grown porous alumina as template material, whose structure can be adjusted by changing the experimental anodization conditions. Then the electrochemical polishing technique was applied to prevent the branching of the NW. By adjusting the parameters of experiments, the length of NW arrays in the range of 100 nm to 5 nm could be produced.

Other nano-restricted templates have also been successfully applied to the synthesis of NW arrays [73]. Yang et al. fabricated a 3D Ni/TiO_2 nanowire network based on 3D acrylic resin (acrylic acid, PAA) template-assisted electrodeposition and ALD process, and the schematic diagram is shown in Figure 3.32a. Side hole structure in the 3D PAA template can eliminate the agglomeration phenomenon in the directionally grown NW arrays, and 3D Ni/TiO_2 NW arrays with staggered structures were obtained as shown in Figure 3.32b–e, which show higher area discharge capability.

3.6.2 Wet Chemical Method

The core-shell NW array design was created by Yu et al. using metal–organic framework nanostructures and nanoporous $NiCo_2O_4$ nanoflakes that were flawlessly painted on a Co_3O_4 NW array [7]. The Ni foam-supported cobalt carbonate

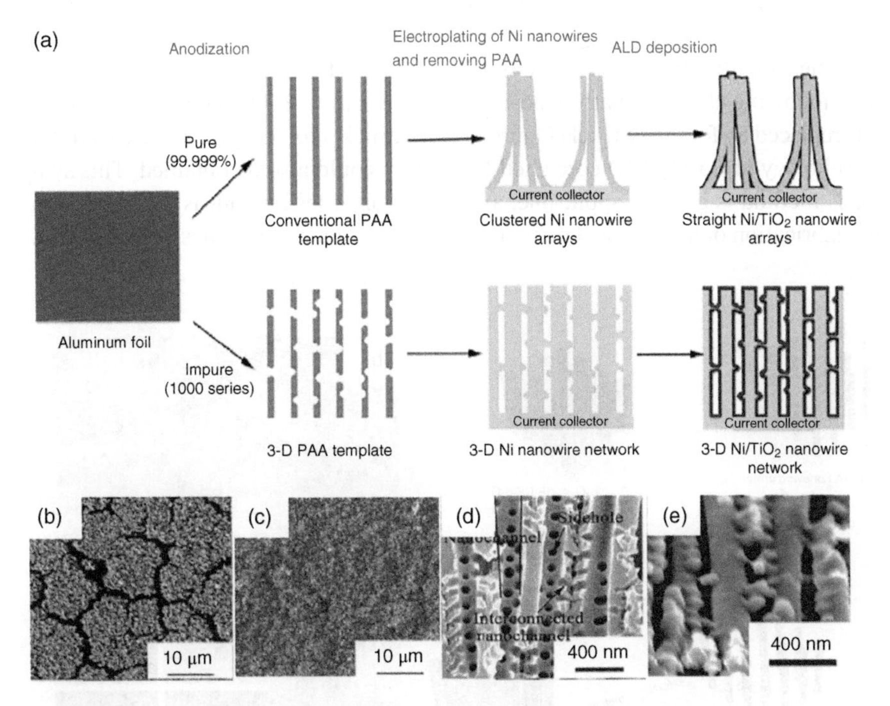

Figure 3.32 Schematic diagram of the synthesis process and morphological characterization of the three-dimensional nanowire array network structure synthesized by the template method. (a, b) SEM image of vertical Ni/TiO_2 nanowire array. (c) SEM image of 3D Ni/TiO_2 nanowire array network. (d) Cross section SEM image of 3D PAA template cross-section. (e) Cross section SEM image between adjacent vertical nanowires in the three-dimensional Ni/TiO_2 nanowire array network. Source: Wang et al. [73] / American Chemical Society.

hydroxide NW array was first prepared by a typical hydrothermal method as the precursor, and then reacted with 2-methylimidazole to convert the surface of the ZIF-67 shell. The shell thickness of the resultant Co-NA@ZIF-67 core-shell NWs (Co@ZIF-67) may be easily regulated by changing the reaction time. The schematic diagram is shown in Figure 3.33a.

The nanoflake structure of Co@ZIF-67 could be observed in SEM image, as shown in Figure 3.33b. From Figure 3.33c, d, the core-shell structure can be distinguished. The holes inside could not only increase the specific surface area but also shorten the ion transport distance, thus providing a path for ion transmission and improving energy storage performance.

Xia et al. fabricated the Co_3O_4/NiO core-shell NW array by a two-step solution method (hydrothermal synthesis and chemical bath deposition) [74]. The synthesis process is shown in Figure 3.34a. Co_3O_4 NWs were grown on a fluorine-doped tin oxide conductive glass substrate through a hydrothermal reaction (Figure 3.34b), then NiO nanosheets were coated on the surface of Co_3O_4 NWs by chemical bath deposition to obtain Co_3O_4/NiO core-shell heterogeneous NW array (Figure 3.34c). The Co_3O_4 NWs had an average diameter of 70 nm, as shown in the SEM images. NiO nanosheets with a thickness of around 10 nm formed on the surface of the Co_3O_4 NWs following chemical bath deposition. Besides, if the FTO conductive substrate was replaced by foamed nickel (Figure 3.34d) or nickel foil (Figure 3.34e), a similar morphology of Co_3O_4/NiO core-shell NW arrays could also be obtained. This synthesis method is general-applicable, and it also provides new ideas for the design and fabrication of hierarchical nano-arrays of transition metal oxides or hydroxides.

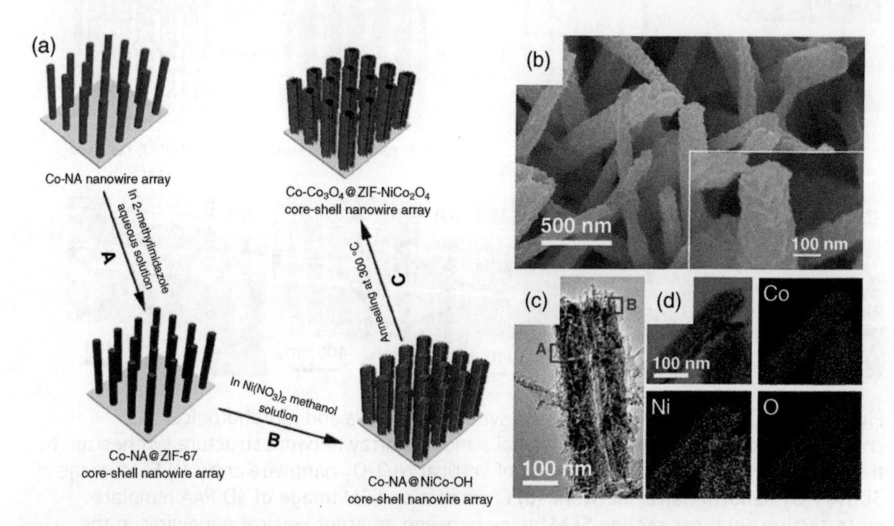

Figure 3.33 Schematic diagram of the synthesis process and morphological characterization of Co@ZIF-67. (a) Schematic illustration of the fabrication process of NW array on Ni foam. (b) SEM image of Co@ZIF-67 core-shell NW array. (c) TEM image and (d) EDX mapping images of Co@ZIF-67. Source: Yu et al. [7] / Royal Society of Chemistry.

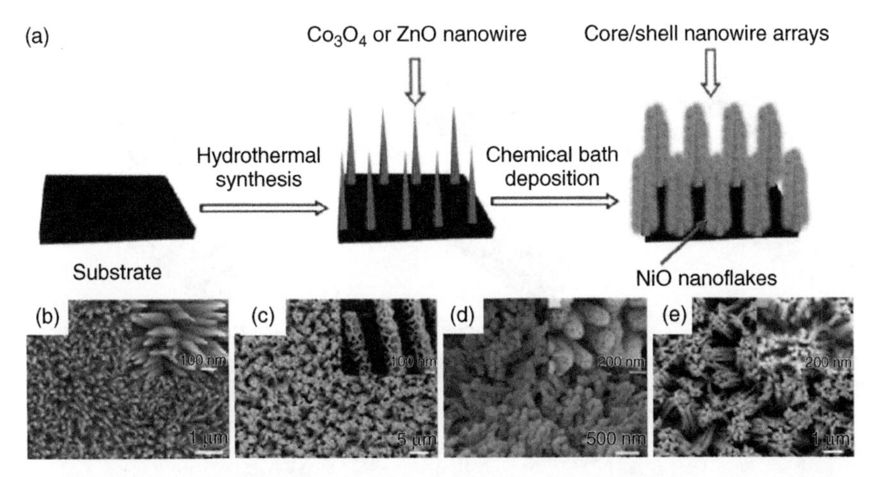

Figure 3.34 Synthesis process and morphology characterization of Co_3O_4/NiO core-shell nanowire arrays. (a) Process flow diagram for synthesis. SEM images of Co_3O_4 NWs on (b) fluorine-doped tin oxide conductive glass, (c) FTO, (d) foamed nickel and (e) nickel foil conductive substrates, with the insets showing their corresponding SEM images under high magnification. Source: Xia et al. [74] / American Chemical Society.

Apart from that, Shen and coworkers successfully fabricated a 3D layered structure of $ZnCo_2O_4$ NW arrays grown on carbon cloth by hydrothermal route and thermal annealing treatment in air [75]. The diameter and length of typical $ZnCo_2O_4$ NWs are in the range of 80–100 nm and 5 μm, respectively. The final sample enhances the electrochemical performance as well as the flexibility of the electrode material [76]. Another Cu_2S NW array on the copper substrate was prepared via the wet chemical route. Cu_2S NWs have uniform diameters and lengths of about 40–120 nm and 700–1400 nm, respectively. And in this way, the regular copper sulfide NW arrays provide compact and uniform coverage of the entire area of the copper substrate.

3.6.3 Chemical Vapor Deposition

CVD on substrates has become a transitional approach, which is inexpensive and suitable for manufacturing [77]. Mai et al. used a simple and unique low-pressure CVD method to synthesize carbon-confined NW arrays. Firstly, 2 methylimidazoles (2 Methylimidazol, 2 MIM) vapor and CoO-NiO NW array reacted at 120 °C and low pressure (about 100 Pa) to obtain a zeolite imidazole ester framework (ZIF) shell-constrained CoO-NiO (CoO-NiO@ZIF) NW array in situ on the surface of the carbon cloth. Then the metal–organic framework was converted into nitrogen-doped carbon at 500 °C and the volume fraction ratio of Ar to H_2 was 95 : 5. Finally, the CoNi@NC mesoporous NW array was obtained on the carbon cloth. The synthesis process and corresponding morphological characterization are shown in Figure 3.35.

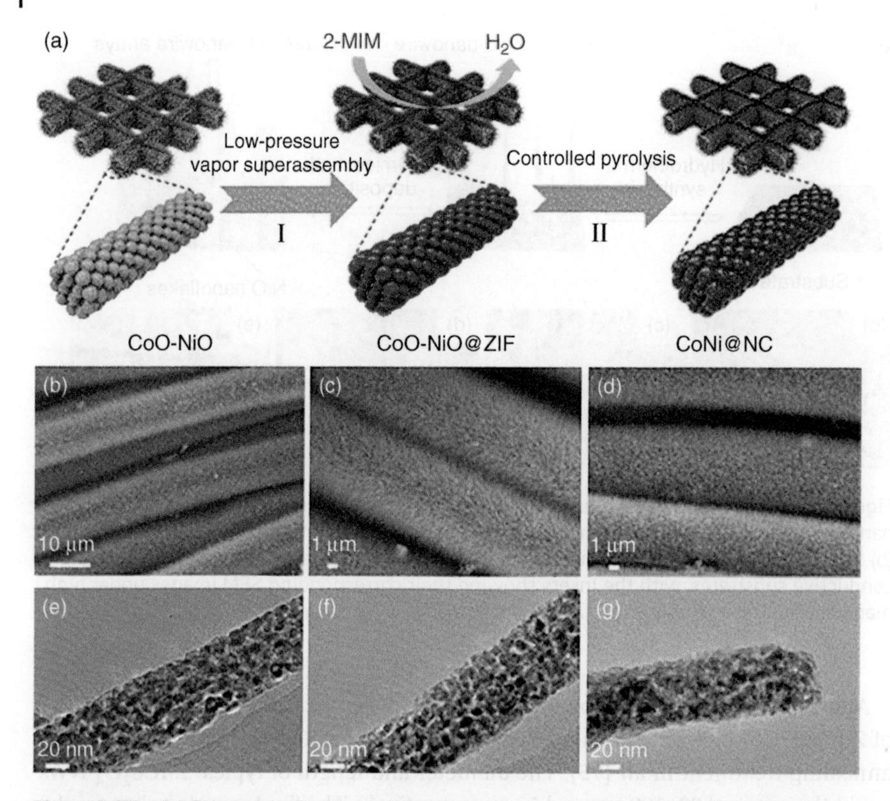

Figure 3.35 (a) Schematic diagram of the synthesis process of CoNi@NC mesoporous nanowire arrays. SEM images of (b) CoO-NiO, (c) CoO-NiO@ZIF and (d) CoNi@NC nanowire arrays. TEM images of (e) CoO-NiO, (f) CoO-NiO@ZIF and (g) CoNi@NC nanowire arrays. Source: Meung et al. [77] / American Chemical Society.

Yang and coworkers used a CVD method with precisely fine-tuned growth parameters to prepare indium phosphide (InP) NW arrays on Si substrates by using trimethylindium (TMIn) and tributyl phosphate (TBP) as raw materials (Figure 3.36a) [78]. Furthermore, the influence of growth time and TBP/TMIn ratio on the morphology of NWs was also explored. The time-dependent growth analysis based on constant TMIN and TBP flow was shown in Figure 3.36b,c. As time increased, it grew linearly and continuously in the axial and radial directions, and finally, multifaceted NWs with larger size generated from the initial needle-like structure. When constant TBP flow rate and a changing TMIN flow rate were applied, the base diameter of the InP NW remained unchanged (Figure 3.36d). While the axial growth rate increased with the increase of the TMIN flow rate and tended to be stable at the flow rate of 15 sccm (Figure 3.36e), this indicated that the TMIN component was the key factor in axial growth. When fixed TMIN flow rate, the growth of the radial sidewall started from the bottom of NW and was exceeded by the axial growth as the TBP flow rate decreased, and finally an asymmetric structure with a wide bottom and fine needle-like tip was obtained (Figure 3.36f).

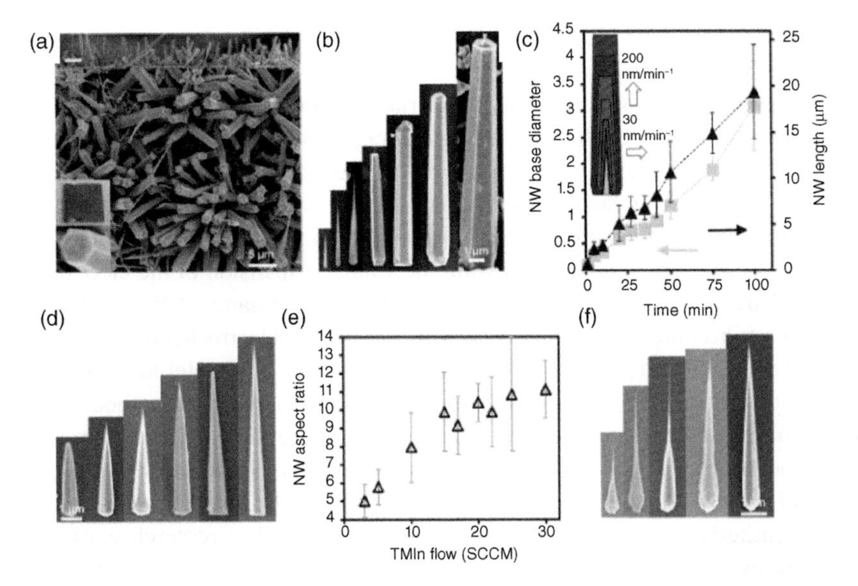

Figure 3.36 Study on the growth state of InP nanowire arrays. (a) The SEM image (top-down) and cross-sectional SEM image (top) of the InP nanowire array. The embedded image is the SEM image of the tip of the nanowire and the corresponding Si chip; (b) and (c) are the SEM diagrams of the nanowires obtained linearly increasing with the growth time and the corresponding diagrams of the relationship between length and diameter, and time; (d) and (e) are the InP nanowires obtained under different flow rates, SEM image, and corresponding aspect ratio statistical graph; (f) InP nanowire SEM graph of TBP flow value change. Source: Kornienko et al. [78] / American Chemical Society.

Here, they classified this growth mechanism as a dual growth mechanism in which coexisting radial epitaxial sidewall development and autocatalytic axial growth were both present.

Lu and coworkers fabricated UV photodetectors (UV PDs) on entangled ZnO NW array grown on SiO_2 substrate by catalyst-free CVD directly [79]. Mechanical and chemical surface treatment of the dielectric is the most critical step in catalyst-free CVD process. In this way, the ZnO NW array was not only impurity-free but also intertwined with each other, which provided an uninterrupted pathway for charge transfer and improved the response efficiency of ZnO NW-based UV PDs.

MOFs and templates with high surface areas were used to prepare orderly layered nanoporous carbon electrodes [80]. The NWs were synthesized by simple hydrothermal method and grown directly on each fiber of the cloth. Then the sample was exposed to 2-MeIm vapor to generate ZnO@ZIF-8 and $Co(CO_3)_{0.5}(OH){\cdot}0.11H_2O$@ZIF-67 NWs. Finally, the samples were carbonized at high temperatures to generate hybrid Co_3O_4/NC NW arrays. By coating raw materials with conformal films of MOF rather than using organic solvents, deposition of continuous nanoporous carbon films could entirely cover the surface of the Co_3O_4 NWs, ensuring less transmission resistance without the use of any polymer binder.

3.7 Summary and Outlook

In recent years, research on various nanomaterials has progressed toward obtaining higher performance electrochemical energy storage devices. Wherein, compared with other nanomaterials, nanowires have unique advantages due to their anisotropy, large specific surface area, excellent tension adaptability, and rapid axial/radial transport of ion and electron, which play an important role in shortening the ion transport path and suppressing the volume expansion of the electrode materials. The research and applications of high-performance NW electrode materials have become one of the hotspots in the field of electrochemical energy storage devices. With the endless preparation methods of NW materials, such as dry/wet chemical method, template method, secondary modification method, NW materials have also developed from the initial simple conventional NWs to porous, hollow, and heterogeneous NWs. This chapter summarizes the latest research progress of nanowire materials, focusing on the structural differences caused by different synthesis methods. Besides, we selectively list typical research work to briefly describe the preparation method, morphology, and corresponding characterization of NWs. Table 3.1 describes the common synthesis methods and growth mechanisms of different types of NWs in detail (conventional, porous, hierarchical, heterogeneous, hollow, and array), as well as their advantages, disadvantages, and influencing factors, respectively.

It can be seen from Table 3.1 that each synthesis method has its own unique advantages and limitations. In fact, in order to prepare ideal nanowire products for practical applications, a combination of multiple methods is usually used to synthesize or modify nanowires with special structures. For example, it is common to calcinate samples after hydrothermal reactions. Researchers should combine the knowledge of chemistry, physics, and materials science to conduct in-depth research on microscopic processes and internal synthesis mechanisms to achieve precise designs. At the same time, in order to make a breakthrough in the synthesis method, it is recommended that the researchers effectively combine the new ideas with the traditional synthesis method, so that the synthesis of nanowires has higher feasibility, an appropriate cost, and good performance.

This chapter provides good theoretical support and guidance for selectively controlling the structure and growth of nanowires to obtain the required performance. However, there is still a long way to go in designing and synthesizing different types of nanowires, as well as in the development of related theories and new preparation technologies to achieve large-scale industrial applications.

Table 3.1 Comparison of advantages, disadvantages, and influencing factors of typical synthesis methods for different types of nanowires.

Synthesis methods	Related reaction	Advantage	Disadvantage	Influencing factors	Types of nanowires
Hydrothermal/ Solvothermal method	Dissolution and recrystallization	Simple equipment, convenient operation, and good crystallization	Steps such as centrifugation and sample washing are required	Temperature, pressure, solvent, and whether the solvent is completely removed	Conventional nanowires, hollow nanowires, and nanowire arrays
Sol–gel method	Hydrolysis and polymerization reactions.	Easy to control the nanostructure	Complex reaction steps, and long preparation cycle	Reaction time, pore size and its distribution	Conventional nanowires
Coprecipitation method	Precipitation reaction	Low cost, simple process, short synthesis cycle, and easy control of preparation conditions	Relatively complicated process; easily affected by external conditions	pH value, temperature, the solubility of the reactants	Conventional nanowires
Ultrasonic spray pyrolysis method	Solvent evaporation, particle thermal decomposition, calcination reaction	Relatively uniform distribution, simple, and low cost	The physical and chemical properties of the reactants are required	Solution concentration, pH value, pyrolysis temperature and time	Conventional nanowires and nanowire arrays
Electrospinning method	Electrostatic stretching, solvent volatilization	Simple device, low cost, diversified raw material selection, and controllable process	Not suitable for preparing small-sized nanowires	Solution uniformity and voltage	Conventional nanowires
High-temperature solid-state method	Solid-state reaction	Low cost, high yield, simple preparation technology, no agglomeration, and good filling property	Difficult to obtain uniform nanowires	Calcination system and feeding ratio	Conventional nanowires
Chemical vapor deposition	Chemically induced nucleation-growth, surface reactions	Low cost and large scale	Strict requirements for working temperature and equipment	Deposition temperature, deposition rate, pressure, gas composition and flow rate	Conventional nanowires, and nanowire arrays
Physical-vapor deposition method	Evaporation, and magnetron sputtering	Cleaner reproducible and efficient method	Chemical impurities are difficult to remove, requiring high target purity	Vacuum conditions, evaporation current, deposition conditions, process temperature, and the degree of vacuum	Conventional nanowires
Template method	Nanoconfinement or porous nanowires	Easy to control shapes, low cost, and easy to obtain	Certain restrictions on the use of templates	Temperature, heating rate, time	Porous nanowires, hollow nanowires, and nanowire arrays
Self-assembly method	Self-assemble	Easy to implement, a wide range of universality	Strict requirements for experimental conditions, material ratio	Material ratio, solvent type, and the structure of components	Porous nanowires, hierarchical nanowires, porous nanowires

References

1 Yu, X., Wang, H., Liu, Y. et al. (2013). One-step ammonia hydrothermal synthesis of single crystal anatase TiO_2 nanowires for highly efficient dye-sensitized solar cells. *Journal of Materials Chemistry A* 1 (6): 2110–2117.

2 Li, W., Gao, X., Xiong, D. et al. (2017). Hydrothermal synthesis of monolithic Co_3Se_4 nanowire electrodes for oxygen evolution and overall water splitting with high efficiency and extraordinary catalytic stability. *Advanced Energy Materials* 7 (17): 1602579.

3 Hu, P., Zhu, T., Wang, X. et al. (2018). Highly durable $Na_2V_6O_{16}\cdot 1.63H_2O$ nanowire cathode for aqueous zinc-ion battery. *Nano Letters* 18 (3): 1758–1763.

4 Wang, X., Wang, X., Huang, W. et al. (2005). Sol–gel template synthesis of highly ordered MnO_2 nanowire arrays. *Journal of Power Sources* 140 (1): 211–215.

5 Cao, H., Xu, Y., Hong, J. et al. (2001). Sol ± gel template synthesis of an array of single crystal CdS nanowires on a porous alumina template. *Advanced Materials* 18: 1393–1394.

6 Jo, Y.R., Myeong, S.H., and Kim, B.J. (2018). Role of annealing temperature on the sol-gel synthesis of VO_2 nanowires with in situ characterization of their metal-insulator transition. *RSC Advances* 8 (10): 5158–5165.

7 Yu, D., Wu, B., Ge, L. et al. (2016). Decorating nanoporous ZIF-67-derived $NiCo_2O_4$ shells on a Co_3O_4 nanowire array core for battery-type electrodes with enhanced energy storage performance. *Journal of Materials Chemistry A* 4 (28): 10878–10884.

8 Vivekchand, S.R.C., Gundiah, G., Govindaraj, A. et al. (2004). A new method for the preparation of metal nanowires by the nebulized spray pyrolysis of precursors. *Advanced Materials* 16 (20): 1842–1845.

9 Greiner, A. and Wendorff, J.H. (2007). Electrospinning: a fascinating method for the preparation of ultrathin fibers. *Angewandte Chemie International Edition in English* 46 (30): 5670–5703.

10 Huang, Z.-M., Zhang, Y.Z., Kotaki, M. et al. (2003). A review on polymer nanofibers by electrospinning and their applications in nanocomposites. *Composites Science and Technology* 63 (15): 2223–2253.

11 Hsu, P.C., Kong, D., Wang, S. et al. (2014). Electrolessly deposited electrospun metal nanowire transparent electrodes. *Journal of the American Chemical Society* 136 (30): 10593–10596.

12 Ren, W., Zheng, Z., Luo, Y. et al. (2015). An electrospun hierarchical LiV_3O_8 nanowire-in-network for high-rate and long-life lithium batteries. *Journal of Materials Chemistry A* 3 (39): 19850–19856.

13 Xiao, X., Yang, L., Zhao, H. et al. (2011). Facile synthesis of $LiCoO_2$ nanowires with high electrochemical performance. *Nano Research* 5 (1): 27–32.

14 Zhang, J., Li, W., Jia, Q. et al. (2015). Molten salt assisted synthesis of 3C–SiC nanowire and its photoluminescence properties. *Ceramics International* 41 (10): 12614–12620.

15 Fu, Q.-G., Li, H.-J., Shi, X.-H. et al. (2006). Synthesis of silicon carbide nanowires by CVD without using a metallic catalyst. *Materials Chemistry and Physics* 100 (1): 108–111.

16 Suzuki, H., Araki, H., Tosa, M. et al. (2007). Formation of silicon nanowires by CVD using gold catalysts at low temperatures. *Materials Transactions* 48 (8): 2202–2206.

17 Iriarte, G.F. (2010). Growth of nickel disilicide nanowires by CVD. *Journal of Non-Crystalline Solids* 356 (23–24): 1135–1144.

18 Shen, L., Cui, Q., and Cheng, T. (2008). Preparation of monocrystal AlN nanowires by direct nitridation method. *Electronic Components and Materials* 27 (10): 78–80.

19 Xue, C., Li, H., Zhuang, H. et al. (2009). Synthesis of GaN nanowires with tantalum catalyst by magnetron sputtering. *Rare Metal Materials and Engineering* 38 (7): 1129–1131.

20 Teng, D., Wu, L., He, W. et al. (2014). High-density silicon nanowires prepared via a two-step template method. *Langmuir* 30 (8): 2259–2265.

21 Ma, M., Pan, Z., Guo, L. et al. (2012). Porous cobalt oxide nanowires: notable improved gas sensing performances. *Chinese Science Bulletin* 57 (31): 4019–4023.

22 Peraca, N.M., Haney, P.M., and Hamadani, B.H. (2019). The effect of luminescent coupling on modulated photocurrent measurements in multijunction solar cells. *Applied Physics Letters* 115 (8): 083506.

23 Bechelany, M., Abou Chaaya, A., Frances, F. et al. (2013). Nanowires with controlled porosity for hydrogen production. *Journal of Materials Chemistry A* 1 (6): 2133–2138.

24 Ran Liu, S.B.L. (2008). MnO_2/poly(3,4-ethylenedioxythiophene) coaxial nanowires by one-step coelectrodeposition for electrochemical energy storage. *Journal of the American Chemical Society* 130 (10): 2942–2943.

25 Li, F., He, J., Zhou, W.L., and Wiley, J.B. (2003). Synthesis of porous wires from directed assemblies of nanospheres. *Journal of the American Chemical Society* 125 (52): 16166–16167.

26 Zhu, T., Wu, H.B., Wang, Y. et al. (2012). Formation of 1D hierarchical structures composed of Ni_3S_2 nanosheets on CNTs backbone for supercapacitors and photocatalytic H_2 production. *Advanced Energy Materials* 2 (12): 1497–1502.

27 Zhou, X., Yu, L., and Lou, X.W. (2016). Nanowire-templated formation of SnO_2/carbon nanotubes with enhanced lithium storage properties. *Nanoscale* 8 (15): 8384–8389.

28 Yuan, C., Zhang, X., Hou, L. et al. (2010). Lysine-assisted hydrothermal synthesis of urchin-like ordered arrays of mesoporous $Co(OH)_2$ nanowires and their application in electrochemical capacitors. *Journal of Materials Chemistry* 20 (48): 10809–10816.

29 Liu, J., Jiang, J., Cheng, C. et al. (2011). Co_3O_4 nanowire@MnO_2 ultrathin nanosheet core/shell arrays: a new class of high-performance pseudocapacitive materials. *Advanced Materials* 23 (18): 2076–2081.

30 Yu, L., Zhang, G., Yuan, C. et al. (2013). Hierarchical $NiCo_2O_4@MnO_2$ core-shell heterostructured nanowire arrays on Ni foam as high-performance supercapacitor electrodes. *Chemical Communications (Cambridge)* 49 (2): 137–139.

31 Vinzons, L.U., Shu, L., Yip, S. et al. (2017). Unraveling the morphological evolution and etching kinetics of porous silicon nanowires during metal-assisted chemical etching. *Nanoscale Research Letters* 12 (1): 385.

32 Hao Tang, L.-G.Z., Zhao, L., Zhang, X. et al. (2012). Carrier dynamics in Si nanowires fabricated by metal-assisted chemical etching. *ACS Nano* 6 (9): 7814–7819.

33 Sandu, G., Avila Osses, J., Luciano, M. et al. (2019). Kinked silicon nanowires: superstructures by metal-assisted chemical etching. *Nano Letters* 19 (11): 7681–7690.

34 Chen, Y., Liu, L., Xiong, J. et al. (2015). Porous Si nanowires from cheap metallurgical silicon stabilized by a surface oxide layer for lithium ion batteries. *Advanced Functional Materials* 25 (43): 6701–6709.

35 Bang, B.M., Kim, H., Song, H.-K. et al. (2011). Scalable approach to multi-dimensional bulk Si anodes via metal-assisted chemical etching. *Energy & Environmental Science* 4 (12): 5013–5019.

36 Veerbeek, J., Ye, L., Vijselaar, W. et al. (2017). Highly doped silicon nanowires by monolayer doping. *Nanoscale* 9 (8): 2836–2844.

37 Zhang, T., Wu, S., Xu, J. et al. (2015). High thermoelectric figure-of-merits from large-area porous silicon nanowire arrays. *Nano Energy* 13: 433–441.

38 Liu, J. and Xue, D. (2008). Thermal oxidation strategy towards porous metal oxide hollow architectures. *Advanced Materials* 20 (13): 2622–2627.

39 Guo, J., Wang, X., Geoyhegan, D.B. et al. (2008). Thermal characterization of multi-wall carbon nanotube bundles based on pulsed laser-assisted thermal relaxation. *Functional Materials Letters* 1 (1): 71–76.

40 Molenda, J. and Marzec, J. (2008). Functional cathode materials for Li-ion batteries—part I: fundamentals. *Functional Materials Letters* 1 (2): 91–95.

41 Xu, J. and Xue, D. (2007). Five branching growth patterns in the cubic crystal system: a direct observation of cuprous oxide microcrystals. *Acta Materialia* 55 (7): 2397–2406.

42 Maity, D., Ding, J., and Xue, J. (2008). Synthesis of magnetite nanoparticles by thermal decomposition: time, temperature, surfactant and solvent effects. *Functional Materials Letters* 1 (3): 189–193.

43 Mai, L., Xu, L., Han, C. et al. (2010). Electrospun ultralong hierarchical vanadium oxide nanowires with high performance for lithium ion batteries. *Nano Letters* 10 (11): 4750–4755.

44 Mai, L.Q., Yang, F., Zhao, Y.L. et al. (2011). Hierarchical $MnMoO_4/CoMoO_4$ heterostructured nanowires with enhanced supercapacitor performance. *Nature Communications* 2: 381.

45 Yang, L., Wang, S., Mao, J. et al. (2013). Hierarchical MoS_2/polyaniline nanowires with excellent electrochemical performance for lithium-ion batteries. *Advanced Materials* 25 (8): 1180–1184.

46 Liu, F. and Xue, D. (2010). Fabrication of Nb_2O_5 nanotrees with controlled branching degrees. *Physica Scripta* T139: 014074.

47 Wu, H., Xu, M., Wang, Y. et al. (2013). Branched Co_3O_4/Fe_2O_3 nanowires as high capacity lithium-ion battery anodes. *Nano Research* 6 (3): 167–173.

48 Zhou, W., Cheng, C., Liu, J. et al. (2011). Epitaxial growth of branched α-Fe_2O_3/SnO_2 nano-heterostructures with improved lithium-ion battery performance. *Advanced Functional Materials* 21 (13): 2439–2445.

49 Cheng, C., Liu, B., Yang, H. et al. (2009). Hierarchical assembly of ZnO nanostructures on SnO_2 backbone nanowires: low-temperature hydrothermal preparation and optical properties. *ACS Nano* 3 (10): 3069–3076.

50 Cao, M., Gao, L., Lv, X. et al. (2017). TiO_2-B@VS_2 heterogeneous nanowire arrays as superior anodes for lithium-ion batteries. *Journal of Power Sources* 350: 87–93.

51 Bi, Y., Hu, H., Ouyang, S. et al. (2012). Selective growth of Ag_3PO_4 submicro-cubes on Ag nanowires to fabricate necklace-like heterostructures for photocatalytic applications. *Journal of Materials Chemistry* 22 (30): 14847–14850.

52 Li, Y., Sun, Y., Xu, G. et al. (2014). Tuning electrochemical performance of Si-based anodes for lithium-ion batteries by employing atomic layer deposition alumina coating. *Journal of the American Chemical Society* 2 (29): 11417–11425.

53 Feng, Y., Cho, I.S., Rao, P.M. et al. (2013). Sol-flame synthesis: a general strategy to decorate nanowires with metal oxide/noble metal nanoparticles. *Nano Letters* 13 (3): 855–860.

54 Jia, Y., Yu, X.Y., Luo, T. et al. (2014). Necklace-like mesoporous MgO/TiO_2 heterojunction structures with excellent capability for water treatment. *Dalton Transactions* 43 (6): 2348–2351.

55 Yu, K., Pan, X., Zhang, G. et al. (2018). Nanowires in energy storage devices: structures, synthesis, and applications. *Advanced Energy Materials* 8 (32): 1802369.

56 Lu, Q. and Gao, F. (2016). Synthesis and property studies of hollow nanostructures. *CrystEngComm* 18 (39): 7399–7409.

57 Mahdi, M.A., Hassan, J.J., Ng, S.S. et al. (2012). Growth of CdS nanosheets and nanowires through the solvothermal method. *Journal of Crystal Growth* 359: 43–48.

58 Jiang, Y., Wu, Y., Zhang, S. et al. (2000). A catalytic-assembly solvothermal route to multiwall carbon nanotubes at a moderate temperature. *Journal of the American Chemical Society* 122 (49): 12383–12384.

59 Yan, M., Wang, F., Han, C. et al. (2013). Nanowire templated semihollow bicontinuous graphene scrolls: designed construction, mechanism, and enhanced energy storage performance. *Journal of the American Chemical Society* 135 (48): 18176–18182.

60 Xia, X., Tu, J., Mai, Y. et al. (2011). Self-supported hydrothermal synthesized hollow Co_3O_4 nanowire arrays with high supercapacitor capacitance. *Journal of Materials Chemistry* 21 (25): 9319–9325.

61 Liqiang, M. (2004). Study on preparation, structure, and properties of low-dimensional vanadium oxide nanomaterials. Doctoral thesis. Wuhan University of Technology.

62 Liang, H.-W., Liu, J.-W., Qian, H.-S. et al. (2012). Multiplex templating process in one-dimensional nanoscale: controllable, synthesis, macroscopic assemblies, and applications. *Accounts of Chemical Research* 46 (7): 1450–1461.

63 Huczko, A. (2000). Template-based synthesis of nanomaterials. *Applied Physics A* 70 (4): 365–376.

64 Liang, H.-W., Liu, S., Gong, J.-Y. et al. (2009). Ultrathin Te nanowires: an excellent platform for controlled synthesis of ultrathin platinum and palladium nanowires/nanotubes with very high aspect ratio. *Advanced Materials* 21 (18): 1850–1854.

65 Lee, J.H., Kim, J.Y., Kim, J.H. et al. (2017). Synthesis and gas sensing properties of membrane template-grown hollow ZnO nanowires. *Nano Convergence* 4 (1): 27.

66 Caruso, R.A., Schattka, J.H., and Greiner, A. (2001). Titanium dioxide tubes from sol–gel coating of electrospun polymer fibers. *Advanced Materials* 13 (20): 1577–1579.

67 Zhou, G., Xu, L., Hu, G. et al. (2019). Nanowires for electrochemical energy storage. *Chemical Reviews* 119 (20): 11042–11109.

68 Niu, C., Meng, J., Wang, X. et al. (2015). General synthesis of complex nanotubes by gradient electrospinning and controlled pyrolysis. *Nature Communications* 6 (1): 7402.

69 Liu, Z., Guo, R., Meng, J. et al. (2017). Facile electrospinning formation of carbon-confined metal oxide cube-in-tube nanostructures for stable lithium storage. *Chemical Communications* 53 (59): 8284–8287.

70 Meng, J., Niu, C., Liu, X. et al. (2016). Interface-modulated approach toward multilevel metal oxide nanotubes for lithium-ion batteries and oxygen reduction reaction. *Nano Research* 9 (8): 2445–2457.

71 Liu, H., Hu, L., Meng, Y.S. et al. (2013). Electrodeposited three-dimensional Ni-Si nanocable arrays as high performance anodes for lithium ion batteries. *Nanoscale* 5 (21): 10376–10383.

72 Forrer, P., Schlottig, F., Siegenthaler, H. et al. (2000). Electrochemical preparation and surface properties of gold nanowire arrays formed by the template technique. *Journal of Applied Electrochemistry* 30: 533–541.

73 Wang, W., Tian, M., Abdulagatov, A. et al. (2012). Three-dimensional Ni/TiO_2 nanowire network for high areal capacity lithium ion microbattery applications. *Nano Letters* 12 (2): 655–660.

74 Xia, X., Tu, J., Zhang, Y. et al. (2012). High-quality metal oxide core/shell nanowire arrays on conductive substrates for electrochemical energy storage. *ACS Nano* 6 (6): 5531–5538.

75 Lai, C., Huang, K., Cheng, J. et al. (2010). Direct growth of high-rate capability and high capacity copper sulfide nanowire array cathodes for lithium-ion batteries. *Journal of Materials Chemistry* 20 (32): 6638–6645.

76 Liu, B., Zhang, J., Wang, X. et al. (2012). Hierarchical three-dimensional $ZnCo_2O_4$ nanowire arrays/carbon cloth anodes for a novel class of high-performance flexible lithium-ion batteries. *Nano Letters* 12 (6): 3005–3011.

77 Meng, J., Liu, X., Li, J. et al. (2017). General oriented synthesis of precise carbon-confined nanostructures by low-pressure vapor superassembly and controlled pyrolysis. *Nano Letters* 17 (12): 7773–7781.

78 Kornienko, N., Gibson, N.A., Zhang, H. et al. (2016). Growth and photoelectrochemical energy conversion of wurtzite indium phosphide nanowire arrays. *ACS Nano* 10 (5): 5525–5535.

79 Zhan, Z., Xu, L., An, J. et al. (2017). Direct catalyst-free chemical vapor deposition of ZnO nanowire array UV photodetectors with enhanced photoresponse speed. *Advanced Engineering Materials* 19 (8): 1700101.

80 Young, C., Wang, J., Kim, J. et al. (2018). Controlled chemical vapor deposition for synthesis of nanowire arrays of metal–organic frameworks and their thermal conversion to carbon/metal oxide hybrid materials. *Chemistry of Materials* 30 (10): 3379–3386.

4

Nanowires for In Situ Characterization

4.1 In Situ Electron Microscopy Characterization

4.1.1 In Situ Scanning Electron Microscopy (SEM) Characterization

Scanning electron microscopy (SEM) is a characterization method that involves transferring electron signals from materials into digital images under the excitement of electron beam. SEM images often use secondary electrons for image formation and have the capability of observing the surface morphology of materials.

In research on lithium-ion batteries, since common ex situ SEM tests require complicated sample treatments, they cannot reflect the true state of electrode material. For example, the material should be collected from a full cell, and electrolytes must be removed. The obtained image might lose the information that researchers required after all the treatment; hence, it is necessary to develop in situ SEM observation.

In situ SEM system is constructed by integrating electrochemical testing device into SEM in order to leave out unnecessary troubles brought by sample treatment. By applying in situ SEM to battery research, the understanding of reasons for failure, structural changes, and reaction mechanisms could be pushed to a new level [1–5].

Strelcov et al. applied in situ SEM in studying the V_2O_5 nanowire during cycles [3]. As schematically shown in Figure 4.1a, the experiment system was constructed by V_2O_5 nanobelts (NBs) growing on palladium as the working electrode and $LiCoO_2$ particles fixed on gold tip as anode. Ionic liquid electrolyte bis(trifluoromethane)sulfonimidate of lithium (LiTFSI) in (1-butyl-1-methylpyrrolidinium bis(trifluoro-methylsulfonyl)imide) was used to build a liquid electrolyte environment in vacuum chamber of SEM. The SEM image of integrated battery system is exhibited in Figure 4.1b. With the help of cyclic voltametric tests, morphology of V_2O_5 in different potentials was observed. The relationship between potential and micromorphology of V_2O_5 was deeply explored. The results demonstrated that there was no obvious shape distortion of V_2O_5 nanowire during cycles, indicating that V_2O_5 possessed a good potential in lithium battery.

Not only cathode, but anodes were also studied by in situ SEM. Si anodes, known as high-capacity Li-ion anodes, suffer from bulky volume expansion throughout

Nanowire Energy Storage Devices: Synthesis, Characterization, and Applications, First Edition.
Edited by Liqiang Mai.

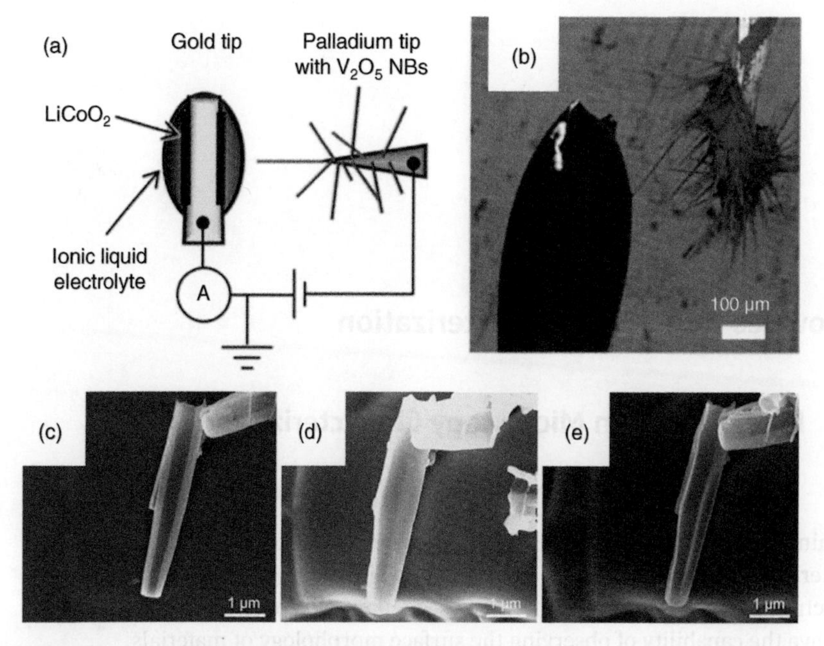

Figure 4.1 (a) Schematic of the experimental setup; (b) backscattered electron detector SEM micrograph of the assembled battery with one V_2O_5 whisker submerged in electrolyte. Source: Strelcov et al. [3]/with permission of Royal Society of Chemistry. In situ observations: SEM images of a C@Si@C nanotube (c) as fabricated, (d) after discharge (lithiation), and (e) after charge (delithiation). Source: Liu et al. [6]/with permission of the American Chemical Society.

lithiation and delithiation (~400%), which causes harmful internal stresses on the electrode. Since in situ SEM could real-time monitor the morphology changes during cycling, it is necessary to utilize in situ SEM to explore effective solutions addressing the serious problem of Si anode. Liu et al. developed a nanotube sandwich structure aimed at easing the mechanical issue [6]. First, ZnO nanorod was deposited on nickel foam by hydrothermal method. Then carbon was coated on ZnO nanorod by polymerization and carbonization of furfuryl alcohol. After chemical vapor deposition (CVD) of Si, the polymerization and carbonization processes were performed again to deposit another carbon layer. In the end, ZnO was etched by acetic acid, forming a C@Si@C sandwich structure. In situ SEM images of C@Si@C nanotube at different states clearly revealed changes of a single nanotube. The thickness and length of as-fabricated C@Si@C nanotube both endure enlargement after Li-ions insert into Si crystal (Figure 4.1c,d). However, the shape nearly reversed to its original state after delithiation (Figure 4.1e), illustrating that the method successfully enhanced reversibility of Si crystal structure.

Shi et al. performed in situ SEM characterization of Si/C anode by placing the battery parallel with electron beam [7]. In this case, captured picture includes information on every layer of battery, including current collector, cathode, separator, and anode. Through a rational design of Si/C anode three-dimensional (3D)-lined-structure, the influence of volume expansion was reduced to a minimum. In situ SEM played a great role in revealing the cycling process of Si/C. SEM images together with videos of unstructured anode and 3D-lined-structure anode showed that 3D-lined-structure with gaps between Si bars could keep good contact between anode and current collector while the volume expansion is induced by lithiation. The unstructured anode not only suffers from interior stress but is also detached from current collector due to volume expansion.

Overall, in situ SEM can instantaneously visually reveal the morphological deviations of battery content during cycles. However, signal from SEM is not abundant enough to further understand the electrochemical process of battery reaction. It is essential to combine in situ SEM with other characterization methods.

4.1.2 In Situ Transmission Electron Microscope (TEM) Characterization

Similar to SEM, TEM also uses an electron beam as excitement source to obtain various signals from materials. TEM has higher accelerating voltage to ensure that electrons have more power to transmit material. The electrons that penetrate through materials could reflect some information about internal structure, which makes TEM suitable for some in-depth research on batteries [8–11].

For in situ transmission electron microscopy, if the morphology of the nanowires is to be demonstrated in situ, it is necessary to make appropriate adjustments to the transmission electron microscopy sample table. In situ transmission in the field of electrochemical energy storage requires secondary battery characterization, which is mainly divided into five types [12]. As shown in Figure 4.2, Figure 4.2a is a solid in situ open cell sample bar design to investigate the anode structure failure (volume change, etc.), which enables the high spatial resolution image. However, its form of electrochemical contact is different from the real environment, which is full of liquid electrolyte. Figure 4.2b depicts a sealed liquid in situ battery sample bar, enabling research on the evolution of SEI and Li dentrite at the interface between electrolyte and electrode. Figure 4.2c describes the in situ heating sample bar, which allows the stability of metal oxide cathode. Figure 4.2d is an in situ open cell based on ionic liquid. Figure 4.2e represents a nanothin film that researches solid electrolyte, in which low ionic conductivity and unstable interface are not major problems to be solved.

In a common case, the traditional nanowire battery in situ testing platform is constructed with single nanowire as cathode, Li metal or $LiCoO_2$ as lithium source, and ionic liquid/Li_2O as electrode. Ionic liquid has flaws in that some reactions of real electrolytes might not happen in it. At the same time, Li_2O solid electrolyte has a great overpotential due to its large resistance. In order to solve these problems, Gu et al. developed a testing system with the usage of 1.0 M $LiClO_4$ in EC:DMC (3 : 7, v/v) [13]. The schematic and microscopic views of the system are shown in

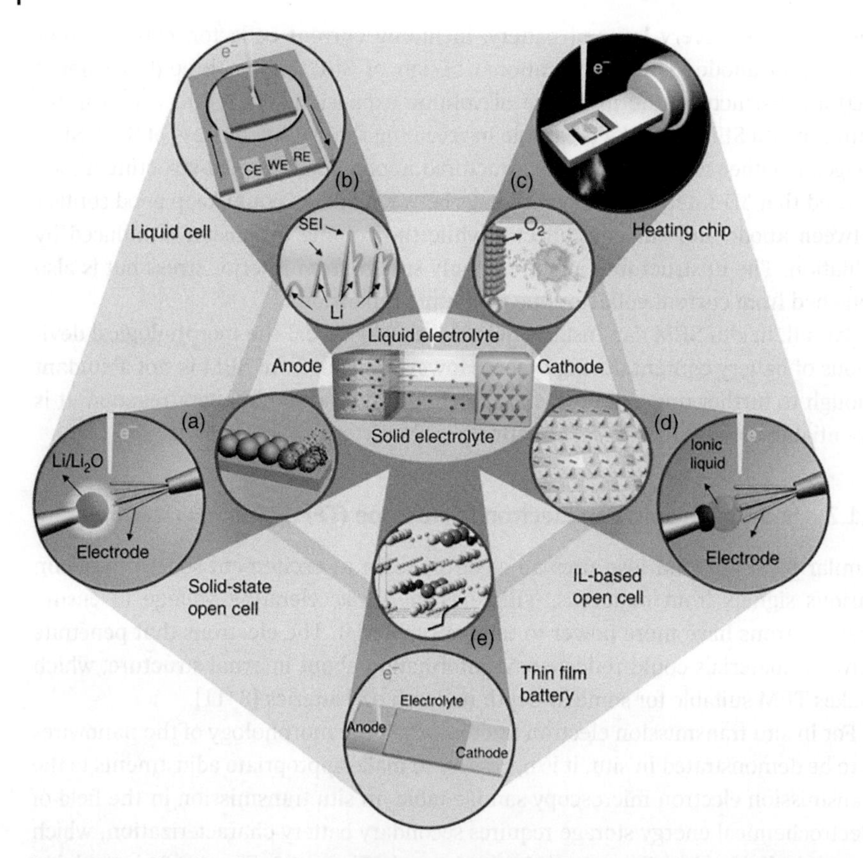

Figure 4.2 Classification of in situ TEM for secondary batteries. (a) solid-state open cell, (b) liquid-cell, (c) in situ heating stage, (d) liquid-based open cell, and (e) nanoscale thin-film battery. Source: Yuan et al. [12], Reproduced with permission from Springer Nature.

Figure 4.3. It indicated that Si NWs can be fully immersed by electrolyte in the testing device, which prevents an uneven lithiation degree caused by longitudinal Li-ion diffusion (Figure 4.3b). The "core-shell" lithiation behavior helped monitor the interface between Li and the unlithiated part. Results illustrated that the total diameter of the wire increased from 100 to 298 nm at 1658 seconds and to 391 nm at 2462 seconds. However, liquid electrolyte makes the difference between SEI layers and Li_xSi hard to be discriminated, indicating room for improvement for design of the whole system.

Luo et al. illustrated the relationship between morphology of Si NW and speed of lithiation by analyzing in situ TEM results [14]. TEM images of lithiated and unlithiated parts of Si NWs have a distinct contrast difference that guides researchers monitoring the process of lithiation. During the lithiation process of alucone-coated Si NW, the further from contact point in axial direction, the less lithiation depth of Si NW. The phenomenon was called "V-shaped" lithiation front. On the contrary, Si NW with ordinary Al_2O_3 coating was "H-shaped" due to the even lithiation in longitude direction. The speed of lithiation was measured by measuring the distance

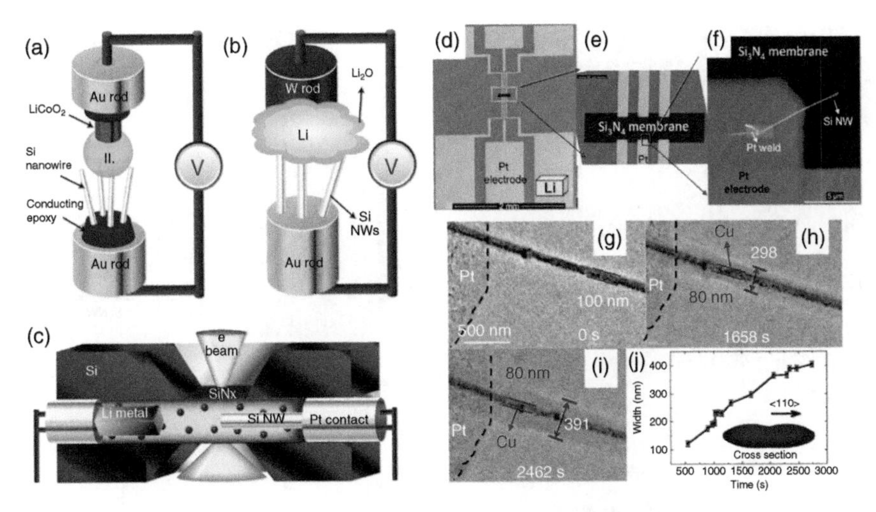

Figure 4.3 (a) Schematic drawing showing the experimental setup of the open-cell approach using ionic liquid as electrolyte; (b) schematic drawing showing the setup of the open-cell approach using Li metal as lithium source and Li_2O as solid electrolyte; (c) schematic drawing showing the setup of the liquid cell battery. (d) SEM image of the inner side of the biasing chip; (e) magnified view of the region labeled by the orange rectangle; (f) SEM image showing the welded Si NW electrode onto the Pt contact. Note that the Li location is labeled by the light blue object in panel (a). In situ liquid-cell TEM observation of the lithiation of the Cu-coated Si (Cu–Si) NW. (g) TEM image showing the pristine state of the Cu–Si NW at 0 seconds; (h) core–shell formation of the Cu–Si NW during lithiation at 1658 seconds; (i) TEM image of the Cu–Si NW at 2462 seconds; (j) plotted width changes of the NW as a function of time. Note that, in all images from (a) to (c), the Pt contact region is labeled by the black lines on the left of the image. The inset in panel (c) illustrates the cross-sectional image after anisotropic swelling of the Si nanowire upon lithium insertion with maximum volume expansion along the ⟨110⟩ direction. Source: Gu et al. [13]/with the permission of American Chemical Society.

that the boundary between Si and Li_xSi moved from its initial state (Figure 4.4). It turned out that Si NWs with alucone possess much faster lithiation kinetics compared to Si NWs without alucone coating, demonstrating successful regulation of Si NW anode. By in situ TEM, the mechanism of lithiation and relationship between mechanical stress and lithiation were explained and comprehended, leading to a new and effective direction of Si NWs modification.

Apart from widely reported Si NWs with poor conductivity, metal Ge attract attention from researchers who aimed at exploring NW for anode of Lithium-ion batteries. Liu et al. studied the Ge NW anode with in situ TEM method [15]. The in situ TEM images clearly showed the morphology changes of different states of lithiation and delithiation process (Figure 4.5a). At initial state, the Ge NW was straight. As time passed, Ge NW became bent due to the elongation of its total length. The increased thickness of Ge NW was also observed at the same time. It can be concluded that the lithiation of Ge NWs was more isotropic compared to the reported Si NWs. In addition, nanopores would form during delithiation rather than lithiation, which helped researchers track the progress of reaction. Figure 4.5b displays the

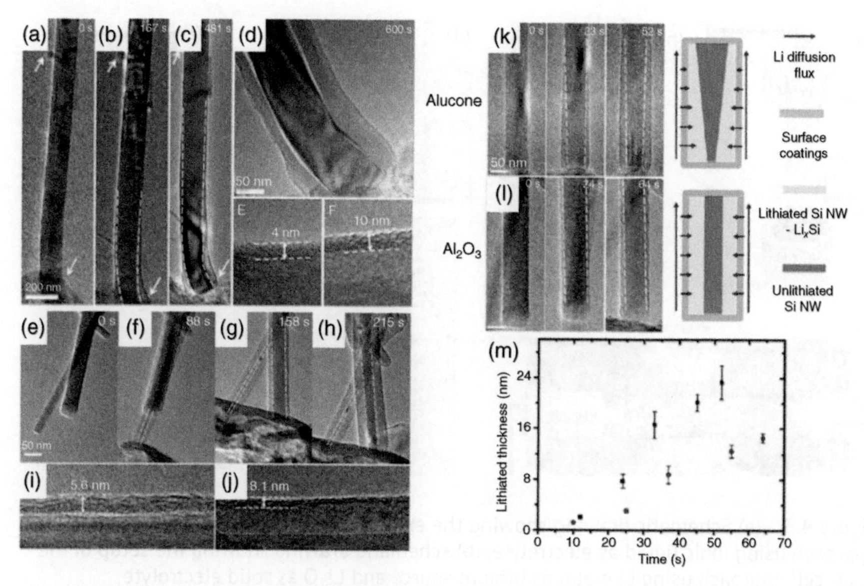

Figure 4.4 (a–c) Time-resolved TEM images of an alucone-coated Si NW under lithiation. White dashed lines indicate the interface that separates lithiated and unlithiated regions. (d) Magnified view of the close end of Si NW after 600 seconds lithiation; (e–h) Time-resolved bright-field TEM images of an Al_2O_3-coated Si NW under lithiation, with white dash lines indicating the interface between lithiated and unlithiated regions; TEM images of the alucone coatings before (i) and after (j) lithiation. (k, l) Time-resolved TEM images show the development of lithiation profiles of the alucone and Al_2O_3 coated Si NWs and schematics of the Li diffusion paths through the Si NWs that dictate the lithiation behavior; (m) average lithiation thickness versus time for the alucone (black square) and Al_2O_3 (red dot) coated Si NWs. Source: Luo et al. [14]/with the permission of American Chemical Society.

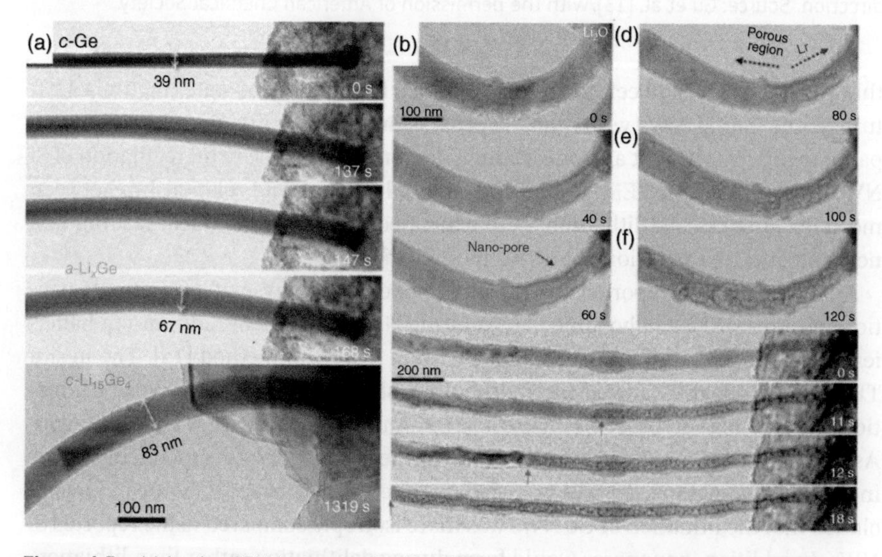

Figure 4.5 (a) In situ TEM images reflecting different states of the Ge NW lithiation process. (b) In situ TEM images of Ge NW delithiation process that analyzed by nanopores. The blue arrows indicate the "delithiation fronts." Source: Liu et al. [15]/with the permission of the American Chemical Society.

gradual morphological transformation of the delithiation process. Nanopores began to form at right site of nanowire, and then formed a distinct "lithiation front," indicating the trend of lithiation. The lithiation front moved leftward, and nanopores filled all the area of Ge NW, indicating an axial directional delithiation. *Pore memory effect* was also founded with in situ TEM. After repeating the cycle of lithiation and delithiation, the nanopores on Ge NW always formed at the same spots. These findings, gained with the assistance of in situ TEM, might give a new understanding of damage mechanism of batteries consisting of Ge anode.

In summary, TEM has the advantage of revealing differences between various materials, such as initial electrode, SEI, and products of lithiation or delithiation, when compared to SEM. However, not every content could be discriminated by difference in contrast. In other words, sometimes shape changes are not always reliable to estimate the reaction of nanowires. Up until now, in situ TEM is mostly applied to simple nanowires like Si NWs and Ge NWs, which involve simple reactions and distinct volume expansions that simplify the analysis of reaction processes. To further push forward the understanding of reaction mechanism, it is crucial to utilize in situ TEM to investigate more kinds of material. In addition, it is difficult to keep the original chemical reaction state in regular TEM devices due to the fact that elements in battery might be sensitive to electron beam. Fortunately, developing cryo-electron microscope technology can greatly enhance research about chemical reactions inside electrode materials. It can be attributed to the property of cryo-electron microscope that it can preserve intrinsic structure of materials and more factually reflect the structural evolution of materials.

4.2 In Situ Spectroscopy Characterization

4.2.1 In Situ X-ray Diffraction

X-ray diffraction is one of the most commonly used characterization methods to analyze the phase or multiple phases of crystal materials. In the field of lithium-ion battery, it can be easily predicted that the time spent on sample treatment might severely affect the state we observed. Through in situ XRD, the phase of battery material at every potential during cycling can be detected. Compared to ex situ XRD, in situ XRD can assist researchers in grasping the process of electrochemical reaction without influence of complicated sample treatment [16–18].

Constructing in situ XRD testing system is a critical matter that affects the reliability of experiment results. The first step is assembling in situ cells. In situ cell models have been proposed and developed in a short period of time. Their construction principle is similar to that of ordinary button batteries, including battery shell, cathode material, anode material, battery diaphragm, electrolyte, and current collector. The biggest difference between normal cell and in situ cell lies in the fact that in situ cell has an X-ray transmitting window and a cavity part with a certain height. In 2001, McBreen et al. in Brookhaven National Laboratory designed a in situ cell accompanying with Mylar film as a window material and two piece of open aluminum sheet

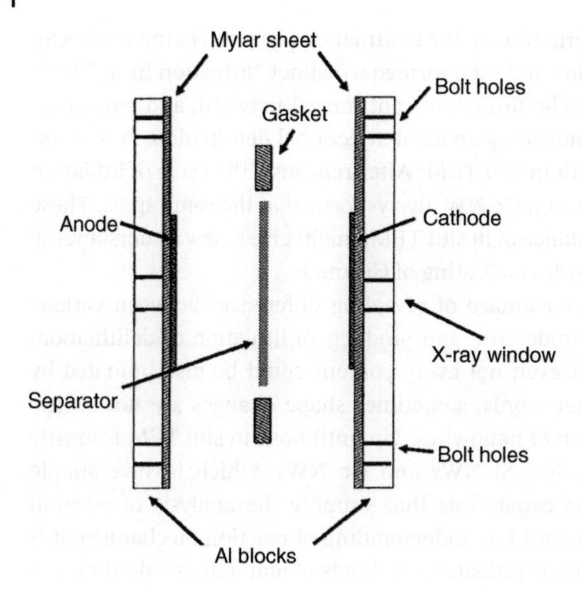

Figure 4.6 The in situ electrochemical cell for XRD. Source: McBreen et al. [19], Reproduced with permission from Elsevier.

as container of battery (Figure 4.6) [19]. Based on this work, Mai's group optimized this in situ model by replacing Mylar film with Be sheet, resulting in integration of three functions in one part – sealing battery, allowing transmission of X-rays, and serving as a current collector.

The second step is collecting data synchronously. After a short standing time, the assembled in situ cell is placed on the sample station with the Be sheet on top. At the same time, the incident spot of X-ray is adjusted to the center of the window. Then the cell connects with electrochemical testing device to conduct corresponding testing program, finally achieving real-time capture of phase-changing and testing data.

Misra et al. studied Si NW by in situ XRD and obtained interesting results that gave new recognition to the Si anode [20]. Through observation of the phase transformation process at different rates, two new Li–Au phases were discovered, and the temperature for CVD and rate for cycling are the key factors controlling the existence of these two phases. In addition, the reason for formation of $Li_{15}Si_4$ phase was discussed. During a complete cycle at $1C$, peaks from $Li_{15}Si_4$ did not show up (Figure 4.7a), while they appeared after enduring a $C/2$ discharging process. Figure 4.7b) demonstrates the dependency between new phases and rate. In addition, a novel new strategy to improve cycling ability was proposed. After 37 cycles, the cells with Si NWs grown at 500 °C retained 33% more than those batteries with Si NWs fabricated at 485 °C (Figure 4.7c). It could be attributed to absence of crystalline phases, resulting in the avoidance of repeated amorphous-crystalline transformation, which is considered harmful to the stability of anode [21]. Inspired by this phenomenon, lowering the cutting-off capacity can significantly prolong the cycling life (Figure 4.7d). Similar in situ XRD analysis was also used in research on Ge NW [22], which is considered a candidate for replacing Si NWs. After rigorous discussion, the transition reaction process could be described as

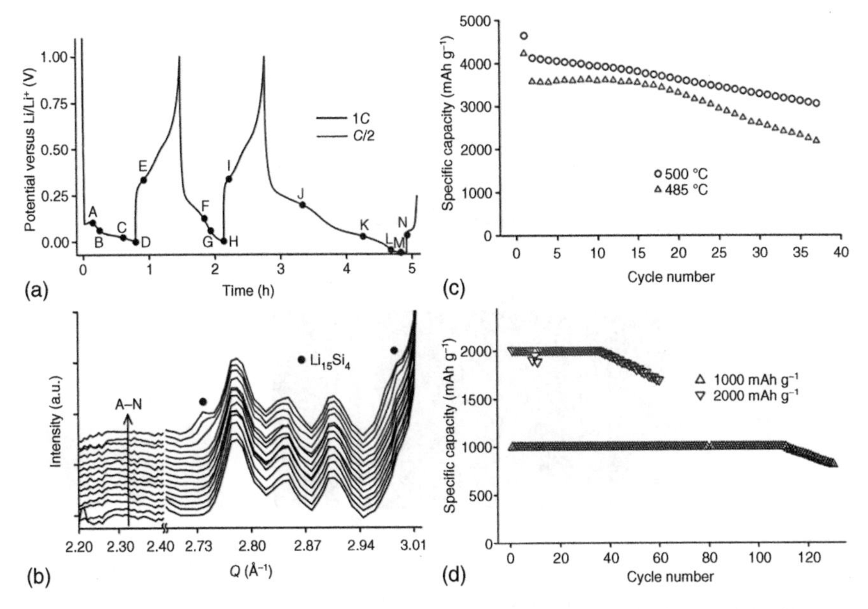

Figure 4.7 In situ XRD results for a stainless sheet-mesh cell under electrochemical cycling: (a) voltage profile for a cell cycled at 1*C* (black) and *C*/2 (blue); (b) XRD patterns showing the appearance of $Li_{15}Si_4$ peaks for corresponding points A–N in (a). Unlabeled peaks are associated with either the polyester pouch or polymer separator. (c) Cycling performance of Si NW samples grown by CVD at 500 and 485 °C. Only lithiation capacities are shown. The Si NWs grown at 500 °C show 33% better capacity retention after 37 cycles compared to Si NWs grown at 485 °C. (d) Constant capacity cycling performance of Si NW anodes cycled at constant capacities of 1000 mAh g^{-1} (blue) and 2000 mAh g^{-1} (dark red); only lithiation capacities are shown. Source: Misra et al. [20]/with the permission of American Chemical Society.

c-Ge → a-Li_xGe → c-Li_xGe (Li_7Ge_2) → c-Li_xGe ($Li_{22}Ge_5$ and $Li_{15}Ge_4$) → c-Li_xGe (Li_7Ge_2) → a-Li_xGe → a-Ge, which is helpful for exploring modifying method.

Aside from graphite and Si anode, Nb-based anode for Li-ion battery has also attracted attention from researchers. For a better understanding of charge storage mechanism, Li et al. studied the process of lithiation and delithiation of $ZrNb_{14}O_{37}$ nanowire [23]. As depicted in Figure 4.8, XRD data was processed by visualization, and a major peak shift was obviously observed by colors. While lithium ion was inserted into $ZrNb_{14}O_{37}$, intermediate phase of $Li_xZrNb_{14}O_{37}$ formed, and the (−111) peak slowly shifted to a lower angle after being gradually discharged to 1.61 V. During discharging between 1.55 and 1 V, the (−111) peak disappeared, and a peak at 22.4° was generated and moved to lower angles. The same phase transformation was processed in a reversed direction, and the initial phase was recovered, indicating a reversible lithiation-delithiation mechanism.

Tang et al. from the same group studied the more complicated reaction of $BiNbO_4$ nanowire during charge–discharge with in situ XRD method [24]. The in situ XRD pattern indicated that small numbers of lithium ions begin to embed and react with $BiNbO_4$ to form Bi nanocrystalline and Li_3NbO_4, and peaks of $BiNbO_4$ diminish.

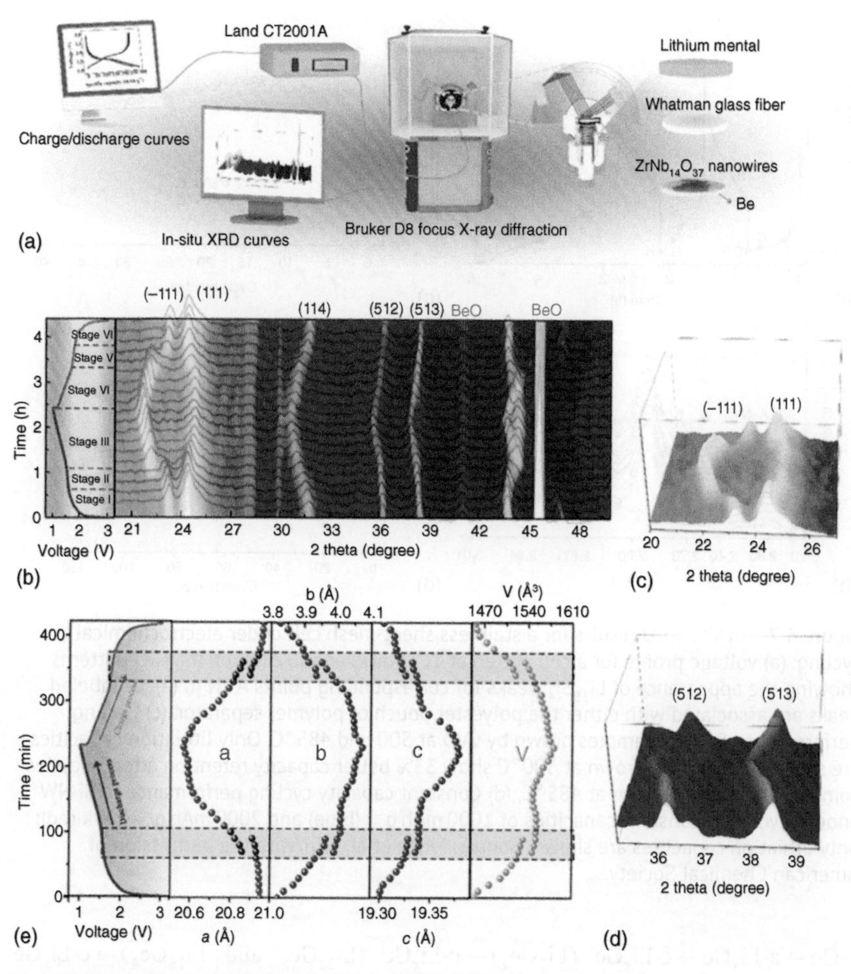

Figure 4.8 (a) Schematic construction of an in situ XRD system; (b) in situ XRD patterns of $ZrNb_{14}O_{37}$ nanowires during the initial cycle; (c, d) change of the (−111), (111), (512), and (513) reflections during cycling; and (e) evolution of lattice parameters during the initial lithiation/delithiation process. Source: Li et al. [23]/with the permission of the American Chemical Society.

Three new reflections finally appeared at 27.1°, 38.6°, and 39.6° referring to (012), (104), and (110) peaks of Bi, respectively, demonstrating the formation of Bi. Due to the prolonged discharging time, Bi nanocrystals grow into large, inactive Bi crystals that have no storage ability for lithium ions. The changes in the intensity of the characteristic diffraction reflections are displayed in detail in Figure 4.9d–f. And a new reflection appears at 38.1°, which can be indexed to (200) diffraction reflection of Li_3Bi, indicating that the Li_3Bi alloy formed after lithium ions were combined with active Bi nanocrystalline. As depicted in Figure 4.9i, peak at 38.1° becomes strong during in situ discharge process while the intensity decreases in the subsequent charging process, which indicates reversible transformation between Bi nanocrystallites and Li_3Bi alloy. Images of the phase transition for $BiNbO_4$ (Figure 4.9g–i. From

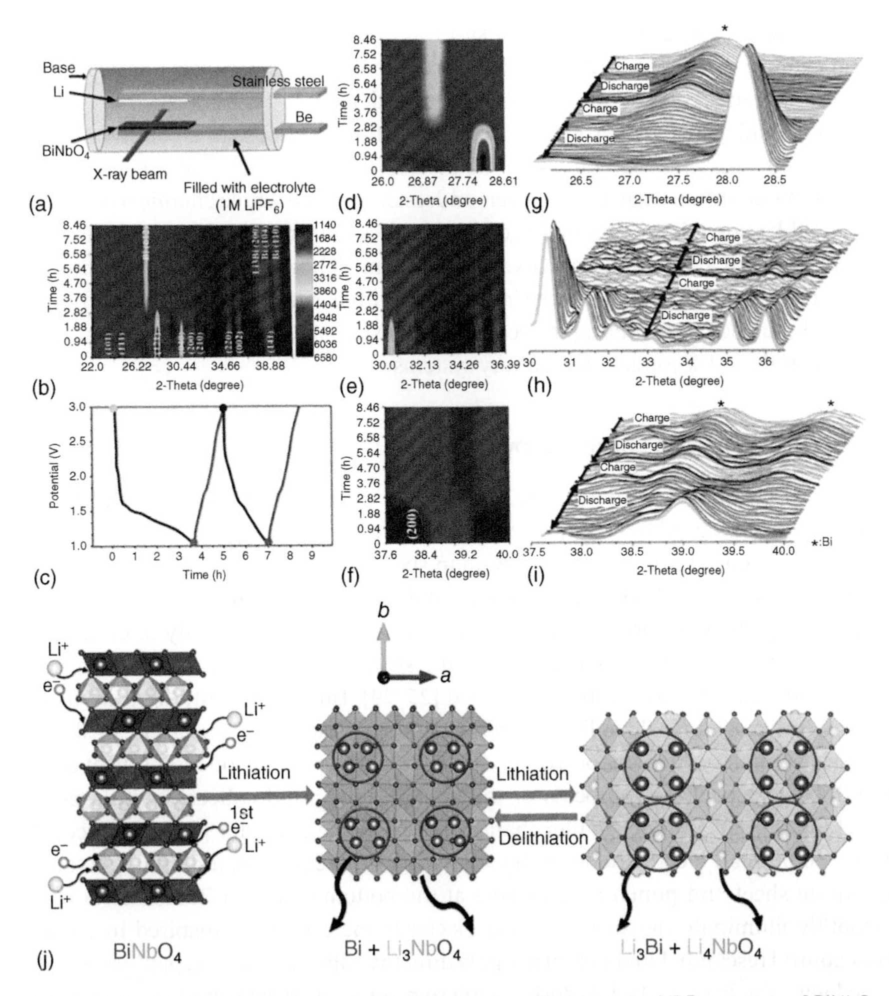

Figure 4.9 (a) The detailed structure of the in situ cell. (b) In situ XRD patterns of $BiNbO_4$ nanowires and (c) corresponding discharge/charge curves. (d–i) Local in-situ reflection images for XRD patterns selected from (d) 26.0° to 28.61°, (e) 30.0° to 36.39°, and (f) 37.6° to 40.0°. (j) Schematic image of structural transformation of $BiNbO_4$ during the first discharge/charge process. Source: Tang et al. [24]/with the permission of Elsevier.

those results, Tang et al. illustrated the energy storage mechanism with schematic image (Figure 4.9j). These results from in situ XRD analysis laid the foundation for researching Bi inserted Nb oxides as anodes.

Despite these remarkable progresses of in situ XRD researches about nanowire-based electrodes, there are still problems restricting further development:

(1) Design of an in situ cell device. There are severe problems with sealing in- situ batteries nowadays, which further affect the reaction inside the electrode. This phenomenon could be more obvious after ultralong testing time (ordinarily a dozen hours). In addition, electrode material and electrolyte could easily react with Be sheet when used as window material.

(2) The accuracy and sensitivity of in situ test. While in situ XRD aims to capture the intermediate products, the ultrashort time that these products may exist for, resulting in the limited time for machines to detect.
(3) Combination with other in situ characterization methods. In a common case, single in situ characterization method is qualitative analysis and reflects information in certain aspects. However, understanding the whole chemical reaction could be possible by rationally combining these means together. For instance, relationship and evolution between phases, morphology, and chemical bonds can be discovered by combining in situ XRD with in situ SEM, TEM, Raman, and FTIR, further contributing to recognizing, designing, and controlling electrode materials and secondary ion battery systems.

4.2.2 In Situ Raman Spectroscopy

Raman spectroscopy can detect the information of molecular vibration and rotation, so it can be used to characterize the valence bonds and functional groups of molecules in samples. As for in situ Raman spectroscopy, it combines the concept of in situ testing with Raman spectrometer, and through real-time monitoring of the position of Raman absorption peak and the change of peak intensity of electrode materials in the electrochemical process, the structure change of materials in the electrochemical process can be understood [25–29]. During the test, there are usually two electrochemical configurations available to ensure that the laser can shine on the electrode surface smoothly. The first is to pre-punch the top metal sheet during the device packaging process or to use a grid-like fluid collector with a pore size 1 order of magnitude smaller than the laser source diameter [30]. The second is to compose an electrochemical device with a fluid collector, a diaphragm, and a lithium sheet and punch a small hole at the bottom so that a Raman laser can smoothly illuminate the surface of the electrode material [31]. Inspired by small flow control research, Liqiang Mai use poly dimethyl siloxane, such as platform combined with single nano device, design a micro reaction pool, can characterize single nanowires in the electrochemical reaction of the structure evolution, and using the in situ Raman spectroscopy and atomic force microscopy in combination to study the electrochemical process of energy storage materials at atomic scale [32].

The typical transition metal layered oxide $Na_{0.5}Ni_{0.25}Mn_{0.75}O_2$, as a cathode material for sodium ion batteries, exhibits obvious deintercalation reaction and is a promising cathode material with high energy density. Zhou et al. from Nanjing University proposed a novel reaction mechanism of co-embedding of cations and/or anions for such layered materials through in situ Raman characterization [33]. Figure 4.10 shows the results of $Na_{0.5}Ni_{0.25}Mn_{0.75}O_2$ in situ Raman test.

Through in situ Raman test, the Raman representation diagram in the open circuit state is first obtained. Figure 4.10b shows the bending vibration peak of O–M–O (about 500 cm^{-1}) in the same plane and the stretching vibration peak of M—O bond in space (about 600 cm^{-1}). As the charging process progresses, the stretching vibration peak of O—O bond and the geometric vibration peak of ClO_4^- appear in the Raman spectrum peak of Figure 4.10c. When the chemical reaction continues and

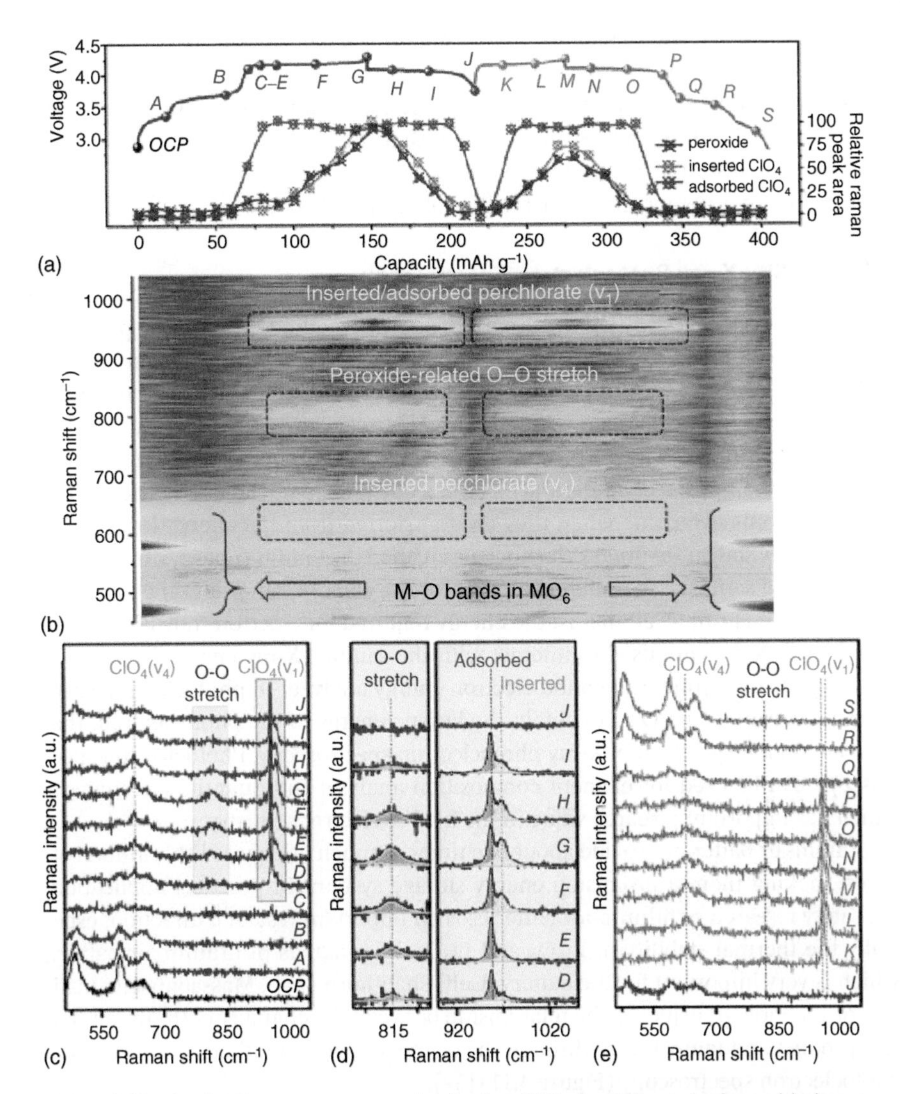

Figure 4.10 In situ Raman spectra study of anion (ClO_4^-) uptake and release (a) charge and discharge curves of the first and second cycles collected during in situ observations. (b) Color intensities of several bands. (c) Capacity-dependent spectra were recorded during the first cycle. (d) Magnified spectral profiles of the specific regions outlined in (c). Peroxo-related (blue) and ClO_4^- related (green) regions are respectively fitted for clarity. (e) In situ spectra collected during the second cycle. Source: Zhou et al. [33], Reproduced with permission from Wiley-VCH Verlag GmbH & Co.

the reaction discharges to point J, the geometric vibration peak of ClO_4^- gradually decreases to disappear. This indicates that, along with the interaction between ClO_4^- and the material crystal, ClO_4^- is also involved in ion embedding and removal during the electrochemical reaction. Through in situ Raman characterization, they successfully explained the high capacity of transition metal layered oxides, proved the electrochemical reaction model of cation-anion co-embedding, and opened up a new idea for the electrochemical mechanism research of cathode materials.

Spectra collected at different charge/discharge stages can be distinguished by specific colors and letters marked in (A). Note that the pair of ClO_4^- related peaks (adsorbed and inserted) and peroxo-related peaks are fitted, and each of their capacity dependence is demonstrated inside the voltage profiles.

4.2.3 In Situ X-ray Photoelectron Spectroscopy

X-ray photon spectroscopy (XPS) is a surface analysis method based on the principle of the interaction between light and matter. It was developed by physicist Sigbarn and his colleagues at Uppsala University in Sweden after two decades of research. As a common characterization method, X-ray photoelectron spectroscopy plays an important role in the characterization of elements and contents of materials [34–38].

Atoms are composed of atoms and electrons outside the nucleus. The energy that bound electrons overcome when they are transferred from the energy level to the lowest energy state unbound by the nucleus is called the binding energy of electrons. For real solid samples, the reference point for the calculation of binding energy is chosen as the Fermi level, that is, the energy required for electron transition to the Fermi level. X-rays are used to interact with the sample. X-ray energy is absorbed by the solid, and through the photoelectron energy analyzer in photoelectron spectroscopy, it can be analyzed to obtain the kinetic energy of the photoelectron, that is, to obtain the spectrum of X-ray photoelectron spectroscopy. Photoelectron spectroscopy can be used for element composition analysis and quantitative analysis, which has important research value in the field of electrochemistry.

Lithium-air batteries provide about five times as much energy as lithium-ion batteries, making them a promising energy storage system. In practical application, the safety research of lithium-air batteries is an important research direction, especially the thermal stability of Li_2O_2 and Li_2O, the products of lithium-air battery, which is very important for the battery itself. Shao-horn et al., Massachusetts Institute of Technology, explored the relationship between the change in lithium-air battery product and temperature through the characterization method of in situ X-ray photoelectron spectroscopy (Figure 4.11) [39].

In situ XPS characterization results show that for Li_2O, the 0.1s orbital changes most obviously with the increase in temperature. That is, when Li_2O_2 is heated, there will be an obvious growth of Li_2CO_3 in the product. However, with the increase in temperature, the growth of Li_2CO_3 in the product was not obvious. In situ characterization results show that Li_2O and Li_2CO_3 will form and grow on the surface of Li_2O_2 with increasing temperature. With the help of characterization, they found that the decomposition temperature of Li_2O_2 is 250 °C.

4.2.4 In Situ XAS Characterization

Unlike XRD, X-ray Adsorption spectroscopy is a characterization method that gains information from the intensity of X-ray beam that passes through sample. By analyzing the relationship between ratio of initial intensity and transmitted intensity.

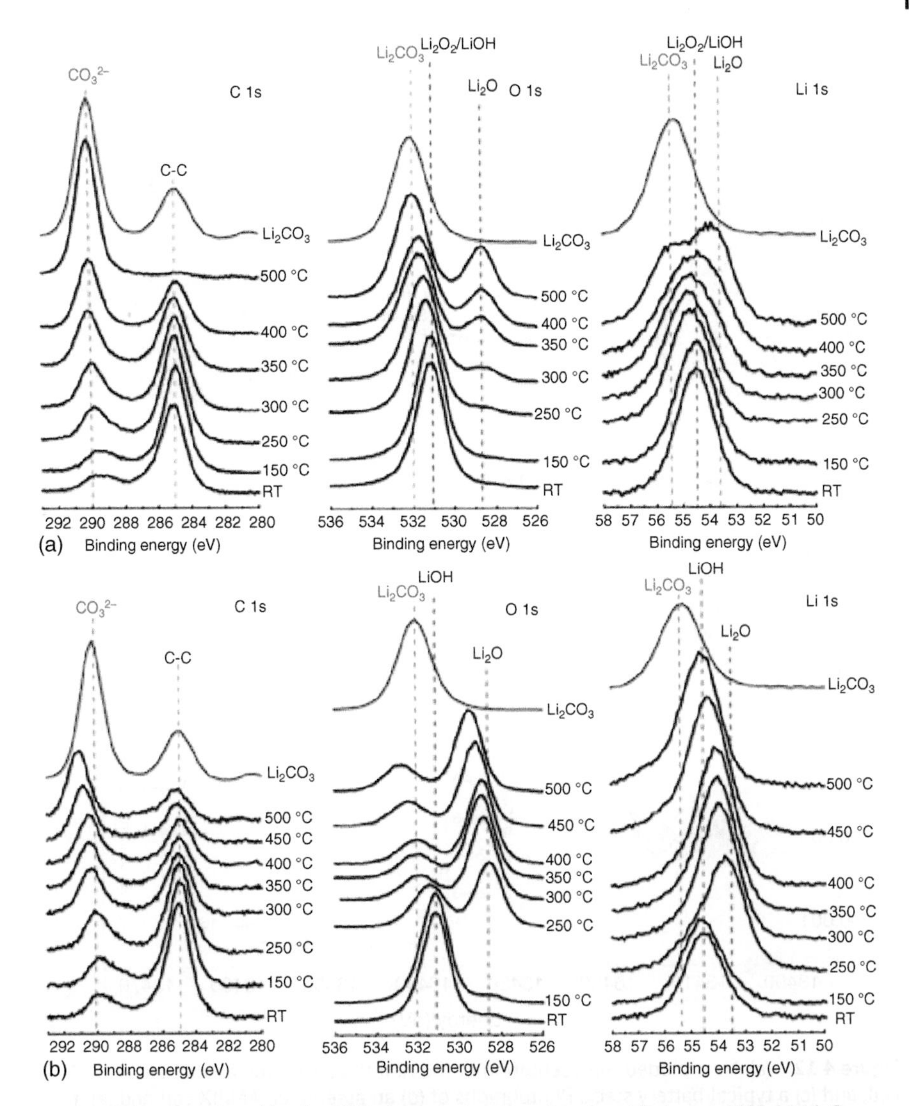

Figure 4.11 Results of in situ XPS characterization of each orbital peak [39]. (a) Li_2O_2; (b) Li_2O. Source: Shao-horn et al. [39]/with the permission of Manchester IOP Publishing.

Following equation can describe the principle of XAS analysis:

$$I_t = I_0^{-\mu(E)t} \tag{4.1}$$

where I_0 is the intensity of incident X-ray, I_t is the intensity of transmitted X-ray, t is the thickness of sample, and E is the absorption coefficient, which is dependent on the photon energy.

Synchrotron radiation (SR) is the electromagnetic radiation emitted by charged particles traveling in curved orbits under the action of electromagnetic fields. Since the discovery of X-rays by Roentgen in 1895, XAS has achieved unprecedented

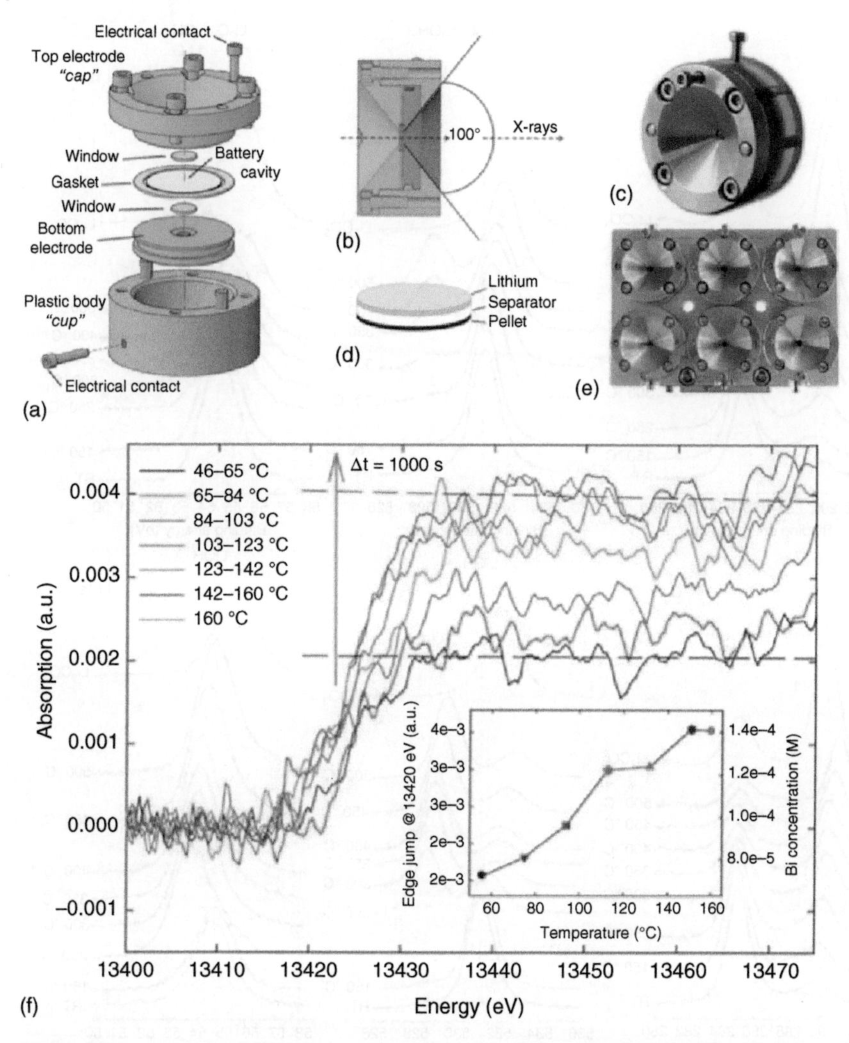

Figure 4.12 (a) An exploded representation of the AMPIX cell, (b) the assembled AMPIX cell, and (c) a typical battery stack. Photographs of (d) an assembled AMPIX cell and (e) a six-cell multisample holder. Source: Borkiewicz et al. [45], Reproduced with permission from JOHN WILEY & SONS, INC. (f) Bi-L_3-edge XANES spectra were recorded in QEXAFS mode (liquid phase). Inset: temperature-dependent Bi concentration. Source: Patzke et al. [46], Reproduced with permission from Wiley-VCH Verlag GmbH & Co.

development since the construction of SR light source, which analyzes material element composition, electronic states, and microstructure by using the changes of X-ray signals before and after radiation.

XAS has some discrete peaks or wave-like fluctuations near the absorption edge and its high-energy end, which are called fine structures (XAFS). By adjusting the energy of X-rays, SR XAFS technology is widely used to study the chemical environment around the atoms of various elements in condensed matter and obtain important information such as local structure, oxidation state, and neighboring coordination atoms of the constituent elements in the material. In addition, by combining the in situ concept with SR technology, information about electrode materials can be obtained in the process of electrochemical energy storage, such as structure evolution, local structural change, and oxidation state of transition metal ions, which has important guiding significance for understanding the precise battery reaction mechanism, method for optimizing the material composition and structure, and the development of new high-performance materials [17, 40–44]. Argonne National Laboratory has designed a multipurpose X-ray in situ battery that can be used at multiple SR line stations to conduct wide-range in situ SR electrochemical X-ray scattering and spectrum analysis [45].

Patzke et al. from the University of Zurich, Switzerland, used X-ray fine structure characterization to confirm that the bulky α-Bi_2O_3 precursor and sulfuric acid nanowires overgrow and form solid–liquid interface continuously (Figure 4.12) [46]. Only 0.4% (in mole ratio) of Bi dissolved, confirming that it takes time to process.

4.3 In Situ Characterization of Nanowire Devices

4.3.1 Nanowire Device

In the research field of micro/nano energy storage devices, micro/nano devices assembled based on single nanomaterials play an important role, such as single nanowires and single nanosheet devices. Nano materials because of its reach nanoscale on two dimensions, which presents the massive materials do not have special properties (for example, the nanowire electrode has a unique anisotropy, as for the single nano-device building provides great convenience, but at the same time has a large specific surface area, continuous axial electron transmission, fast radial ion diffusion, etc).

By assembling single nanowire micro/nano devices, researchers can extract materials from complex and mixed systems and intuitively present principles that are difficult to explain by conventional means through clever device design, including accurate measurement of basic physical property parameters (including electricity, heat, and magnetism [47–49]). This provides an effective method to study the evolution of material properties at nanometer scale. With the continuous maturity of semiconductor-controlled precision machining technology, the research on single nanodevices has gradually expanded from the basic frontiers of physics to the application of electrochemical energy storage.

4.3.2 Nanowire Device Characterization Example

A lithium-ion battery is like a black box. The study of capacity attenuation usually requires dismantling the device, which is easily disturbed by external water and oxygen, leading to the inability to detect its real state in situ in real time. On the other hand, active materials, conductive additives, binders, and fluid collectors in energy storage devices are mixed together, and the interface is complex, which seriously hinders the direct observation of active materials.

Therefore, Wuhan university of science and technology Liqiang Mai in 2010 to create a single nanowires device such as universal new model of in situ characterization of electrochemical process, design and assembly of the international the first single nanowires on solid-state electrochemical devices, the proposed single nanowires electrochemical device for the first time micro-nano circuit with precise integration of energy storage devices, high precision, strong signal collection[49]; At the same time, a monochromatic laser was introduced to detect the material structure information to realize multifield real-time monitoring of electrical/electrochemical/spectroscopy (Figure 4.13). The device model takes the lead in extracting the electrode material – fluid collector interface and electrode material – electrolyte interface from the mixed interface, which overcomes the problems of the interaction between complex interfaces and the difficulty of resolving collected signals. The intrinsic relationship between electrode structure, transport, charging and discharging states, and capacity decay was established. With the charging and discharging processes, the Raman spectra show that the nanowire structure deteriorates gradually, and the voltammetry curve shows that the electrode–collector contact changes from ohmic contact to Schottky contact, accompanied by a continuous decrease in electrical conductivity. It is found that structural deterioration, the change of electrode – collector contact, and the decrease of electrical conductivity are the intrinsic factors leading to the capacity decay.

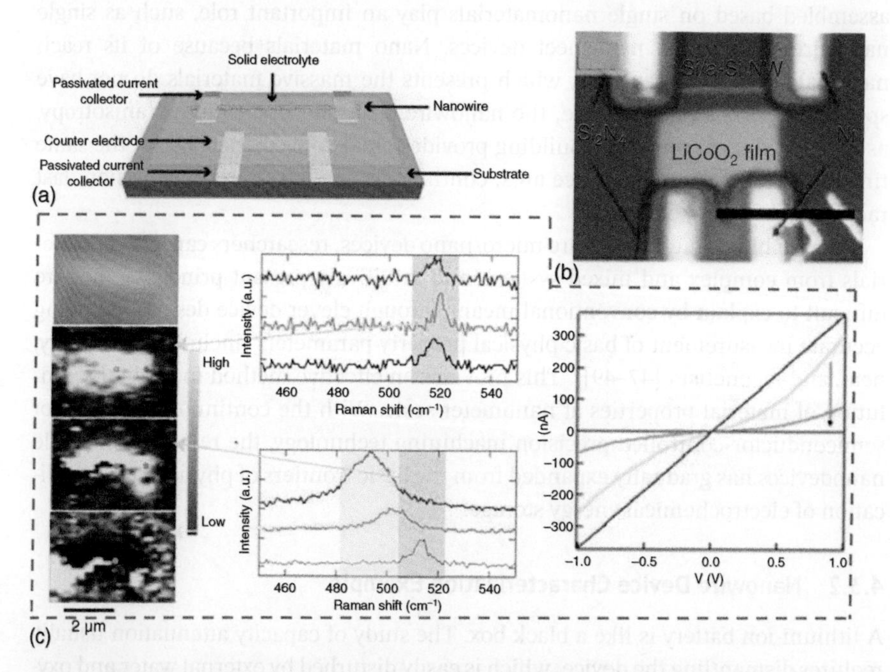

Figure 4.13 In situ characterization of a single nanowire electrochemical device. (a) Schematic diagram of a single nanowire lithium ion energy storage device, (b) Optical micrograph of a single Si nanowire Li-ion energy storage device, (c) In situ Raman spectra and *I–V* curves of a single silicon nanowire Li-ion energy storage device. Source: Mai et al. [49]/with the permission of the American Chemical Society.

Therefore, based on the solid-state single nanowires devices in situ study of this credit platform, we can build the nanowires electric transport, structure, and state of charge and discharge the direct link so as to reveal the intrinsic electrochemical behavior of nanowires and capacity attenuation, the nature of performance degradation, and to solve the energy storage material capacity attenuation problem with fast, low conductivity. The key problem is of great significance. Based on the research results, Mak and others wrote a paper titled "Track Autonomous in Real Rime" at the invitation of *Nature*.

Subsequently, Mai et al. conducted a series of researches on the optimization mechanism of composite nanolimited energy storage materials based on single nanowire devices [50]. In order to explore the mechanism by which graphene optimizes the performance of nanowire energy storage materials, three single nanowire devices, namely pure MnO_2, MnO_2/rGO, and MnO_2/porous graphene, were assembled and electrochemical tests were performed (Figure 4.14). The results show that the coating of rGO can significantly improve the conductivity of MnO_2 and reduce the ion diffusion rate. Porous graphene, on the other hand, provides a gap for ions to enter while improving the conductivity, thus maintaining a high ion diffusion rate.

By studying the optimization mechanism of the midline structure of semi-hollow double-continuous V_3O_7/graphene scrolls, the team pointed out that the specific capacity, cyclic stability, and multiplier performance of VGS are better than those of V_3O_7 nanowires, and the charge-transfer resistance (32 Ω) of VGS is much smaller

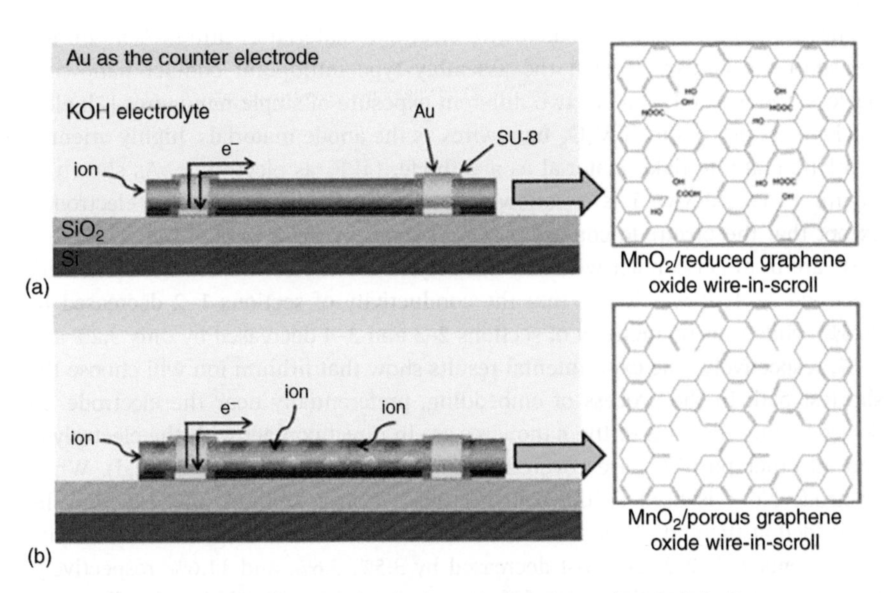

Figure 4.14 Schematic diagram of a single nanowire electrochemical device. (a) Schematic representation of the MnO_2/rGO single-wire nanowire device and rGO, where ions are transported only through the gap between graphene and MnO_2 nanowire, (b) Schematic representation of MnO_2/pGO single nanowire devices, where porous GO ions are transported not only through the gap between graphene and MnO_2 nanowire but also through pores in graphene. Source: Hu et al. [50]/with the permission of the American Chemical Society.

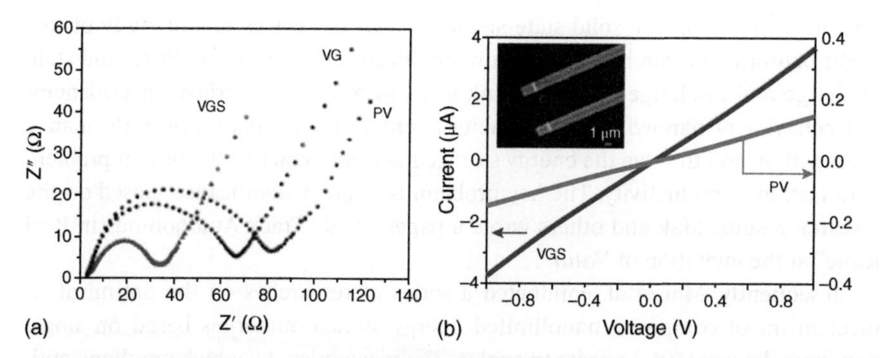

Figure 4.15 The midline conductivity characteristics of a semi-hollow double-continuous V_3O_7/Graphene roll are illustrated. (a) Impedance test of VGS, VG, and PV in initial state, (b) Transmission characteristics of single nanowire of VGS and PV (Check figure: SEM image of single nanowire device of VGS). Source: Yan et al. [51]/with the permission of the American Chemical Society.

than that of pure V_3O_7 nanowires (81 Ω) (Figure 4.15). Based on the electrochemical performance test, a single device was further assembled, and current–voltage (I–V) tests were carried out on pure V_3O_7 nanowire and VGS, respectively. The experimental results showed that the conductivity of VGS (1056 S m^{-1}) was 27 times higher than that of pure V_3O_7 nanowire, revealing the essence of its electrochemical performance improvement [51].

The team and to study the nanowire electrode material of lithium/sodium ion transport mechanism, design the assembly type multipoint contact nanowires devices, the research state of two different exposure of single nanowires [2]: electrochemical process to $H_2V_3O_8$ nanowires as the anode materials, highly oriented pyrolytic graphite flake material as a cathode, $LiPF_6$ as electrolyte. As shown in Figure 4.16a,b, device 1 is a nanowire material connected with four electrodes, except that the electrode contact part is covered by SU-8 photoresist. The other parts are in direct contact with the electrolyte. When electrodes were tested with electrode No.1, it was found that the conductivity of sections 1–2 decreased by 96.5%, while the conductivity of sections 2–3 and 3–4 decreased by only 9.2% and 1.9%, respectively. The experimental results show that lithium ion will choose the shortest path in the process of embedding, preferentially near the electrode. In device 2, only the ends of the nanowires are in direct contact with the electrolyte, and all other positions are covered by SU-8 photoresist (Figure 4.16c,d). When electrode No 1 is used as the test electrode, lithium/sodium ions can only be embedded at the end of the nanowires. After Li-ion embedding, the conductivity of segments 1–2, 2–3, and 3–4 decreased by 3.5%, 5.6%, and 11.6%, respectively. Due to the high migration rate of electrons along the nanowire's axial direction, it is the low axial migration rate of lithium ions that limits the electrochemical reaction kinetics. The final results show that the farther away from the end, the less structural damage and the less decrease in electrical conductivity. The exquisiteness of the single nanowire micro-cell provides great convenience for studying the basic reaction principle of electrode materials in electrochemical reactions and deepens the understanding of the reaction mechanism.

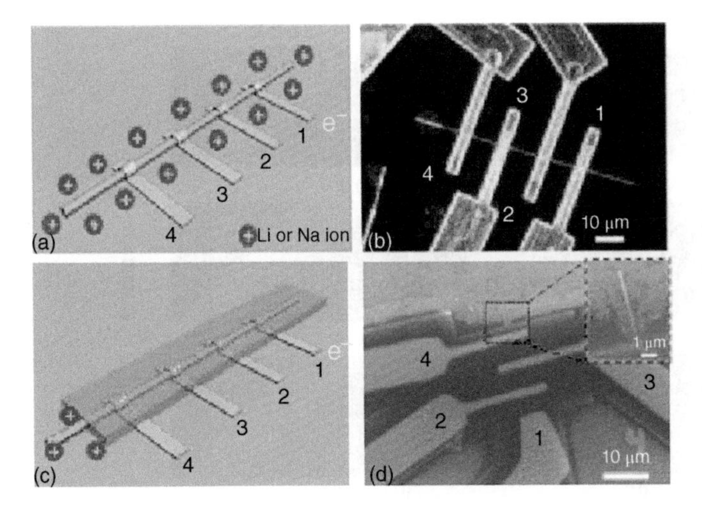

Figure 4.16 A single nanowire electrochemical device with two different exposure states. (a) Schematic diagram of a fully exposed $H_2V_3O_8$ single nanowire device, (b) SEM diagram of the device corresponding to the fully exposed state, (c) Schematic diagram of a single $H_2V_3O_8$ nanowire device with exposed end, (d) SEM diagram of the device corresponding to the end-exposed state. Source: Xu et al. [2]/with the permission of the American Chemical Society.

4.4 Other In Situ Characterization

4.4.1 In Situ Atomic Force Microscopy Characterization

When the material is characterized by AFM, the slight changes of the microcantilever can be detected by the sensor, and then the surface information of the material can be obtained. Therefore, in situ AFM can be used to detect the surface changes of materials, which is of great significance for some difficult-to-detect material interfaces, such as the SEI changes of nanowire materials during electrochemical processes [52–56].

Wan Lijun et al., Institute of Chemistry, Chinese Academy of Sciences, studied the interfacial electrochemistry of the electrode through in situ AFM based on a single silicon nanowire and observed the morphology of the silicon negative electrode surface and the change of Young's modulus during the growth of SEI [57]. They monitored the samples in a specially designed AFM fluid cell under an argon atmosphere. For in situ monitoring, they connect it to an external Autolab workstation, which controls the voltage of the fluid battery. In addition, they removed the SEI from the surface of the nanowire in contact mode, applying a scanning force slightly greater than that of contact-mode imaging. Subsequently, quantitative nanomechanical experiments were performed.

As shown in Figure 4.17, the formation process of SEI on the negative electrode of silicon nanowires in 1 $mol\,l^{-1}$ $LiPF_6$/PC was characterized by various operating modes of in situ AFM. Three distinct stages were observed in SEI formation at different potentials. The primary film formed on the negative electrode of the silicon nanowire is above 0.4 V, and when the potential is reduced to 0.33 V, the surface is

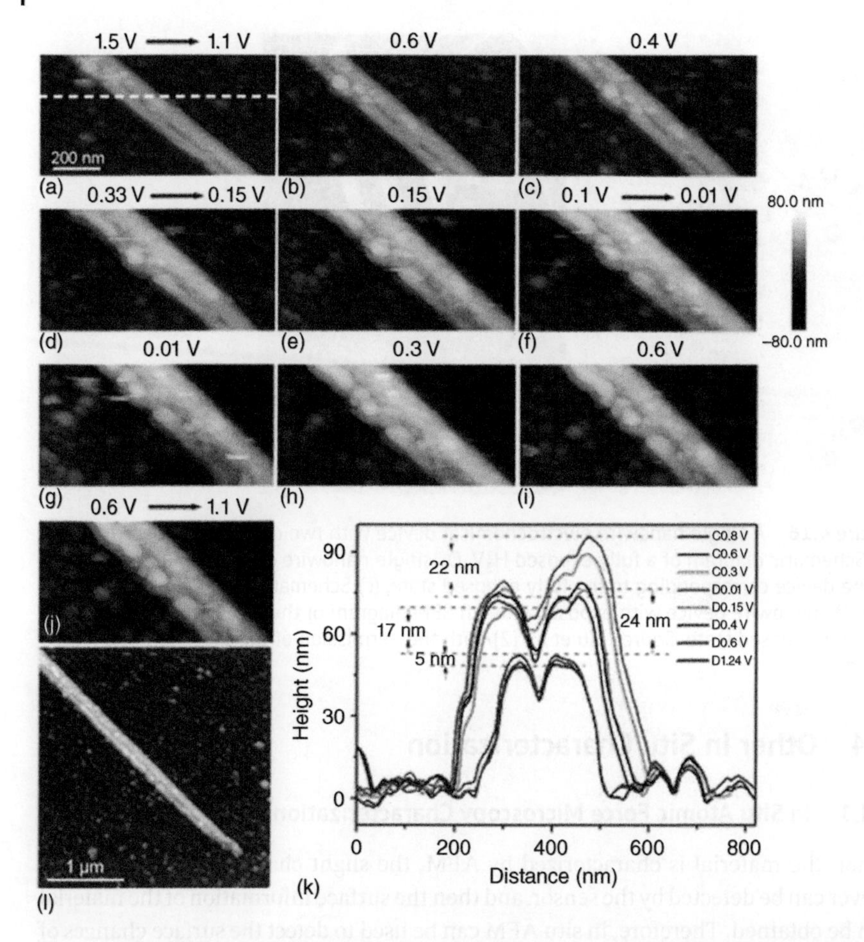

Figure 4.17 Schematic diagram of in situ AFM characterization results of SEM on the Si negative electrode. (a–j) The in situ AFM height images of a single Si NW anode during the first cycle at different discharge/charge states. The scan direction is from bottom to top, and the scan area is $1 \times 0.5\ \mu m^2$. (k) Cross-section analysis of the height of the location marked by a white dashed line during the cycle. (l) The AFM image of the Si NW after the first cycle. Source: Liu et al. [57]/with the permission of the American Chemical Society.

quickly covered by thick granular SEI. When the temperature is lower than 0.1 V, SEI continues to grow during the evolution. After the first cycle, the SEI formed on the Si nanowire was rough and heterogeneous, with a thickness of (28 ± 10) mm. The SEI film remained intact after the first cycle, showing the ability to withstand large volume changes during the first lithium and delithium cycles.

As shown in Figure 4.18, the statistical values of Young's modulus of intact SEI in the range of 50–400 MPa are consistent with the compositional properties of SEI film. In addition, with the decrease in voltage at both ends of the material, the SEI formed.

Based on the above analysis, the team proposed a mixed-filled structure model of the SEI film on the negative electrode of silicon nanowire, indicating the in situ AFM of the single-atom electrochemical process.

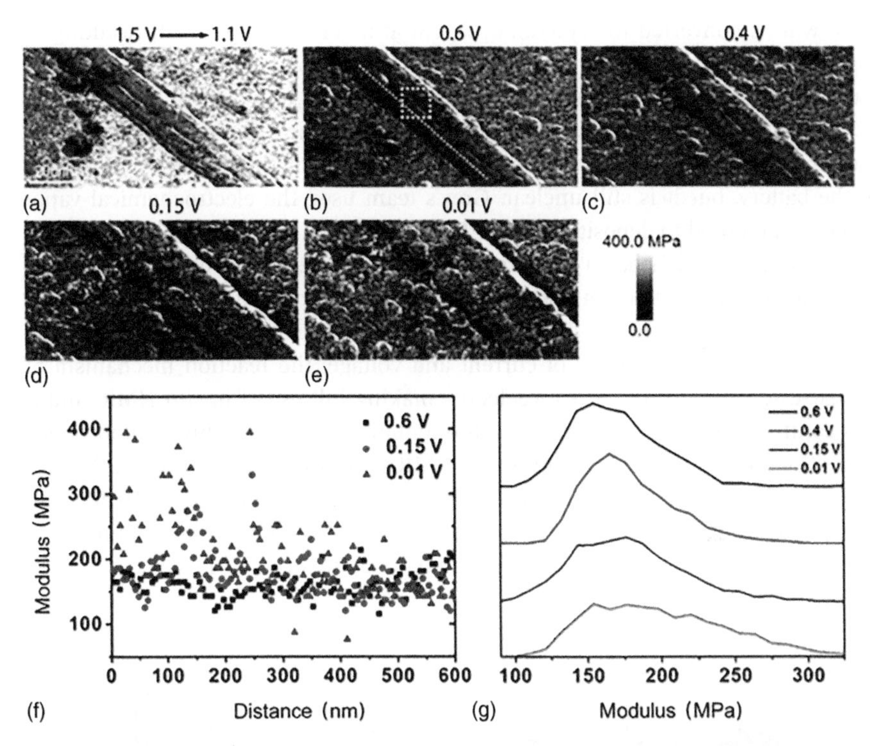

Figure 4.18 In situ AFM characterization of the negative electrode of a single silicon nanowire. (a–e) The AFM images of Young's modulus mapping of a single Si NW anode in the first discharge process. The scan direction is from bottom to top, and the scan area is $1 \times 0.5\ \mu m^2$. (f) The distribution of Young's modulus along the location marked with a dashed yellow line. (g) The statistic values of Young's modulus of the area marked with a dashed white box. Source: Liu et al. [57]/with the permission of American Chemical Society.

4.4.2 In Situ Nuclear Magnetic Resonance

Nuclear magnetic resonance (NMR) technology has a wide range of applications in physics, medicine, chemistry, biology, food, and other fields. It can provide molecular dynamics and molecular chemical structure information and has become a conventional technical means to analyze the molecular structure and characterize the physical and chemical properties of substances [58–61].

Based on in situ NMR technology [62], Grey et al., University of Cambridge, UK, characterized silicon nanowire electrodes at the atomic level during battery cycle, revealing the complex kinetic and thermodynamic processes involved in electrochemical reactions and material structure rearrangement [63]. When Si is used as the anode material of lithium-ion battery, its capacity is 10 times that of commonly used graphite material. However, the huge volume expansion of 300% will cause the destruction of electrode structure and the attenuation of capacity. A practical solution is to use a Si/graphite (with a small proportion of Si) composite material that combines the cycling properties of graphite with the high capacity of Si, thus exhibiting substantially increased capacity. In order to fully utilize the capacity of graphite material, it is often discharged to a very low potential, but the amorphous

Li_xSi will be converted to crystalline $Li_{3.75}Si$ at a very low potential, resulting in a great overpotential in the process of negative electrode delithium. Researchers have demonstrated that crystalline Si is converted to c-$Li_{3.75}Si$ or $Li_{4.4}Si/Li_{4.2}Si$ upon initial discharge below 70 mV. The study of the reaction mechanism in the second and subsequent cycles is very important for the actual cycle process of the battery, but it is still unclear. Grey's team used the electrochemical vapor deposition method to deposit and grow Si nanowires on the carbon fiber-based gas diffusion layer, which has a high crystallinity and a length of more than 50 μm. The diameter is about 60 nm. Based on this material model, further in the process of constant current charge–discharge and constant potential charge–discharge, through the accurate control of current and voltage, the reaction mechanism of charge–discharge process was revealed by making full use of in situ NMR, and it was confirmed that $Li_{3.75}Si$ with crystal orientation grew unevenly in amorphous phase (Figure 4.19). Electrochemical and NMR signals show that the overlithiation process of c-$Li_{3.75}Si$ in the reaction is released below the potential of 50 mV, and small clusters will be formed in $Li_{3.75}Si$ phase under charging state (Figure 4.20). Combining these results with density functional theory (DFT) calculations, it

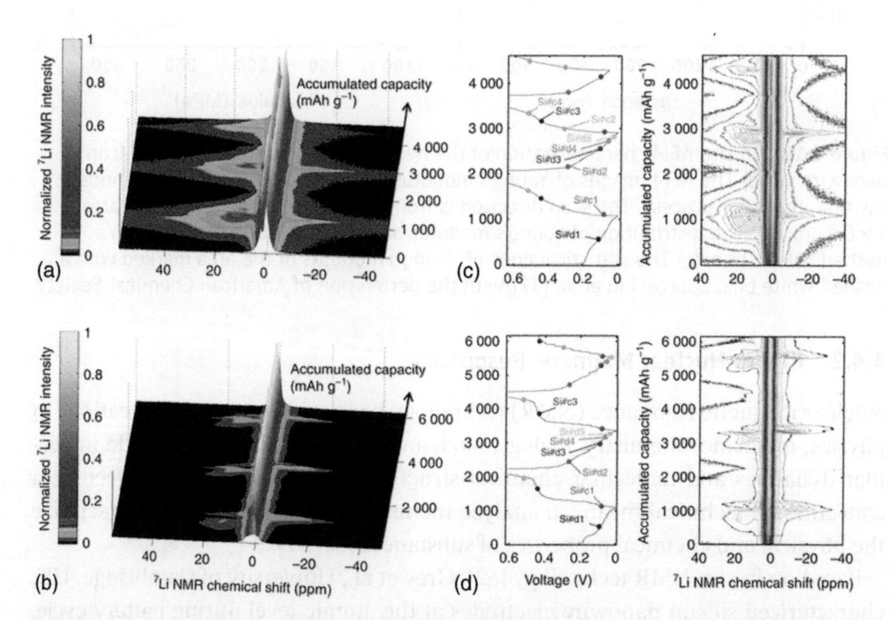

Figure 4.19 Si nanowire - carbon fiber-based vapor diffusion matrix (Si NW-CFGDL). (a) Acquired with a *C*/30 galvanostatic experiment with potential limits of 2–0 V for the first and second cycles and 50 mV cutoff for the third discharge, (b) acquired with a potentiostatic experiment with potential limits of 2–0 V for the first and second cycles and 30 mV cutoff for the third discharge. The corresponding contour and Li/Li^+ voltage versus accumulated specific capacity plots are shown for the galvanostatic and the potentiostatic experiments in (c) and (d), respectively. The ^7Li resonances at around 20–10, 10–0 and −10 ppm are labeled as P1 (Li nearby small Si clusters), P2 (isolated Si anions, c-$Li_{3.75}Si$ and extended Si networks), and P3 (the over-lithiated crystalline phase, c-$Li_{3.75+\delta}Si$), respectively. Source: Key et al. [63]/with the permission of the American Chemical Society.

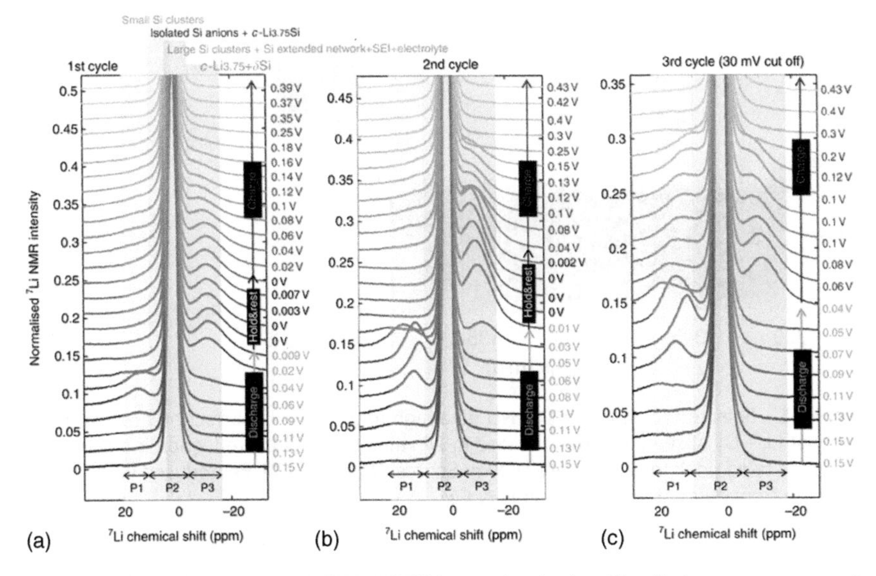

Figure 4.20 LiNMR spectra of Si NW-CFGDL complex in situ. The discharge was stopped at 0 V for the (a) first and (b) second cycles and (c) 30 mV for the third cycle. P1–3 are assigned to the ^{7}Li environments, as explained in the legend to Figure 4.19.

is shown that the phase formed by delithiation strongly depends on the rate of lithiation, which can rationalize the large voltage lag phenomenon in these systems.

4.4.3 In Situ Neutron Diffraction

The principle of neutron diffraction is similar to that of X-ray diffraction, which is mainly used to study the crystal structure of materials and is also applicable to the Bragg formula. Its application is complementary to X-ray diffraction. When the X-ray or electron current meets the material and scatters, the scattering center is mainly the electron in the atom, and its scattering power increases with the increase in the atomic number of the material and with the diffraction angle of 2θ increase and decrease. In neutron diffraction, the neutron flow is uncharged. When it interacts with matter, it mainly interacts with the atomic nucleus to produce isotropic scattering. This scattering power has no relationship with the atomic number of the material, so the diffraction spectrum avoids the influence of the atomic number of the element. Therefore, neutron diffraction is more reliable for the characterization of some substances containing elements with large atomic numbers [64–68].

Sharma of the Prague Research Institute in Australia and others explored the reaction mechanism of Li ion deinking and phase transition of $Li_4Ti_5O_{12}/TiO_2$ cathode material during charging and discharging using in situ neutron diffraction technology (Figure 4.21) [69]. In the 475–490 minutes time range of in-situ detection, the diffraction signal of (112) crystal plane of Li_xTiO_2 disappears, while the (112) crystal plane of TiO_2 displays. Therefore, it was proved by neutron diffraction that a two-phase reaction occurred in this electrochemical process.

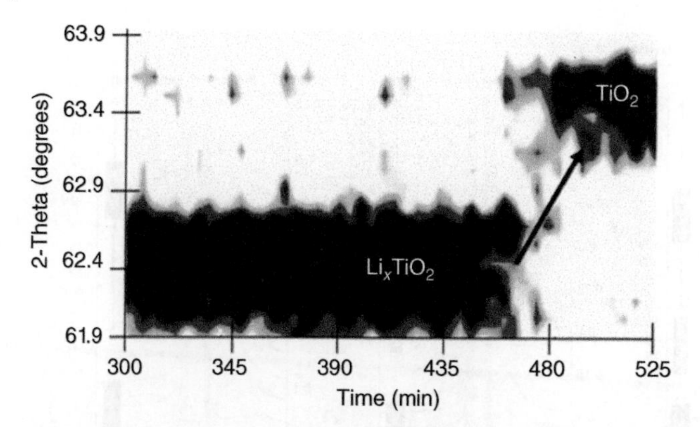

Figure 4.21 Selected region of in situ ND data highlighting the Li extraction process in ~80 mg of Li_xTiO_2 observed by the increase in 2θ value of the (112) reflection indicated by the arrow. Scaled peak intensities are shown. Source: Du et al. [69], Reproduced with permission from Wiley-VCH Verlag GmbH & Co.

Wang et al. reported an in situ neutron diffraction study of a large pocket cell [70]. The continuity of the Li-graphite intercalated phase is completely captured at 1*C* charging and discharging (that is, filling to full capacity within one hour). However, the lithiation and dilithiation pathways are quite different, and unlike the slow charging experiments where the Li-graphite phase diagram was established, no LiC_{24} phase was found during charging at 1*C* rate. Approximately 75 mol. % of the graphite is converted to LiC_6 at full charge, and up to 4% lattice expansion is observed during the charge–discharge cycle. Our work demonstrates the potential for in situ, time-and space-resolved neutron diffraction studies of the dynamic chemistry and structural changes of "real-world" batteries under real cycling conditions, which will provide microscopic insights into the important role of degradation and diffusion dynamics in energy storage materials. For accurate phase analysis, neutron diffraction patterns were measured at full charge (4.2 V) and discharge (2.7 V) for a counting time of approximately one hour. The results are shown in Figure 4.22a, along with the Rietveld improvements for each mode. Four phases are observed under complete discharge conditions: NMC cathode and graphite anode, as well as Al and Cu collectors. Under charged conditions, most of the graphite phase has been converted to LiC_6 and LiC_{12}, and only a small amount of graphite phase remains. An additional peak was found at ~2.05 A that could not be identified with any of the above but remained constant during charging and discharging. In situ neutron diffraction patterns collected during the cycle are shown in Figure 4.22b. Neutron diffraction data were collected continuously using an event-mode data acquisition system for approximately 20 hours, during which time the cell went through seven cycles of charge–discharge sequences. The data were divided into diffraction patterns corresponding to a duration of 2.5 minutes and strobe averaged over seven charging and discharging cycles. More details on event-based data acquisition and strobe averaging are given in the Supplementary material. Despite averaging over seven cycles (i.e. a total count time of 17.5 minutes), the statistical quality of each

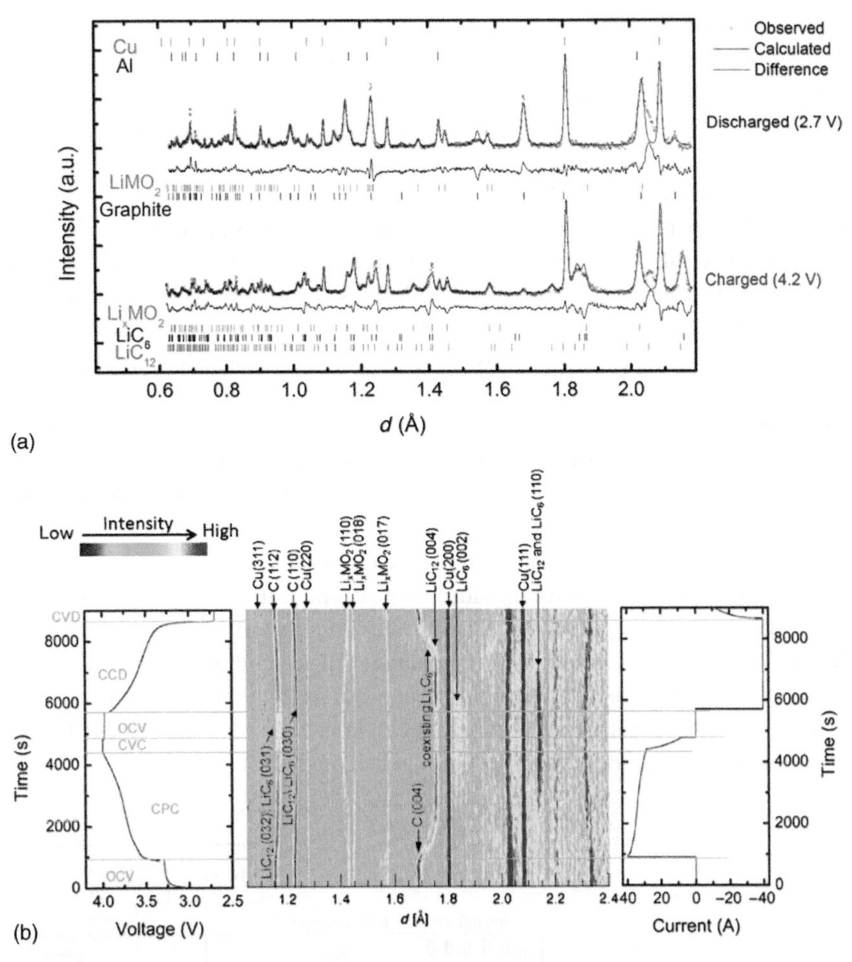

Figure 4.22 (a) Neutron diffraction patterns obtained at charged (4.2 V) and discharged (2.7 V) conditions along with the results of Rietveld refinement. (b) In situ neutron diffraction pattern during a charge and discharge cycle. The in situ diffraction data, averaged over seven cycles, are binned into 2.5-minutes histograms. The voltage (current) is plotted in the side panel to the left (right) of the diffraction data. The acronyms used in the figure are as follows: constant voltage discharge (CVD), constant current discharge (CCD), open circuit voltage (OCV), constant voltage charge (CVC), and constant power charge (CPC). Source: Wang et al. [70], Reproduced with permission from Springer Nature.

dataset is not sufficient for full Rietveld refinement of complex diffraction patterns. One possible reason for the lack of data is that the data may not have been averaged over the same seven periods.

4.4.4 In Situ Time-of-Flight Mass Spectrometry

The mass analyzer of the in situ time-of-flight mass spectrometer is an ion drift tube in which the ions produced by the ion source are first collected, and all ions become velocity zero in the collector. It is accelerated using a pulsed electric field into a field-free drift tube and flies to the ion receiver at a constant speed. The larger the ion mass, the longer it takes to reach the receiver; The smaller the mass of ions, the shorter the time it takes to reach the receiver. According to this principle, ions of different masses can be separated according to their m/z values. A time-of-flight mass spectrometer has a wide range of molecular weights, fast scanning speed, and simple instrument structure. The main disadvantage of traditional time-of-flight mass spectrometer is its low resolution because ions with the same mass-to-charge ratio have a certain time distribution to reach the detector due to their different initial energies when leaving the ion source, resulting in a decrease in resolution. Improvement is one of the methods in linear detector in front of the mirror, plus a set of electrostatic field will fly free ions, push back, big initial energy ions due to the initial speed, long into the electrostatic field reflecting mirror, return journey is long, and small initial energy ions return journey is short, and the return journey will focus on a certain position. Thus, the resolution of the instrument is improved. This time-of-flight mass spectrometer with an electrostatic field mirror is called a reflective time-of-flight mass spectrometer.

As shown in Figure 4.23, the mass spectrometer needs to operate in a vacuum to protect the detector and improve measurement accuracy. In the instrument shown, gas samples are first sampled through microwells and then arrive at the ion source, where pulsed electric fields are fed into the time-of-flight module. The ions are then accelerated using a pulsed electric field perpendicular to the direction of delivery. The main purpose of this is to determine that all ions have no initial velocity in the horizontal direction. After the V or W flight, reach the sensor.

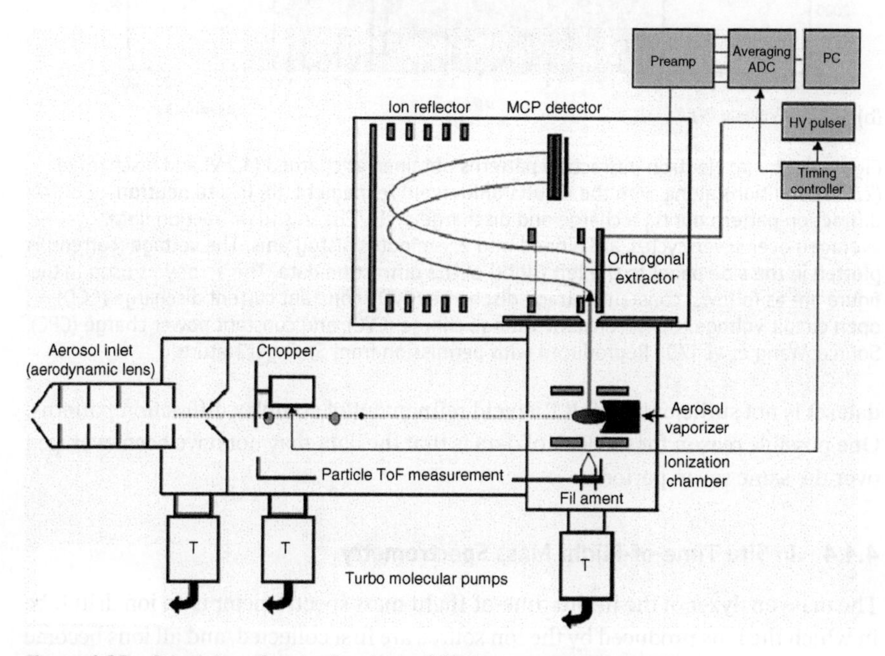

Figure 4.23 Principles of an in situ time-of-flight mass spectrometer.

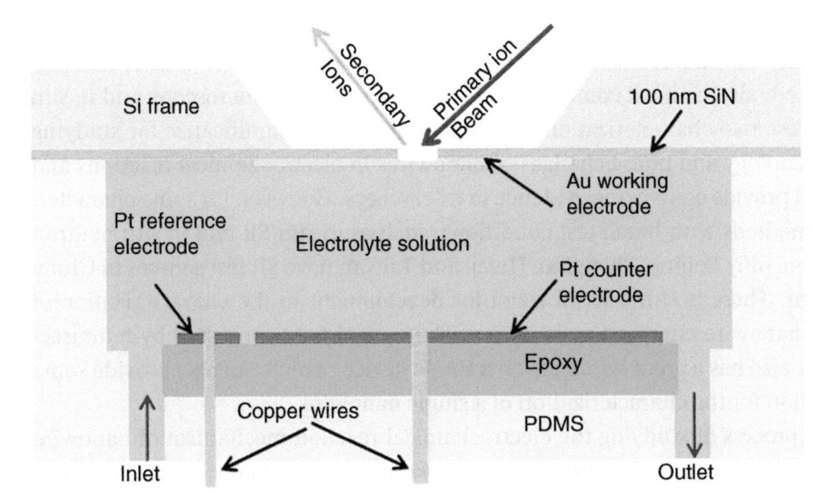

Figure 4.24 Schematic illustration of the microfluidic device for real-time and in situ ToF-SIMS analysis of the EEI. Source: Liu et al. [71]/with the permission of the Royal Society of Chemistry.

Different ions arrive at the sensor at different times, thus pushing out m/z. The conventional assumption is that an ion has only one charge, so the resulting signal corresponds directly to the relative atomic mass of the ion detected, so in most mass spectrometric diagrams, the X-axis is in atomic mass units.

Time-of-flight mass spectrometry (TOFMS) is also a commonly used method to characterize elements. The collector first gathers the ions produced by the ion source. The ions with zero initial velocity are accelerated by the pulsed electric field, then enter the field-free drift tube and fly to the ion receiver at a constant speed. The larger the ion mass is, the longer it takes to reach the receiver; The smaller the ion mass, the shorter the time it takes to reach the receiver. According to this principle, ions with different masses can be separated according to the ratio of mass to charge. Secondary ion mass spectrometry (SIMS, Figure 4.24) is one of the useful methods in high surface mass spectrometry (MSI) technology [72]. The horizontal resolution of SIMS can be reduced to below 50 nm. It uses a focused primary ion beam to bombard the sample surface and ionize it, and then generates characteristic secondary ions to point to the mass spectrometer. Due to the high sensitivity of the surface and the high resolution of the depth profile, ToF SIMS can be used to study problems related to the electrode–electrolyte interface in the battery system, such as the composition difference of the relevant interface before and after the electrochemical reaction [71, 73–77].

4.5 Summary and Outlook

Modern scientific research is inseparable from the development of modern testing technology, and the in situ characterization of the unique 1-D structure of nanowires is a hot topic in current research. The related research results have been reported many times in *Nature*, *Science*, and other journals. Among the various characterization methods, the most common ones are in situ electron microscopy and in situ XRD. These two characterization techniques are of great significance for studying the morphology and phase changes of nanowires in electrochemical reactions and can often provide convincing evidence to researchers. However, for some characterization methods with harsh test conditions, such as in situ SR and in situ neutron diffraction, only Beijing, Shanghai, Hefei, and Taiwan have SR test sources in China at present. There is still a lot of room for development in the characterization of a single nanowire energy storage device. The optical field generated by femtosecond laser also has a great influence on a single device, which can also provide some information for the characterization of a single nanowire.

In the process of studying the electrochemical reaction mechanism of nanowire materials, it is often necessary to explain a variety of test results, so that the conclusion can be more convincing. By using more in situ osmoelectrochemical technology in the characterization technology, real-time monitoring of the changes of nanowire materials in the entire electrochemical reaction will undoubtedly enable people to more accurately understand the electrochemical energy storage mechanism of nanowire and explore new theories.

References

1 Wang, C.M., Xu, W., Liu, J. et al. (2010). In situ transmission electron microscopy and spectroscopy studies of interfaces in Li ion batteries: challenges and opportunities. *Journal of Materials Research* 25: 1541–1547.

2 Xu, X., Yan, M., Tian, X. et al. (2015). In situ investigation of Li and Na ion transport with single nanowire electrochemical devices. *Nano Letters* 15: 3879–3884.

3 Strelcov, E., Cothren, J., Leonard, D. et al. (2015). In situ SEM study of lithium intercalation in individual V_2O_5 nanowires. *Nanoscale* 7: 3022–3027.

4 Unocic, R.R., Sun, X.-G., Sacci, R.L. et al. (2014). Direct visualization of solid electrolyte interphase formation in lithium-ion batteries with in situ electrochemical transmission electron microscopy. *Microscopy and Microanalysis* 20: 1029–1037.

5 Kennedy, T., Mullane, E., Geaney, H. et al. (2014). High-performance germanium nanowire-based lithium-ion battery anodes extending over 1000 cycles through in situ formation of a continuous porous network. *Nano Letters* 14: 716–723.

6 Liu, J., Li, N., Goodman, M.D. et al. (2015). Mechanically and chemically robust sandwich-structured C@Si@C nanotube array Li-ion battery anodes. *ACS Nano* 9: 1985–1994.

7 Shi, H., Liu, X., Wu, R. et al. (2019). In situ SEM observation of structured Si/C anodes reactions in an ionic-liquid-based lithium-ion battery. *Applied Science* 9: 956.

8 Fan, Z., Zhang, L., Baumann, D. et al. (2019). In situ transmission electron microscopy for energy materials and devices. *Advanced Materials* 31: 1900608.

9 Mukherjee, A., Ardakani, H.A., Yi, T. et al. (2017). Direct characterization of the Li intercalation mechanism into α-V_2O_5 nanowires using in-situ transmission electron microscopy. *Applied Physics Letters* 110: 213903.

10 Li, J., Johnson, G., Zhang, S. et al. (2019). In situ transmission electron microscopy for energy applications. *Joule* 3: 4–8.

11 Gregorczyk, K.E., Liu, Y., Sullivan, J.P. et al. (2013). In situ transmission electron microscopy study of electrochemical lithiation and delithiation cycling of the conversion anode RuO_2. *ACS Nano* 7: 6354–6360.

12 Yuan, Y., Amine, K., Lu, J. et al. (2017). Understanding materials challenges for rechargeable ion batteries with in situ transmission electron microscopy. *Nature Communications* 8: 15806.

13 Gu, M., Parent, L.R., Mehdi, B.L. et al. (2013). Demonstration of an electrochemical liquid cell for operando transmission electron microscopy observation of the lithiation/delithiation behavior of Si nanowire battery anodes. *Nano Letters* 13: 6106–6112.

14 Luo, L., Yang, H., Yan, P. et al. (2015). Surface-coating regulated lithiation kinetics and degradation in silicon nanowires for lithium ion battery. *ACS Nano* 9: 5559–5566.

15 Liu, X.H., Huang, S., Picraux, S.T. et al. (2011). Reversible nanopore formation in Ge nanowires during lithiation-delithiation cycling: an in situ transmission electron microscopy study. *Nano Letters* 11: 3991–3997.

16 Kim, S.-W., Lee, H.-W., Muralidharan, P. et al. (2011). Electrochemical performance and ex situ analysis of $ZnMn_2O_4$ nanowires as anode materials for lithium rechargeable batteries. *Nano Research* 4: 505–510.

17 Silberstein, K.E., Lowe, M.A., Richards, B. et al. (2015). Operando X-ray scattering and spectroscopic analysis of germanium nanowire anodes in lithium ion batteries. *Langmuir* 31: 2028–2035.

18 Li, A., Song, H., Wan, W. et al. (2014). Copper oxide nanowire arrays synthesized by in-situ thermal oxidation as an anode material for lithium-ion batteries. *Electrochimica Acta* 132: 42–48.

19 Balasubramanian, M., Sun, X., Yang, X.Q. et al. (2001). In situ X-ray diffraction and X-ray absorption studies of high-rate lithium-ion batteries. *Journal of Power Sources* 92: 1–8.

20 Misra, S., Liu, N., Nelson, J. et al. (2012). In situ X-ray diffraction studies of (de)lithiation mechanism in silicon nanowire anodes. *ACS Nano* 6: 5465–5473.

21 Beaulieu, L.Y., Hatchard, T.D., Bonakdarpour, A. et al. (2003). Reaction of Li with alloy thin films studied by in situ AFM. *Journal of the Electrochemical Society* 150: A1457.

22 Liu, H., Wu, T., Zhang, L. et al. (2022). Germanium nanowires via molten-salt electrolysis for lithium battery anode. *ACS Nano* 16: 14402–14411.

23 Li, Y., Zheng, R., Yu, H. et al. (2019). Observation of $ZrNb_{14}O_{37}$ nanowires as a lithium container via in situ and ex situ techniques for high-performance lithium-ion batteries. *ACS Applied Materials & Interfaces* 11: 22429–22438.

24 Tang, G., Zhu, H., Yu, H. et al. (2018). Ultra-long $BiNbO_4$ nanowires with hierarchical architecture exhibiting reversible lithium storage. *Journal of Electroanalytical Chemistry* 823: 245–252.

25 Krause, A., Tkacheva, O., Omar, A. et al. (2019). In situ Raman spectroscopy on silicon nanowire anodes integrated in lithium ion batteries. *Journal of the Electrochemical Society* 166: A5378.

26 Li, J., Xie, W., Zhang, S. et al. (2021). Boosting the rate performance of Li–S batteries under high mass-loading of sulfur based on a hierarchical NCNT@Co-CoP nanowire integrated electrode. *Journal of Materials Chemistry A* 9: 11151–11159.

27 Stancovski, V. and Badilescu, S. (2014). In situ Raman spectroscopic–electrochemical studies of lithium-ion battery materials: a historical overview. *Journal of Applied Electrochemistry* 44: 23–43.

28 Shang, H., Zuo, Z., Li, L. et al. (2018). Ultrathin graphdiyne nanosheets grown in situ on copper nanowires and their performance as lithium-ion battery anodes. *Angewandte Chemie, International Edition* 57: 774–778.

29 Stokes, K., Flynn, G., Geaney, H. et al. (2018). Axial Si–Ge heterostructure nanowires as lithium-ion battery anodes. *Nano Letters* 18: 5569–5575.

30 Lanz, P., Villevieille, C., and Novák, P. (2013). Electrochemical activation of Li_2MnO_3 at elevated temperature investigated by in situ Raman microscopy. *Electrochimica Acta* 109: 426–432.

31 Gross, T., Giebeler, L., and Hess, C. (2013). Novel in situ cell for Raman diagnostics of lithium-ion batteries. *The Review of Scientific Instruments* 84: 073109.

32 Liu, Q., Hao, Z., Liao, X. et al. (2019). Langmuir-Blodgett nanowire devices for in situ probing of zinc-ion batteries. *Small* 15: 1902141.

33 Li, Q., Qiao, Y., Guo, S. et al. (2018). Both cationic and anionic co-(de) intercalation into a metal-oxide material. *Joule* 2: 1134–1145.

34 Lu, Y.-C., Crumlin, E.J., Veith, G.M. et al. (2012). In situ ambient pressure X-ray photoelectron spectroscopy studies of lithium-oxygen redox reactions. *Scientific Reports* 2: 1–6.

35 Endo, R., Ohnishi, T., Takada, K. et al. (2020). In situ observation of lithiation and delithiation reactions of a silicon thin film electrode for all-solid-state lithium-ion batteries by X-ray photoelectron spectroscopy. *Journal of Physical Chemistry Letters* 11: 6649–6654.

36 Li, H., Shen, L., Ding, B. et al. (2015). Ultralong $SrLi_2Ti_6O_{14}$ nanowires composed of single-crystalline nanoparticles: promising candidates for high-power lithium ions batteries. *Nano Energy* 13: 18–27.

37 Cai, X., Yan, H., Zheng, R. et al. (2021). $Cu_2Nb_{34}O_{87}$ nanowires as a superior lithium storage host in advanced rechargeable batteries. *Inorganic Chemistry Frontiers* 8: 444–451.

38 Wang, Y., Cao, L., Li, J. et al. (2020). Cu/Cu_2O@Ppy nanowires as a long-life and high-capacity anode for lithium ion battery. *Chemical Engineering Journal* 391: 123597.

39 Yao, K.P., Kwabi, D.G., Quinlan, R.A. et al. (2013). Thermal stability of Li_2O_2 and Li_2O for Li-air batteries: in situ XRD and XPS studies. *Journal of the Electrochemical Society* 160: A824.

40 Liu, D., Shadike, Z., Lin, R. et al. (2019). Review of recent development of in situ/operando characterization techniques for lithium battery research. *Advanced Materials* 31: 1806620.

41 Lin, D., Li, K., and Zhou, L. (2021). Advanced in-situ characterizations of nanocomposite electrodes for sodium-ion batteries – a short review. *Composites Communications* 25: 100635.

42 Lu, J., Wu, T., and Amine, K. (2017). State-of-the-art characterization techniques for advanced lithium-ion batteries. *Nature Energy* 2: 1–13.

43 Li, H., Guo, S., and Zhou, H. (2021). In-situ/operando characterization techniques in lithium-ion batteries and beyond. *Journal of Energy Chemistry* 59: 191–211.

44 Liu, X., Tong, Y., Wu, Y. et al. (2021). In-depth mechanism understanding for potassium-ion batteries by electroanalytical methods and advanced in situ characterization techniques. *Small Methods* 5: 2101130.

45 Borkiewicz, O.J., Shyam, B., Wiaderek, K.M. et al. (2012). The AMPIX electrochemical cell: a versatile apparatus for in situ X-ray scattering and spectroscopic measurements. *Journal of Applied Crystallography* 45: 1261–1269.

46 Zhou, Y., Grunwaldt, J.D., Krumeich, F. et al. (2010). Hydrothermal synthesis of $Bi_6S_2O_{15}$ nanowires: structural, in situ EXAFS, and humidity-sensing studies. *Small* 6: 1173–1179.

47 Xu, S., Qin, Y., Xu, C. et al. (2010). Self-powered nanowire devices. *Nature Nanotechnology* 5: 366–373.

48 Ghosh, S., Bao, W., Nika, D.L. et al. (2010). Dimensional crossover of thermal transport in few-layer graphene. *Nature Materials* 9: 555–558.

49 Mai, L., Dong, Y., Xu, L. et al. (2010). Single nanowire electrochemical devices. *Nano Letters* 10: 4273–4278.

50 Hu, P., Yan, M., Wang, X. et al. (2016). Single-nanowire electrochemical probe detection for internally optimized mechanism of porous graphene in electrochemical devices. *Nano Letters* 16: 1523–1529.

51 Yan, M., Wang, F., Han, C. et al. (2013). Nanowire templated semihollow bicontinuous graphene scrolls: designed construction, mechanism, and enhanced energy storage performance. *Journal of the American Chemical Society* 135: 18176–18182.

52 Becker, C.R., Strawhecker, K.E., McAllister, Q.P. et al. (2013). In situ atomic force microscopy of lithiation and delithiation of silicon nanostructures for lithium ion batteries. *ACS Nano* 7: 9173–9182.

53 Huang, S., Cheong, L.-Z., Wang, S. et al. (2018). In-situ study of surface structure evolution of silicon anodes by electrochemical atomic force microscopy. *Applied Surface Science* 452: 67–74.

54 Tokranov, A., Sheldon, B.W., Li, C. et al. (2014). In situ atomic force microscopy study of initial solid electrolyte interphase formation on silicon electrodes for Li-ion batteries. *ACS Applied Materials & Interfaces* 6: 6672–6686.

55 Zhang, L., Yang, T., Du, C. et al. (2020). Lithium whisker growth and stress generation in an in situ atomic force microscope–environmental transmission electron microscope set-up. *Nature Nanotechnology* 15: 94–98.

56 Liu, X., Wang, D., and Wan, L. (2015). Progress of electrode/electrolyte interfacial investigation of Li-ion batteries via in situ scanning probe microscopy. *Scientific Bulletin* 60: 839–849.

57 Liu, X.-R., Deng, X., Liu, R.-R. et al. (2014). Single nanowire electrode electrochemistry of silicon anode by in situ atomic force microscopy: solid electrolyte interphase growth and mechanical properties. *ACS Applied Materials & Interfaces* 6: 20317–20323.

58 Harks, P.P.R.M.L., Mulder, F.M., and Notten, P.H.L. (2015). In situ methods for Li-ion battery research: a review of recent developments. *Journal of Power Sources* 288: 92–105.

59 Lee, H.-W., Li, Y., and Cui, Y. (2016). Perspectives in in situ transmission electron microscopy studies on lithium battery electrodes. *Current Opinion in Chemical Engineering* 12: 37–43.

60 Zhang, Y., Li, Y., Wang, Z. et al. (2014). Lithiation of SiO_2 in Li-ion batteries: in situ transmission electron microscopy experiments and theoretical studies. *Nano Letters* 14: 7161–7170.

61 Krachkovskiy, S., Trudeau, M.L., and Zaghib, K. (2020). Application of magnetic resonance techniques to the in situ characterization of Li-ion batteries: a review. *Materials* 13: 1694.

62 Ogata, K., Salager, E., Kerr, C. et al. (2014). Revealing lithium–silicide phase transformations in nano-structured silicon-based lithium ion batteries via in situ NMR spectroscopy. *Nature Communications* 5: 1–11.

63 Key, B., Bhattacharyya, R., Morcrette, M. et al. (2009). Real-time NMR investigations of structural changes in silicon electrodes for lithium-ion batteries. *Journal of the American Chemical Society* 131: 9239–9249.

64 Lai, S.Y., Knudsen, K.D., Sejersted, B.T. et al. (2019). Silicon nanoparticle ensembles for lithium-ion batteries elucidated by small-angle neutron scattering. *ACS Applied Energy Materials* 2: 3220–3227.

65 Qian, D., Ma, C., More, K.L. et al. (2015). Advanced analytical electron microscopy for lithium-ion batteries. *NPG Asia Materials* 7: e193–e193.

66 Srinivasan, R.M., Chandran, K.R., Chen, Y. et al. (2022). In-operando neutron diffraction investigation of structural transitions during lithiation of Si electrode in Li-ion battery. *Journal of the Electrochemical Society* 169: 100545.

67 Zhou, H., An, K., Allu, S. et al. (2016). Probing multiscale transport and inhomogeneity in a lithium-ion pouch cell using in situ neutron methods. *ACS Energy Letters* 1: 981–986.

68 Didier, C., Pang, W.K., Guo, Z. et al. (2020). Phase evolution and intermittent disorder in electrochemically lithiated graphite determined using in operando neutron diffraction. *Chemistry of Materials* 32: 2518–2531.

69 Du, G., Sharma, N., Peterson, V.K. et al. (2011). Br-doped $Li_4Ti_5O_{12}$ and composite TiO_2 anodes for Li-ion batteries: synchrotron X-ray and in situ neutron diffraction studies. *Advanced Functional Materials* 21: 3990–3997.

70 Wang, X.-L., An, K., Cai, L. et al. (2012). Visualizing the chemistry and structure dynamics in lithium-ion batteries by in-situ neutron diffraction. *Scientific Reports* 2: 1–7.

71 Liu, B., Yu, X.-Y., Zhu, Z. et al. (2014). In situ chemical probing of the electrode–electrolyte interface by ToF-SIMS. *Lab on a Chip* 14: 855–859.

72 Lu, J., Hua, X., and Long, Y.-T. (2017). Recent advances in real-time and in situ analysis of an electrode–electrolyte interface by mass spectrometry. *The Analyst* 142: 691–699.

73 Bordes, A., De Vito, E., Haon, C. et al. (2016). Multiscale investigation of silicon anode Li insertion mechanisms by time-of-flight secondary ion mass spectrometer imaging performed on an in situ focused ion beam cross section. *Chemistry of Materials* 28: 1566–1573.

74 Liu, Z., Lu, P., Zhang, Q. et al. (2018). A bottom-up formation mechanism of solid electrolyte interphase revealed by isotope-assisted time-of-flight secondary ion mass spectrometry. *Journal of Physical Chemistry Letters* 9: 5508–5514.

75 Yulaev, A., Oleshko, V., Haney, P. et al. (2018). From microparticles to nanowires and back: radical transformations in plated Li metal morphology revealed via in situ scanning electron microscopy. *Nano Letters* 18: 1644–1650.

76 Wu, S.-C., Ai, Y., Chen, Y.-Z. et al. (2020). High-performance rechargeable aluminum–selenium battery with a new deep eutectic solvent electrolyte: thiourea-$AlCl_3$. *ACS Applied Materials & Interfaces* 12: 27064–27073.

77 Pereira-Nabais, C., Światowska, J., Chagnes, A. et al. (2013). Interphase chemistry of Si electrodes used as anodes in Li-ion batteries. *Applied Surface Science* 266: 5–16.

5

Nanowires for Lithium-ion Batteries

5.1 Electrochemistry, Advantages, and Issues of LIBs Batteries

5.1.1 History of Lithium-ion Batteries

Rechargeable batteries (secondary batteries) have been widely used in many applications, including consumer electronics, electric tools, and more recently, rechargeable cells are being developed as a power source for electric and hybrid electric vehicles and grid storage [1]. The basic components of each battery include two electrodes, the anode and the cathode, and an electrolyte (separator) to separate them [2]. Electric power is stored in a battery as chemical energy between two electrodes. Among batteries operated by various types of electrochemistry, rechargeable batteries based on lithium intercalation chemistry are the most promising candidate and lithium-ion batteries (LIBs) have achieved great success in the market [3]. The history of rechargeable lithium batteries began in 1960s, when chemists in Europe explored the intercalation chemistry of lithium into layered transition-metal sulfides. In 1976, Stanley Whittingham created a rechargeable lithium battery composed of a layered titanium disulfide (TiS_2) cathode, a lithium metal anode, and organic electrolytes with an operation potential of 2.0 V and good initial performance [4]. However, during repeated charging process, Li dendrites forms on the anode surface, which grows across the electrolyte to the cathode to create an internal short circuit. Consequently, the large amount of heat generated during the short circuit would ignite the flammable electrolyte, raising serious safety hazards [5]. In addition, the operation voltage of TiS_2 cathode is rather limited to <2.5 V due to the overlap of the Ti^{4+}:3d band with the top of the S^{2-}:3p band. Such an overlap results in the introduction of holes or removal of electrons from the S^{2-}:3p band and the formation of molecular ions such as S_2^{2-}, rather than the accessibility of higher states of M^{n+} ions, i.e. higher voltages.

Recognizing the instability of sulfides, John B. Goodenough's group at Oxford University focused on layered oxide cathodes during the 1980s. They demonstrated reversible Li insertion and extraction in a layered lithium cobalt oxide ($LiCoO_2$) compound [6]. Over half of the lithium could be extracted reversibly at a high voltage of

Nanowire Energy Storage Devices: Synthesis, Characterization, and Applications, First Edition.
Edited by Liqiang Mai.

~4 V versus Li^+/Li, showing significantly improved energy density compared with TiS_2 cathode [7]. Based on this innovation, Akira Yoshino, then at the Asahi Kasei Corporation, made the first rechargeable LIB by combining the $LiCoO_2$ cathode with a graphitic-carbon anode, which was commercialized by Sony Corporation in 1991 to power the first portable phone [8]. After more than two decades of development, the commercial LIBs based on "intercalation chemistry" offer a gravimetric energy density of >250 Wh kg^{-1} and a volumetric energy density of >700 Wh l^{-1}, which is almost five times higher than that of the Pb–acid batteries [9]. For now, the LIBs have dominated the portable electronics market and are now actively expanding their applications into emerging markets including electric vehicles and grid storage. Given the huge success and great influence of LIBs, the 2019 Nobel Prize in Chemistry was jointly given to John B. Goodenough, Stanley M. Whittingham, and Akira Yoshino, recognizing their pioneering work on lithium intercalation materials for LIBs.

5.1.2 Electrochemistry of Lithium-ion Batteries

The cell configuration and operation principle of a typical rechargeable LIB-based $LiCoO_2$ cathode and graphitic carbon anode are shown in Figure 5.1 [10]. During the charge process, electrons leave the cathode via an external circuit accompanied with Co^{3+} being oxidized to Co^{4+} and flows toward to anode, where carbon is reduced. To retain charge neutrality in electrodes, Li ions (guests) are released from the layered host (CoO_6 octahedral layer) to electrolyte, and the working cation of the electrolyte carries positive charge to the anode, finally inserting into the graphite host structure. This process is reversed during discharge process. The cell electrochemical reactions

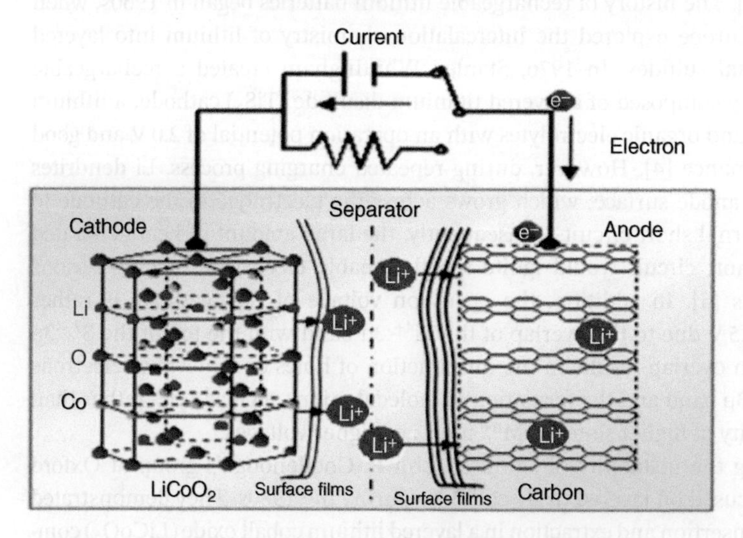

Figure 5.1 Schematic illustration of the configuration and working principle of a typical Li ion battery based on "rocking chair" chemistry, which employs carbon as anode and lithium transition metal oxide as cathode. Source: Etacheri et al. [10].

during charge process are described as follows [2, 11].

Cathode: $LiCoO_2 \rightarrow Li_{1-x}CoO_2 + xe^- + xLi^+$

Anode: $C + ye^- + yLi^+ \rightarrow Li_xC$

Overall reaction: $LiCoO_2 + (x/y)C \rightarrow x/yLi_yC + Li_{1-x}CoO_2$

There are several important factors to evaluate the performance of a LIB, including operation potential, theoretical/practical specific capacity of electrode materials and cells, and theoretical/practical specific energy of the cell, since they determine the theoretical energy stored in the cell upon one charge. We will discuss them in the following section in detail.

5.1.2.1 Theoretical Operation Potential

The open-circuit voltage V_{OC} of such a lithium cell is given by the difference in the lithium chemical potential between the cathode (μ_C) and the anode (μ_A) in the following equation, where E_A and E_C represent the theoretical electrode potential for anode and cathode, respectively; F is the Faraday constant ($F = 96\,485$ C mol^{-1}) [11, 12].

$$V_{OC} = E_C - E_A = (\mu_A - \mu_C)/F \quad (5.1)$$

For example, E_C of a $LiCoO_2$ cathode is ~4.0 V versus Li^+/Li and E_A of a graphite anode is ~0.2 V versus Li^+/Li, giving a full cell with operation voltage of 3.8 V [12].

5.1.2.2 Theoretical Specific Capacity of Electrode Materials and Cells

Faraday's law could be employed to calculate the theoretical specific capacity (Q) of a certain electrode material:

$$Q = (n \times F)/(3600 \times M_W) \text{ (unit : Ah g}^{-1}\text{)} \quad (5.2)$$

where n is the number of charge carriers, F is the Faraday constant (96 485 C mol^{-1}), and M_W is the molecular weight of active materials used in the electrode. The theoretical specific capacity of $LiCoO_2$ could be calculated in case a complete extraction of Li occurs during charge:

$$Q_{theoretical} = (1 \times 96\,485)/(3600 \times 97.87) = 0.274 \text{ Ah g}^{-1}$$

Nevertheless, in a practical cell, the specific capacity of electrode during operation is usually different from the theoretical value, which can be calculated according to the voltage-time profile from the galvanostatic cycling test:

$$Q_{practical} = (I \times A \times t)/(3600 \times Mw) \text{ (unit: Ah g}^{-1}\text{)} \quad (5.3)$$

where I is the current density (A m^{-2}), A is the electrode area (m^2), and t is the time to reach the cut-off potential. Taking $Li_{1-x}CoO_2$ as an example, it has a high theoretical specific capacity of 0.274 Ah g^{-1}, but the cathode evolves oxygen release once removing Li^+ beyond $x = 0.55$ because of the pinned Co^{3+}/Co^{4+} redox couple at the top of the O:2p bands. As a result, the reversible delivered capacity of $LiCoO_2$ cathode is limited to ~150 mAh g^{-1} [2].

The theoretical specific capacity of an electrochemical cell (Q_{cell}) is determined by the theoretical specific capacities of both electrodes and can be calculated as follows:

$$Q_{cell} = Q_{theoretical-a} \times Q_{theoretical-c} / (Q_{theoretical-a} + Q_{theoretical-c}) \quad (5.4)$$

where $Q_{theoretical-a}$ denotes the theoretical specific capacity of the anode while $Q_{theoretical-c}$ denotes the theoretical specific capacity of the cathode.

5.1.2.3 Theoretical Specific Energy Density of an Electrochemical Cell

To calculate the theoretical specific energy (unit: Wh kg^{-1}) of an electrochemical cell, we can multiply the theoretical specific capacity (Q_{cell}) and theoretical voltage output (V_{oc}) of the cell. The energy stored in a practically fully charged cell may be obtained by measuring the time Δt for its complete discharge at a constant current $I_{dis} = dq/dt$ (where q represents the state of charge). The energy density of a practical cell can be calculated by the following equation, where wt is the weight of the cell [11].

$$\text{Energy density} = \int_0^{\Delta t} IV(t)dt/wt = \int_0^{Q} V(q)dq/wt \quad (5.5)$$

5.1.3 Key Materials for Lithium-ion Batteries

According to the above discussion, it is clear that the gravimetric specific energy of a rechargeable battery could be improved by increasing the capacity, increasing the operation voltage, or both. To achieve such a goal, it is essential to use battery materials with properly designed structure and compositions. The key materials for LIBs include cathode, anode, electrolyte, and separators, which are discussed in the following section.

5.1.3.1 Cathode

To realize practical LIBs with high energy output and long lifespan, the cathode must meet the following critical requirements:

(i) A high specific capacity and a high electrode potential (or low chemical potential μ_C).
(ii) Highly reversible lithium insertion/extraction reaction with minimal change in structure.
(iii) Mixed conduction capability, i.e. good electronic and ionic conductivity.
(iv) Low-cost, environmentally benign, and thermally and chemically stable.
(v) High tap/press density to ensure high volumetric density.

In the search for advanced electrode materials for high-energy LIBs, cathode materials based on intercalation chemistry are still promising candidates due to their stable cycling performance, high tap density, and high working potential [13]. Insertion-type cathode materials are classified into two major categories: layered oxides and polyanionic compounds [7]. The electrochemical-related features of some typical insertion-type cathodes are summarized in Table 5.1.

Table 5.1 Electrochemical-related features of some typical insertion-type cathodes for LIBs.

Material	Specific capacity ($mAh\,g^{-1}$)	Average potential (V versus Li+/Li)	Pros.	Cons.
$LiCoO_2$	155	3.9[a)]	High volumetric energy density	Expensive
$LiMn_2O_4$	100–120	4.0[a)]	Low price, safer electrochemistry	Poor high temperature durability
$LiFePO_4$	160	3.4[a)]	Low price, high thermal stability	Low volumetric energy density, poor electronic and ionic conductivity
$LiNi_{0.8}Co_{0.1}Mn_{0.1}O_2$	200	3.8[a)]	High energy density, low price	Poor thermal stability, high air sensitivity
$LiNi_{0.8}Co_{0.15}Al_{0.05}O_2$	200	3.73[a)]		
$Li_3V_2(PO_4)_3$	197[b)]	~3.8[a)]	High structure stability and thermal stability	High price of V, poor electronic and ionic conductivity

a) Midpoint voltage.
b) Theoretical value.
Source: Data obtained from Refs. [14, 15].

5.1.3.2 Anode

As for the anode side, the intercalation chemistry has been dominated by graphite and its derivatives, which enable very stable and reversible lithium intercalation/extraction. Given the limited theoretical capacity of insertion-type graphite anode (372 $mAh\,g^{-1}$), new anode materials having higher energy densities are desired. Opportunities have been found in advanced materials operated by new storage mechanism, such as interfacial/nanopore-storage-type porous carbons, alloy-type anodes and their derivatives, conversion-type oxides, and the plating/stripping type Li metal [16–19]. For instance, the Li metal anode has a large theoretical capacity of 3860 $mAh\,g^{-1}$. This value is about ten times greater than the present graphite anode. The electrochemical-related features of some typical anode materials are summarized in Table 5.2.

5.1.3.3 Electrolyte

The electrolyte in LIBs must satisfy several key requirements such as:

(i) A wide electrochemical stable voltage window E_g to ensure high energy density.
(ii) A high Li^+ conductivity σ_{Li} of $>10^{-4}\,S\,cm^{-1}$ and a low electronic conductivity of $\sigma_e < 10^{-10}\,S\,cm^{-1}$ over the temperature range of battery operation.

Table 5.2 Summary of some typical anode materials for LIBs.

Charge storage mechanism	Material	Specific capacity (mAh g^{-1})	Comments
Insertion-type	Graphite	372[a]	Stable cycling, high Coulumbic efficiency, low energy density
	$Li_4Ti_5O_{12}$	175[a]	"Zero-strain" structure, high safety, poor electronic/ionic conductivity, low energy density
Interfacial/ nanopore-storage-type	Nanostructured carbonaceous materials	<1750	High in capacity but poor in Coulombic efficiency, severe side reactions with electrolyte
Alloy-type	Si	4200[a]	Extremely high capacity. Low operation potential. Huge volume variations. Poor electrochemical stability. Low price
	P	2596[a]	
	Sn	960[a]	
Conversion-type	Metal oxides	375–1170	High in capacity but large potential polarizations. Poor electrochemical stability
	Metal sulfides	447–1230	
	Metal phosphides	700–1800	
Stripping/ plating-type	Li metal	3860[a]	Extremely high energy density. Extremely low electrode potential. Poor Coulombic efficiency and cycling stability, safety hazards

a) Theoretical value.
Source: Data obtained from Refs [20–23].

(iii) A transference number $\sigma_{Li}/\sigma_{total} \approx 1$, where σ_{total} includes conductivities by other ions in the electrolyte.
(iv) Good chemical and electrochemical compatibility with both positive and negative electrodes, e.g. the capability to form passivating solid/electrolyte-interface (SEI) layer.
(v) High flash point, high boiling point, and preferably nonflammable and nonexplosive to ensure a safe battery.
(vi) Low toxicity and low cost.

5.1.3.4 Separator

A separator is located between the positive and negative electrodes to avoid short circuit caused by physical contact between two electrodes and works as an electrolyte reservoir [12]. To achieve such a goal, the separator needs to be mechanically

Table 5.3 General requirements for separators used in LIBs.

Parameter	Requirement
Chemical and electrochemical stabilities	Stable for a long period of time
Wettability	Wet out quickly and completely
Mechanical property	>1000 $kg\,cm^{-1}$(98.06 MPa)
Thickness	20–25 μm
Pore size	<1 μm
Porosity	40–60%
Permeability (Gurley)	<0.025 $s\,\mu m^{-1}$
Dimensional stability	No curl up and lay flat
Thermal stability	<5% shrinkage after 60 min at 90 °C
Shutdown	Effectively shut down the battery at elevated temperatures

Source: Wu et al. [24].

and chemically (electrochemically) stable, i.e. does not directly involve any electrochemical reactions. In addition, an ideal separator should meet the following requirement: (i) high wettability and good permeability toward electrolyte for efficient ion transport; (ii) high thermal stability or thermal shutdown ability to avoid short-circuiting at elevated temperatures. The property of separator, including structure, morphology, porosity, and wettability toward electrolytes, has a great influence on the cell kinetics (Table 5.3).

5.1.4 Advantages and Issues of Lithium-ion Batteries

The development of electrochemical energy storage (EES) systems can be traced back to the year 1800. During the next two centuries, EES systems will penetrate almost every aspect of modern civilization, having a great influence on people's daily lives. Throughout the history of EES development, from Pb-acid batteries (1850s), Ni–Cd batteries (1890s), and Ni–MH cells (1960s) to, finally, LIBs at present [25], LIBs outperform other EES systems in terms of high gravimetric/volumetric energy density, high power density, and long cycle life. For now, the gravimetric energy densities of commercial LIBs based on "intercalation chemistry" and non-aqueous liquid electrolytes are almost five times higher than those of the Pb–acid batteries. However, the desire to improve energy density is always the driving force behind the progress of advanced technology. In the future, a target energy density of >300 $Wh\,kg^{-1}$ or even 500 $Wh\,kg^{-1}$ is desired to power electric vehicles with a 300-mile driving range. Nevertheless, the energy density of state-of-the-art LIBs is rather limited by traditional intercalation chemistry due to the relatively small number of crystallographic sites for storing charge-carrier ions. For this reason, materials that operate on new chemistry for Li^+ storage, such as alloying and conversion, are emerging as promising candidates because of the possibility that they will surpass the energy densities of intercalation-type electrodes [26]. Systems governed by new electrochemistry are noted as "post-LIBs" hereafter [27].

The enhanced energy density of post-LIBs makes them promising next-generation energy storage systems; nevertheless, irreversible phase transitions and unstable electrode/electrolyte interfaces are still two major challenges to conquer [27]. In addition, the rapidly developing electronic industry gives rise to diverse function requirement for EES systems, such as flexible batteries and micro-batteries [25]. Careful materials structure optimization and electrode design are essential for post-LIBs to take advantage of their material superiority and eventually fulfill their mission of surpassing state-of-the-art LIBs. Nanowire (NW) materials, thanks to their unique physical/chemical properties, offer new opportunities for addressing the challenges of advanced EES systems [28]. In this section, we will discuss how LIBs benefit from NW-structured materials in terms of cathode, anode, separator, and electrolyte.

5.2 Unique Characteristic of Nanowires for LIBs

As mentioned in the previous section, the intrinsic characteristics of electrode materials directly affect the capacity and life of LIBs, so the present development bottleneck of LIBs is mainly limited by the development of high-performance electrode materials. In recent years, the rapid development of nanotechnology and the wide application of nanomaterials have made the choice of electrode materials of LIBs no longer limited to the combination of different elements [29]. As a typical nanostructure, NW materials possess unique electrochemical characteristics for the application of LIBs, which is not able to provide with traditional bulk materials. The unique features of NW electrode materials include the following [30–33]:

5.2.1 Enhancing the Diffusion Dynamics of Carriers

LIBs are also known as "rocking-chair" batteries because lithium ions shuttle back and forth between cathode and anode during the charge and discharge processes. For conventional bulk electrodes, the slow diffusion kinetics of lithium ions within them is the main reason limiting the rate capability of the battery. Compared with bulk materials, NWs with diameter in the nanometer scale can greatly shorten the diffusion distance of ions in the radial diffusion of ions along NWs, and the electrons can be continuously transmitted along the axis, thus enhancing the diffusion kinetic properties of carriers and improving the rate capability of LIBs.

5.2.2 Enhancing Structural Stability of Materials

The ionic batteries are based on the ion intercalation/extraction reactions in the electrode material. The insertion of charged ions into electrode material will lead to the internal stress of the lattice, resulting in the destruction of the material structure and even the fracture or pulverization of the electrode, and then the rapid decay of the specific capacity. Because of their radial position at the nanometer scale, NW materials have good stress release properties, which can effectively accommodate the volume changes for electrodes during lithiation/delithiation, and improve the structural stability of the materials compared with bulk materials.

5.2.3 Befitting the In Situ Characterization of Electrochemical Process

A deep understanding of the electrochemical reaction mechanism of electrode materials is of great significance to optimizing the electrochemical performance of the battery. At present, although the practical application of LIBs is basically mature, the details of electrochemical reaction of electrode materials are very little known, which is precisely the key to solving the current capacity bottleneck of LIBs. The in situ characterization of electrode materials during charge and discharge is an effective method to observe and understand their intrinsic electrochemical properties. In virtue of their high aspect ratio and robust mechanical properties, NWs have unique advantages over other materials in building in situ electrochemical devices for the in site characterization of electrochemical.

5.2.4 Enabling the Construction of Flexible Devices

With the development of light and miniaturization of modern electronic devices, wearable and flexible energy storage devices have become one of the most competitive potential development directions in the future. The high aspect ratio of NWs makes them possess good bending and robust mechanical properties, which cannot only construct three-dimensional (3D) free-standing and metallic current collector-free electrodes but also interweave with each other to form a flexible NW network. This makes NW materials unique in the fabrication of flexible electrodes to construct flexible devices.

5.3 Nanowires as Anodes in LIBs

5.3.1 Alloy-Type Anode Materials (Si, Ge, and Sn)

5.3.1.1 Lithium Storage in Si Nanowires

The capacity of the full battery depends on the capacities of matched cathode and anode (only considering positive and negative materials), and nowadays, the research on cathode materials has approached its theoretical capacity, while the capacity of full batteries is mainly limited by the anode materials [34]. Thus, there are also a lot of efforts devoted to finding other effective anodes with higher capacity and lower price compared with commercial graphite systems.

Silicon, which possesses a surprising high theoretical capacity (4200 mAh g^{-1}, ten times over the theoretical capacity of Li metal), is one of the most promising anode materials for next-generation LIBs [35, 36]. In addition to its satisfactory capacity, Si is the second most abundant element in the earth's crust resulting in low cost compared to Li. Although its lithiation voltage (0.4 V versus Li/Li^{+}) is higher than that of graphite (0.1 V versus Li/Li^+), it is still a nice choice as anode compared with most other alloys and metal oxides [37]. However, there are some bulk rocks lying on the road to the commercial application of Si-based anodes. Besides the low electronic conductivity, the huge volume expansion (over 300%) of Si-based electrodes during lithiation/delithiation procedure is the most severe problem, which causes an unstable SEI layer, destruction, and electronic isolation of electrode, finally leading to poor cycle stability and rate performance [38, 39].

Nanotechnology should be an effective way to solve the above problems, and there is already evidence showing that the intrinsic physical and chemical properties of Si can be significantly improved by nanocrystallization. Among nano-sized materials, from 0D materials (Si nanoparticles) to 3D materials (nanosphere), the one-dimensional Si-nanowires (SiNWs) have attracted researcher's devotion due to the following reasons: [40] (i) SiNWs can provide rapid electronic and ionic transport pathways; (ii) SiNWs with small diameters can accommodate large volume variation; (iii) void space constructed by adjacent NWs allows for fast penetration and storage of electrolyte; and (iv) SiNWs can offer high material utilization due to a high surface-to-volume ratio.

Of course, pure SiNWs used as anode materials are also not able to solve all the problems. In fact, SiNWs still have to face the structural attenuation caused by the volume change during the cycle and the continuous growth of SEI films, which are obviously weaker than that of other morphological materials. Thus, incorporating additives into the electrode is a necessary strategy for further improvement of electrochemical performance.

A suitable doping material that will not change the crystalline structures of substrate materials after doping procedures can effectively influence the intrinsic chemical and physical properties of base materials. There are two common categories in which SiNWs are growing: (i) bottom-up methods, most of which are based on vapor–liquid–solid (VLS) mechanism, such as chemical vapor deposition (CVD), molecular beam epitaxy (MBE), laser ablation, and supercritical-fluid-liquid–solid (SFLS) method; (ii) top-down methods, the most popular method of which is metal-assisted chemical etching (MACE) [41].

As for doping elements applied in these methods, B and P have been extensively investigated. As it is well-known, the doped B will get an electron from Si when it forms a bond, and an excess hole will be generated accordingly, which is called p-type doping; And P will provide an excess electron, which is called n-type doping. These two doping methods can both greatly increase the carrier concentration in the silicon material to improve its conductivity.

Ge et al. prepared a B-doped porous SiNWs by using MACE method (Figure 5.2a) [42]. They directly etched the B-doped Si wafer with $AgNO_3$, which is possible to adjust the pore size of this porous NWs through concentration during synthesis process. According to this work, the B element cannot only work on improving the intrinsic properties of SiNWs, but also help form its surface structure. The obtained sample possessed an average pore diameter of 8 nm, proving numbers of sites for contacting with electrolyte. Besides, because of this kind of porous structure, shorter diffusion paths of ions/electrons and enough space for volume expansion during charge/discharge processes facilitate the chemical performance

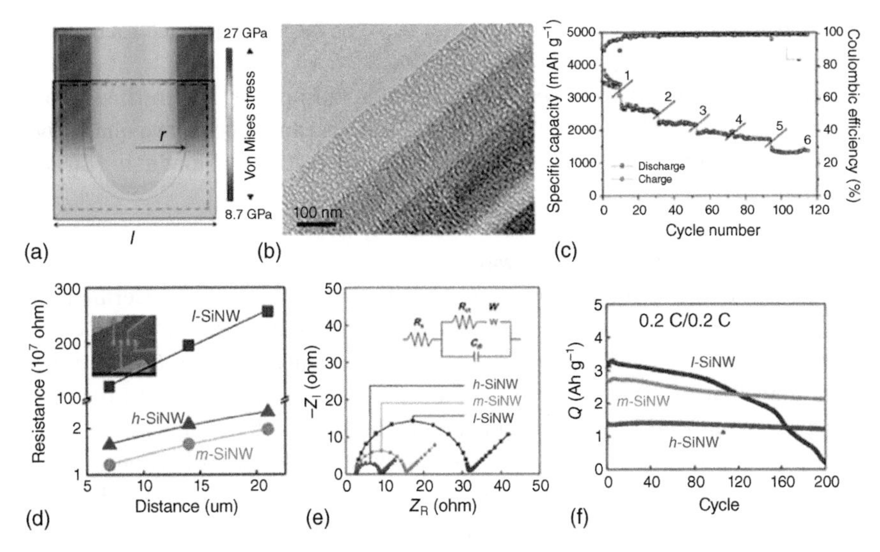

Figure 5.2 (a) One unit of the porous structure is used for theoretical simulation and analysis. (b) TEM images of porous Si nanowires etched with 0.02 M $AgNO_3$. (c) Charge/discharge capacity and Coulombic efficiency of porous silicon nanowire electrodes at current rates of 0.6, 1.2, 2.4, 3.6, 4.8, and 9.6 $A\,g^{-1}$. (d) Electric resistances of SiNWs measured as a function of distance by a probe station. Inset: A SEM image of a single SiNW contacting to four metal contacts with a gap between the nearest contacts = 7 μm. (e) Impedance spectra of coin-type cells of SiNW | lithium metal at fully charged states by electrochemical impedance spectroscopy (EIS). (f) Capacity retention of SiNWs during repeated cycles of lithiation and delithiation at 0.2 C.

of SiNWs-based anodes (Figure 5.2a). Under the synergistic effect between composition and structure, the discharge capacity of this anode is over 1900 $mAh\,g^{-1}$ at the current density of 3.6 $A\,g^{-1}$ during the beginning several cycles (Figure 5.2c). And the Coulombic efficiency (CE) of this cell reaches above 99.5% after 20 cycles with a stable capacity of around 1600 $mAh\,g^{-1}$ at current density of 4 $A\,g^{-1}$ after 250 cycles. In addition to analyzing the influence of element B on SiNWs from a qualitative perspective, other researchers have also studied it from a quantitative perspective. Hwang et al. fabricated B-doped SiNWs with different orders of doping concentration including 1.2×10^{12}, 2.7×10^{15}, $5.7\times10^{19}\,cm^{-3}$, which are named as I-, m-, h-Si NWs respectively [43]. The BET testing shows that the surface areas of the l-, m- and h-SiNWs are 16, 16 and 260 $m^2\,g^{-1}$, respectively. According to the formula $\sigma = ne\mu$, the electron transport ability should be proportional to the doping level. However, the fact is that m-SiNWs has the highest conductivity among the three samples. That's because the h-SiNWs possesses too much excessive surface areas, in which lots of metal oxides are generated, leading to reduction in the electrical conductivity. It is noteworthy that the conductivity of h-SiNWs is slightly lower than that of m-SiNWs and both of them are 100 times higher than that of I-SiNWs (Figure 5.2d,e). Additionally, the property possessing superabundant pore of h-SiNWs leads to the lowest specific capacity, though its stability is the best compared with other two samples during full cell texts. As a consequence, the m-SiNWs-based cathode, having appropriate electronic conductivity and porosity, is a more satisfactory anode for LIBs than other two types according to electrochemical performance (Figure 5.2f).

Similar to B doping, P doping can also adjust the morphology and intrinsic properties of Si NWs with similar mechanisms, and it is also analyzed from a qualitative and quantitative perspective [44, 45]. Other common methods of modifying SiNWs are summarized in Table 5.4.

5.3.1.2 Lithium Storage in Ge Nanowires

Because the Li-ion storage properties are similar to those of Si, Germanium (Ge), which has a faster Li diffusivity (400 times, 6.51×10^{-12} cm^2 s^{-1} for Ge and 1.41×10^{-14} cm^2 s^{-1} for Si) and better electrical conductivity than Si (10^4 times higher than that of Si), is another candidate for anode materials for LIBs [57]. During charge process, each Ge atom can obtain 4.4 Li^+, delivering a theoretical capacity of 1624 mAh g^{-1}, which is second only to that of Si [58]. Though Ge is more expensive than Si, the isotropic lithium behavior of Ge makes it less fracture failure of Ge-based anode, leading to possibility of better rate performance and cycle stability. In addition to advantages above, the content of Ge is as rich as Si in the crust; thus, this project of Ge-based anode in LIBs has also attracted some attention. However, Ge-based anodes face same problem as Si-based anodes, which suffer from dramatic volume variations during the lithiation/delithiation process.

Fortunately, the strategy of nanoization to alleviate the massive volume expansion of Ge is also effective, while Ge nanowires (GeNWs) also show promising performance among them because of the problem of particle pulverization. Cui's group firstly reported a Ge NWs-based anode for LIBs directly grown on stainless steel foil covered with Au catalyst via CVD method [59]. Compared with previous studies, this method is easier and faster. And direct contact between NWs and collectors avoids reducing electronic contact due to pulverization and insufficient binding power of the polymer (Figure 5.3a,b). This simple but effective operation brought a great Ge-based anode with a stable cycling life and great rate performance; its discharge capacity is 600 mAh g^{-1} at 2 C rate (Figure 5.3c). Yuan et al. use a one-step dodecanethiol monolayer passivation process to obtain an alkanethiol-passivated GeNW in a much lower temperature than that of many fabrication methods of C-based Ge composites (Figure 5.3d) [60]. There is a 1 nm-thick passivation layer on the surface of GeNWs. It is precisely because of the existence of the passivation layer that this anode material exhibits higher initial CE and cycle stability (Figure 5.3e). Moreover, the full cell with $LiFeO_4$ as cathode and this alkanethiol-passivated GeNWs as anode had a capacity retention of 90% after 30 cycles (Figure 5.3f).

Some researchers want to combine the advantages of silicon's high capacity with the advantages of germanium's high electronic and ionic conductivity. Then, by adjusting the ratio of Si to Ge in the NWs, the corresponding anode materials can be obtained according to the demands. For example, Kennedy et al. fabricated bind-free GeNWs composited with Si branches by a vapor–solid–solid (VSS) mechanism [61]. The authors have studied three anodes with different Si-to-Ge ratios (1 : 2, 1 : 3, and 1 : 4). And results showed that heterostructured anodes with higher

Table 5.4 Summary of some modifying methods of Si NWs-based LIB anode and their electrochemical performance.

Method	Si NWs-based anodes	Anode performance		References
		Performance of high current density	Cycle performance	
Doping materials	B-doped Si NWs	18 A g^{-1} 1100 mAh g^{-1} > 250 cycles	2 A g^{-1} 2000 mAh g^{-1} > 250 cycles	[42]
	P-doped Si NWs	16 A g^{-1} 586.7 mAh g^{-1} > 5000 cycles	1.5 A g^{-1} 2061.1 mAh g^{-1} >1000 cycles	[44]
Si NWs decorated with carbon materials	Amorphous carbon-coated SiNWs		0.15 A g^{-1} 1326 mAh g^{-1} 40 cycles	[46]
	Si NWs@G@RGO	8.4 A g^{-1} 500 mAh g^{-1}	2.1 A g^{-1} 1600 mAh g^{-1} 100 cycles	[47]
	SiNWs/MWCNTs		1/15 C 1695 mAh g^{-1} 100 cycles	[48]
Si NWs decorated with metallic materials	Cu-coated Si NWs	0.5 C 1750 mAh g^{-1} 50 cycles	0.05 C 2138 mAh g^{-1} 30 cycles	[49]
	Al-coated Si NWs		0.1 C 1300 mAh g^{-1} 100 cycles	[50]
	Sn-coated Si NWs		0.1 C 1865 mAh g^{-1} 100 cycles	[51]

(Continued)

Table 5.4 (Continued)

Method	Si NWs-based anodes	Anode performance		References
		Performance of high current density	Cycle performance	
Si NWs decorated with metal oxides	Al_2O_3-coated Si NWs	2 C 965 mAh g^{-1} 120 cycles	1/16 C 2760 mAh g^{-1} 30 cycles	[52]
	Si NWs/SnO_2		0.1 C 1200 mAh g^{-1} 100 cycles	[53]
Si NWs decorated with polymers	PEDOT-coated Si NWs		0.2 C 2510 mAh g^{-1} 100 cycles	[54]
	SiNWs/PDMS		0.5 C 83.4% retention rate 350 cycles	[55]
Si NWs decorated with Si oxide	SiO_x-coated SiNWs	7.2 A g^{-1} 282 mAh g^{-1}	0.6 A g^{-1} 1503 mAh g^{-1} 560 cycles	[56]

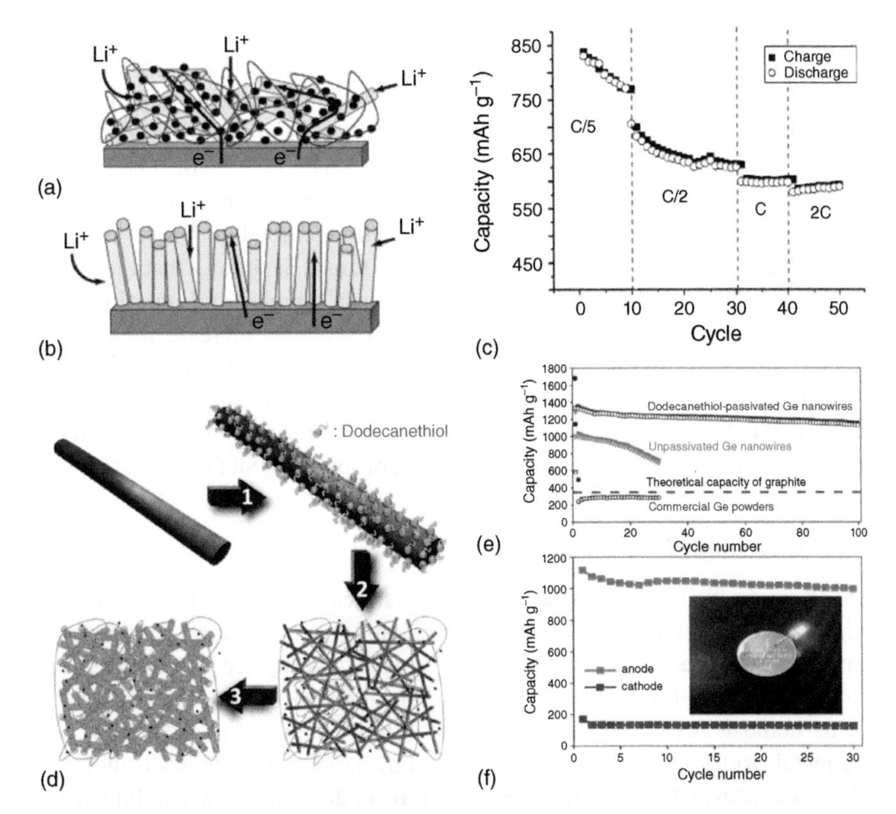

Figure 5.3 (a) A traditional battery electrode consisting of a composite of active material (blue rods). (b) NW electrode design consisting of NWs (blue rods) grown directly on the current collector. (c) Cycling at different discharge rates. (d) Scheme of dodecanethiol-passivated Ge nanowires as anode materials for lithium-ion batteries. (e) Charge/discharge cycle performance of three-type sample at a rate of 0.1 C. (f) Capacity versus cycle number of a coin full cell between 2.0 and 3.8 V with dodecanethiol-passivated Ge nanowire as the anode and $LiFePO_4$ as the cathode.

Si ratio deliver higher capacity that the richest one is about 1800 mAh g^{-1}, while the richest Ge-based electrode exhibited the best rate capability (~800 mAh g^{-1} at high rate of 10 C).

5.3.1.3 Lithium Storage in Sn Nanowires

As an element of the fourth main group, Sn can also be used for constructing the negative electrode of LIBs. The theoretical gravitational capacity of Sn is 990 mAh g^{-1}, which is lower than that of Si and Ge, but its volumetric capacity (2020 mAh cm^{-3}) can be comparable with that of Si (2400 mAh cm^{-3}) [62, 63]. In addition, Sn possesses a higher conductivity and is less fragile than Si so that it's regarded as another promising anode material for LIBs. And a similar large volume expansion problem (up to 259% during lithiation/deithiation) also occurs in Sn-based anodes, while researchers adopted similar approaches.

Generally speaking, carbonous materials are usually used in electrode fabrication because of their abundant surface area, high electrical conductivity, and excellent chemical stability. Zhong et al. encapsulated Sn nanowires with amorphous carbon nanotubes named as Sn@CNTs via in situ CVD growth, while the thickness of carbon nanotubes are able to be accurately controlled by growth time [64]. A large number of core–shell NWs with an average diameter of about 170 nm and a length of 4 to 7 μm are grown together to form a wire-mounted morphology with a larger diameter. And according to electrochemical results, it showed that the thickness of the carbonous shells can obviously influence the electrochemical performance of Sn@CNTs. As the thickness of the outer CNT shells increase, the specific discharge capacity of Sn@CNTs gradually decrease while the retentions increase accordingly. It may contribute to the stabilization properties of outer CNTs. This may be attributed to the stabilization property of outer CNTs and their ability to cushion volume expansion, while the capacity is mainly contributed by Sn ratio in electrodes.

5.3.2 Metal Oxide Nanowires

There are many well-known metal oxides that can react with Li^+ via a conversion mechanism, such as SnO_2, Fe_2O_3, Fe_3O_4, Co_3O_4, and so on. These metal oxides that can provide high specific capacity have been broadly investigated as candidates for next-generation anodes in LIBs. Most of the metal oxides possess the advantages of being nontoxic, easy to obtain, and stable at room temperature [65]. However, these metal oxides have the same issues as alloy-based materials when they are applied as anodes, which include poor electrical conductivity, slow ion diffusivity, easy structure destruction, and similar severe volume expansion during Li^+ insertion/extraction, leading to unsatisfactory chemical performance. In the case of solving these issues, a number of different NWs-based structures have been applied to accelerate reaction kinetics. Improve conductivity and release stress/strain.

Among metal oxides, SnO_2 can be regarded as one of the most promising materials because, after reducing SnO_2 to Sn, it can continue the alloying process with Li^+. Thus, SnO_2 delivers a relative higher capacity than other metal oxides. Kumar et al. fabricated SnO_2 NWs deposited with Sn nanoparticles on the surface via microwave plasma method [66]. The SnO_2 backbone provides efficient electron transport, and the Sn nanoparticles with a specific spacing are believed to be key factors holding the structure stable during cycles. This unique design provided a high reversible capacity, which maintained 814 mAh g^{-1} after 100 cycles.

Inspired by the reported works on alloy-type anode materials, it is foreseeable that building NWs with hollow or porous structures is an effective way to solve the above problems because these kinds of structures show good adaptability to volume expansion and avoid the rapid destruction of the structure after only a few cycles. For instance, Li et al. grew mesoporous Co_3O_4 NW arrays on Ti foil through a simple template-free method, as obtaining composite can be directly used as anode [67]. This particular mesoporous NW array architecture effectively shorten the transmission distance of electron/ion and provide enough volume to sustain the stress caused by expansion during chemical reaction (Figure 5.4a). Finally,

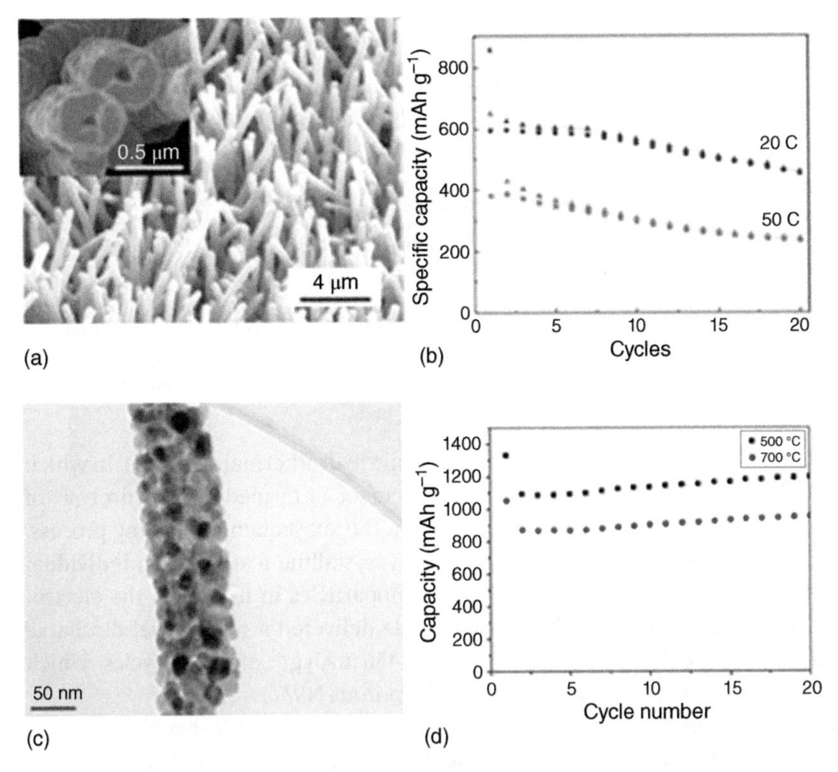

Figure 5.4 (a) SEM images of Co_3O_4 NW arrays growing on Ti foil. (b) Specific capacity of the NW arrays at 20 C (violet) and 50 C (orange) as a function of the cycle number. (c) TEM image of $ZnCo_2O_4$ nanowire. (d) Discharge capacity versus cycle number for the electrodes made from the porous $ZnCo_2O_4$ nanowires synthesized at 500 and 700 °C.

this anode exhibited high capacity and good rate capability, that it can maintain a capacity of about 350 mAh g^{-1} even at 50 C (Figure 5.4b). A porous $ZnCo_2O_4$ NW reported by Du et al. is another example that benefits from its porous structure [68]. It's obvious that the high porosity and large surface area lead to an improvement in the ability to deal with volume variation (Figure 5.4c). As a result, this $ZnCo_2O_4$ NW obtained by heating a sacrificial template showed a high reversible capacity of 1331.5 mAh g^{-1} and maintained at 1197 mAh g^{-1} after 20 cycles, which is much higher than that of $ZnCo_2O_4$ nanoparticle-based anode (Figure 5.4d).

Other promising anodes among metal oxides are iron-oxide-based anodes (Fe_2O_3 and Fe_3O_4), which have a promising theoretical capacity of about 1000 mAh g^{-1} [32]. Taking ferric oxide as an example, the conversion reaction process is as follows: $Fe_2O_3 + 6Li^+ + 6e^- \leftrightarrow 3Li_2O + 2Fe$ [69]. From this equation, we can learn that the formation of Fe^0 from Fe^{3+} involves multiple electron transfer per metal atom, leading to a high theoretical lithium storage capacity. In addition to their high-capacity characteristics, other advantages, including rich content, low synthesis cost, corrosion resistance, and environmental friendliness have attracted a lot of attention in the large-scale application of iron-oxide-based anodes. Liu et al. a

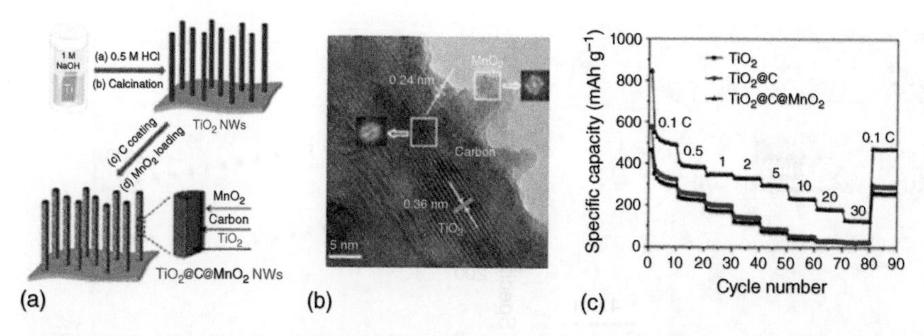

Figure 5.5 (a) Schematic illustration for the fabrication of TiO_2-C/MnO_2 core–shell NW arrays on Ti foil. (b) TEM image of the TiO_2-C/MnO_2 core–shell NW. (c) Rate performance of TiO_2, TiO_2-C, and TiO_2-C/MnO_2 core–shell NW anodes at various current densities from 0.1 to 30 C.

α-Fe_2O_3-NW-based anode to LIBs through a facile hydrothermal synthesis, in which the one-dimensional long-chain polymer precursor is formed after hydrolysis of Fe^{3+} and the final product α-Fe_2O_3 is formed in the subsequent annealing process. According to images of TEM, it shows the polycrystalline feature of an individual NW, which may be caused by the α-Fe_2O_3 nanoparticles in it. And in the electrochemical testing, the α-Fe_2O_3-NW-based anode delivered a high initial discharge capacity of 1303 mAh g^{-1}. But it decreased to 456 mAh g^{-1} after 100 cycles, which may due to the structural disintegration of the porous NWs.

Core–shell structures play an important role in alloy-type NW-based anodes, as they do in metal-oxide-based anodes. The outer shell materials protect inner core materials from agglomeration and excess side reactions, further release strain from volume variation, and sustain structure of anodes in a stable state. For instance, Liao et al. directly grew a TiO_2–C/MnO_2 core–shell NW array on a Ti current collector through a layer-by-layer deposition process (Figure 5.5a,b) [70]. This design integrates the advantages of these three materials: TiO_2 possesses good stability, MnO_2 provides high capacity, and the graphitic carbon layer has excellent electrical conductivity. Furthermore, this unique core–shell structure gives this TiO_2–/MnO_2 anode additive-free capability, which can improve energy density of the whole cell system. And as an anode in half-cell tests, it showed a capacity of 865 mAh g^{-1} at 0.1 C and excellent rate performance of 130 mAh g^{-1} at 30 C, which is obviously more excellent than that of absolute TiO_2 or MnO_2 NWs and TiO_2-C NWs (Figure 5.5c). In addition to preparing electrode materials with a core–shell structure, coating the surface of transition metal oxide NWs with a layer of carbon material or highly conductive metal is also a good way to solve these shortcomings, such as in reported NGA-Co_3O_4 composite [71], MoO_2/Mo_2C/carbon hybrid NWs, [72] and so on.

5.3.3 Carbonaceous Anode Materials

As carbonaceous materials have the characteristics of good mechanical strength, stable chemical properties, low price, abundant content, easy access, excellent reversibility during the insertion and extraction of lithium ions, and most importantly, high electronic conductivity, they have been widely applied in electrode fabrication of LIBs. Although the most successful commercialized anode for LIBs at this stage is carbon material, the low specific capacity of pure carbon materials can no longer meet the growing needs of people. So now carbon materials are usually used as modified materials in combination with other materials to show new effects in anodes.

As mentioned above, the application of SiNWs/GeNWs in anodes of LIBs suffers from their low conductivity and volume expansion problems, which makes their commercialization progress slow. In addition to doping with elements of the third and fifth main groups to improve conductivity, coating with a carbon layer is also a good method. Bogart and co-workers used SFLS method, which is an easy method to realize scalable synthesis, to construct SiNWs coated with a conductive carbon shell (Figure 5.6a) [73]. This anode exhibited a really highly reversible capacity of over 2000 mAh g^{-1} for 100 cycles at current rate of 1/10 C, and when it cycled at 1 C, it could still show a capacity of over 1200 mAh g^{-1}, which is much higher than the performance of bare SiNW-based anodes (Figure 5.6b). These inspiring results obtained with low-cost are attributed to the conductive carbon coatings with increased electrical conductivity, reduced construction and aggregation of NWs, stabilized SEI, and sufficient space for volume variation during cycles. Another example is hierarchical carbon-sheathed GeNWs (C–Ge NWs) through a self-catalytic growth process, which was reported by Choi et al. [74] They adopted chemical reduction reactions involving GeO_2 particles and subsequent self-catalytic growth, at the same time, the carbon layer is uniformly coated on the surface of the GeNWs (Figure 5.6c,d). Obtained C–Ge NW anode showed capacity retention of 40% at 5 C compared with the capacity tested at 0.2 C, while the comparative pure GeNWs cannot even reach 5% under the same conditions (Figure 5.6e).

The method of using carbon materials for modification is also applicable to anode materials constructed of metal oxide NWs, as the anodes consisting of metal oxide NWs face similar problems with alloy types. Hu et al. reported ordered mesoporous carbon/Fe_2O_3 NW composite electrodes through three kinds of template approaches, including soft, hard, and soft–hard dual-template approaches, in order to investigate the influence of templates on the structure and electrochemical performance of electrodes (Figure 5.6f) [75]. This work revealed the synergetic benefits triggered by the soft–hard dual-template, which optimizes the advantages of other two single-template methods and results in more appropriate morphology and much better electrochemical performance (Figure 5.6g). In this special structural design, the in situ grown carbon shells sufficiently endured the stress strain of self and external NWs, sustained the stability of NWs, and prevented them from construction during continual insertion/extraction of Li^+, leading to a highly reversible capacity over 1200 cycles and rate capability of ~400 mAh g^{-1} at a current density of 2 A g^{-1} (Figure 5.6h).

On the other hand, carbonous materials with high specific surface area (SSA) are often used as components of coaxial electrodes to promote the reaction of more active materials because they can provide a large number of active sites for lithium-ion transmission. Chen et al. achieved an additive-free ZGO@C/Cu hybrid anode via simple single-step CVD method [76]. The novel coaxial Zn_2GeO_4@carbon

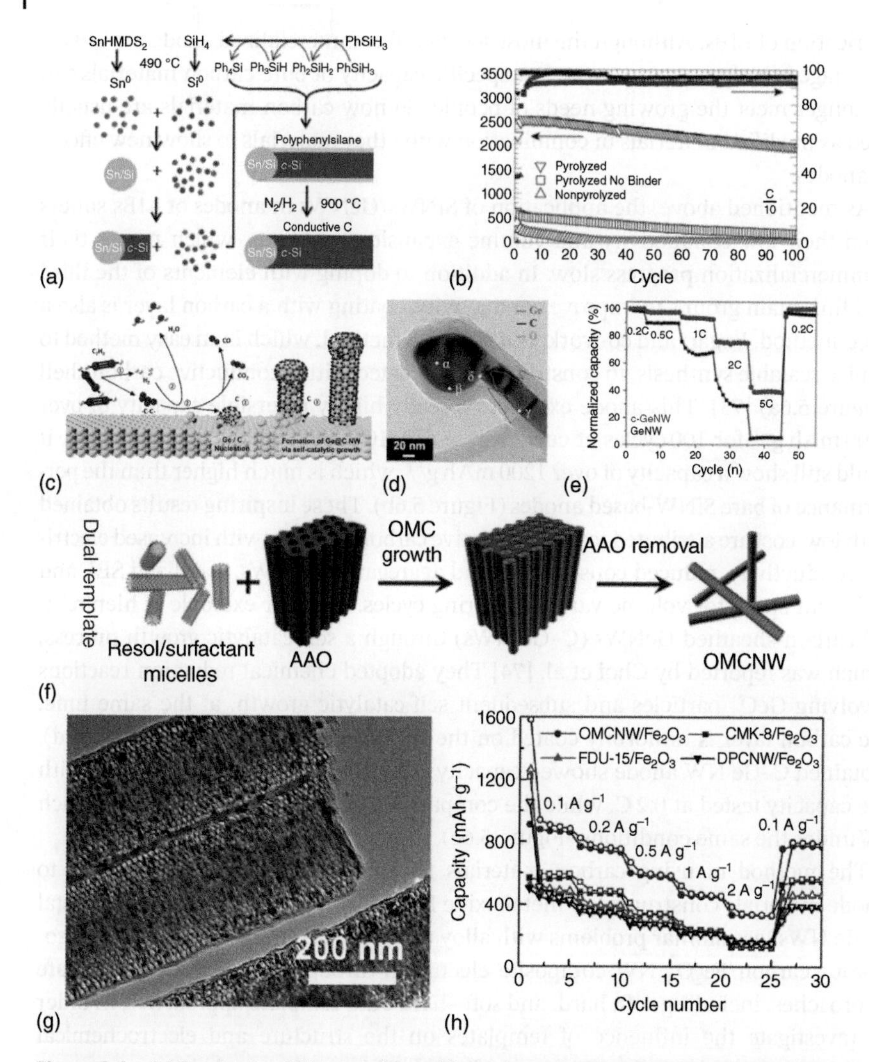

Figure 5.6 (a) Synthesis of Si nanowires with conformal carbon coatings. (b) Batteries were cycled at 1/10 C for 100 cycles. (c) Schematic illustration showing the synthetic route of the C – Ge NW from reaction between decomposed C_2H_2 and GeO_2. (d) TEM image of the carbon-coated Ge NW. (e) Rate capabilities of C – Ge and Ge NW electrodes. (f) Schematic illustration of the proposed dual-template method for the synthesis of porous carbon NW frameworks. (g) TEM image of an ordered mesoporous carbon/Fe_2O_3 NW composite using the soft–hard dual-template approach. (h) Rate capability of different carbon/Fe_2O_3 composites.

NWs were directly grown on the cheap Cu collector. This direct growth cannot only make each individual NW contacts closely with the substrate, providing favorable electron transfer; but also ensure that there is enough space between adjacent NWs to accommodate volume expansion and rapid ion transport channels. Therefore, this ZGO@C/Cu electrode, which is obviously capable of maintaining structural stability and rapid ion transport characteristics during electrochemical cycles, exhibits excellent cycling life, and rate performance. Specifically, at a current density of 2 A g^{-1}, the electrode maintains a discharge- specific capacity of 790 mAh g^{-1} with almost no noticeable decay after 100 cycles, while it can still have a capacity of 465 mAh g^{-1} at a current density of 10 A g^{-1}.

5.4 Nanowires as Cathodes in LIBs

5.4.1 Transition Metal Oxides

The transition metal oxide cathode of traditional LIBs mainly include $LiCoO_2$ with layered structure, $LiMn_2O_4$ with spinel structure, and $LiFePO_4$ with an olivine structure. Layered $LiCoO_2$ has a high theoretical specific capacity of 274 mAh g^{-1}, but the crystal structure will be unstable when more than 70% Li^+ is extracted [77]. Furthermore, the $Li_{1-x}CoO_2$ under high cutoff voltage will promote the occurrence of surficial side reaction, resulting from the continuous decomposition of electrolyte. Therefore, improving the structure and interface stability of $LiCoO_2$ (LCO) is critical to improve its electrochemical performance. At present, the modification strategies mainly include doping of metal ion and coating inert compounds, such as Al_2O_3, ZrO_2, and $AlPO_4$ [78–80] Li et al. synthesized LCO NWs through hydrothermal method and calcination at 500 °C. The obtained $Co(CO_3)_{0.35}C_{10.2}(OH)_{1.1}$ nanowire bunches precursors are composed of radial NWs with a diameter of 150–200 nm and a length of about 10 μm (Figure 5.7a). The micromorphology of the NWs remains intact after annealing, which piles up with smaller nanoparticles. The LCO NWs shows satisfactory rate performance and long-term cycling, where the specific discharge capacity stabilized at 100 mAh g^{-1} at a current density of 1 A g^{-1} after 100 cycles (Figure 5.7b). The favorable electrochemical performance of the LCO NWs may originate from fast Li^+ transportation in their 1D nanostructure [81]. Xia et al. carried out a new approach for fabrication of 3D $LiCoO_2$ arrays as cathode for LIBs. A two-step hydrothermal method was used to synthesize the low-temperature LCO arrays and it was further converted into chain-like high-temperature $LiCoO_2$ arrays through calcination (Figure 5.7c,d). The layered $LiCoO_2$ NW arrays can deliver high areal capacity up to 0.27 mAh cm^{-2}, while maintaining good cycling stability (90%, 50 cycles) and rate capability (103 mAh g^{-1}, 10 C), which makes it promising for application in micro-batteries (Figure 5.7e,f). Except for the wide application of LCO in power batteries, consumer batteries and other end products, its high cost, toxicity, and limitation in resources of cobalt drive the exploration of other cathodes [82].

Spinel $LiMn_2O_4$ is one of representative materials with the advantages of low cost, abundance, and reasonable capacity (theoretical capacity of 148 mAh g^{-1}), which is considered one of promising cathode candidate for LIBs. Although $LiMn_2O_4$ displays many advantages, it also shows fatal shortcomings in commercial applications, mainly in the serious capacity decay during cycling. It can generally be summarized as follows: (i) the dissolution of manganese leads to the collapse of the spinel structure; (ii) When $LiMn_2O_4$ is in deep discharge, especially tetrahedral $Li_2Mn_2O_4$ will be formed on the surface of spinel, undergoing Jahn–Teller effect and lattice

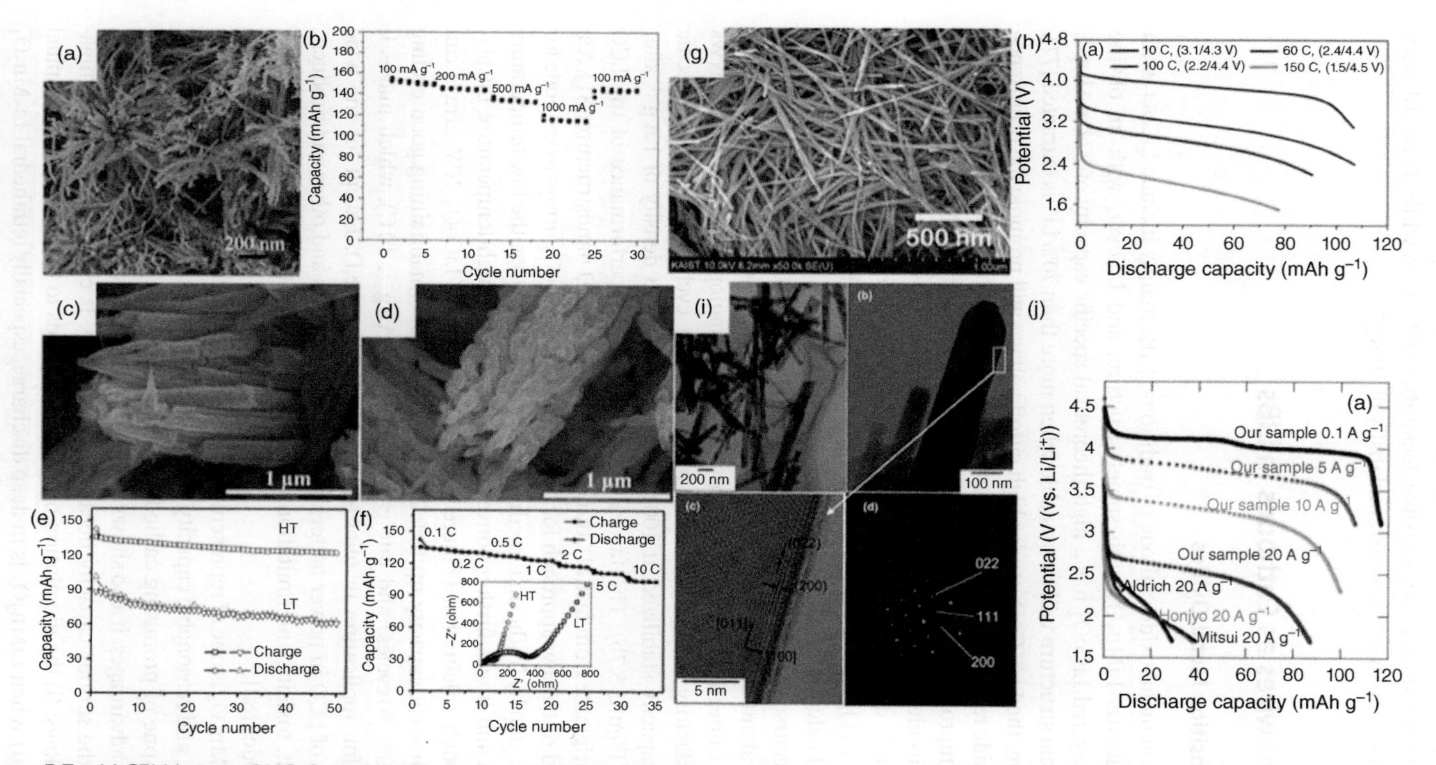

Figure 5.7 (a) SEM image of $LiCoO_2$ nanowires. (b) Rate capability of $LiCoO_2$ nanowires at different rates. SEM images of (c) LT-$LiCoO_2$ nanowires and (d) HT-$LiCoO_2$ nanowires. (e) Cycling performance and (f) Rate performance of LT-$LiCoO_2$ and HT-$LiCoO_2$ nanowire. (Inset is the Nyquist plot for the LT-$LiCoO_2$ and HT-$LiCoO_2$ nanowire. (g) SEM image of $LiMn_2O_4$. (h) Discharge curves of $LiMn_2O_4$ at different rates. (i) TEM, HRTEM and ED patterns of the single crystalline spinel $LiMn_2O_4$ nanowires. (j) Discharge curves of the single crystalline spinel $LiMn_2O_4$ nanowires and commercial $LiMn_2O_4$.

distortion; (iii) The organic electrolyte decomposes at high voltage, which destroys the spinel structure. Generally, the common strategies used in modification were doping to partially replace Mn or O, surface modification, control of the morphology, particle size, or pore structure to stabilize the microstructure, and shorten the ions transport path. Lee et al. obtained the ultrathin $LiMn_2O_4$ NWs by a solvothermal reaction with α-MnO_2 NW followed by solid-state lithiation. (Figure 5.7g) $LiMn_2O_4$ NWs deliver 100 and 78 mAh g^{-1} at 60 and 150 C (Figure 5.7h), respectively. It is due to both the favorable morphologies and the high-crystal NWs [83]. Hosono et al. synthesized high-quality crystalline cubic $LiMn_2O_4$ NWs based on a self-template method with the precursors of $Na_{0.44}MnO_2$ NWs (Figure 5.7i). These single crystalline NWs possess high thermal stability after heating at 800 C and excellent specific capacity of 88 mAh g^{-1} at high rate 20 A g^{-1} (Figure 5.7j). It comes from the microstructures of NWs and the high quality of the crystal, which decreased both the Li^+ and electron diffusion lengths [84].

Li-rich layered manganese oxides (LLMO) are foreground cathode material in LIBs. Its general structural formula is $x\ Li_2MnO_3 \cdot (1-x)\ LiMO_2$ $(0<x<1$, M = Ni, Co, Mn). The Li-rich cathode material can provide a specific capacity of 250 mAh g^{-1} when charged to 4.5 V. There are several thermodynamic and kinetic shortcomings that need to be tackled, including low initial coulombic efficiency, poor rate performance, voltage attenuation, and surface structure transformation. Fabricating lithium-rich oxide NWs are proposed to efficiently overcome these issues. Lee et al. applied $Co_{0.4}Mn_{0.6}O_2$ NWs and $LiNO_3$ as precursors to synthesize Li-rich layered $Li_{0.88}[Li_{0.18}Co_{0.33}Mn_{0.49}]O_2$ NWs via a one-step hydrothermal approach (Figure 5.8a). This NWs provided a high reversible capacity of 230 mAh g^{-1} at 1 C between 2 – 4.8 V and a high-rate performance (220 mAh g^{-1} at 15 C) (Figure 5.8b) [85]. Yu et al. obtained the $Li_4Mn_5O_{12}$ heterostructured NWs composed of small nanoparticles, and the spinel phase is embedded into the layered structure (Figure 5.8c,d), which prevents the voltage from fading under 2 – 4.6 V. The heterostructured structure exhibited a high initial capacity of 290.1 mAh g^{-1} at 0.1 C and maintained 94.4% after 200 cycles at 0.5 C (Figure 5.8e,f) [86].

Besides, other transition metal oxides, such as LiV_3O_8, have been studied as cathode candidate. Xu et al. adopted a general two-step method to fabricate by topotactic lithiation of $H_2V_3O_8$ NWs (Figure 5.8g). A specific capacity of 176 mAh g^{-1} was able to achieve at 1.5 A g^{-1}, and it maintains 160 mAh g^{-1} after 400 cycles (Figure 5.8h). It has a low charge transfer resistance, stable structure, and high crystallinity, which confirmed LiV_3O_8 NWs as cathode for high-rate and long-life LIBs [87]. Ren et al. designed hierarchical LiV_3O_8 NWs by electrospinning through a PEO crosslinking strategy. Its unique structure achieves an effective electrical contact area, low resistance, and improved stability, which offer suitable energy storage in high-rate and long-life LIBs [88].

5.4.2 Vanadium Oxide Nanowires

Vanadium oxides are considered an alternative material for high-energy cathodes for LIBs owing to their high capacity, low-cost, rich natural resources, and simple

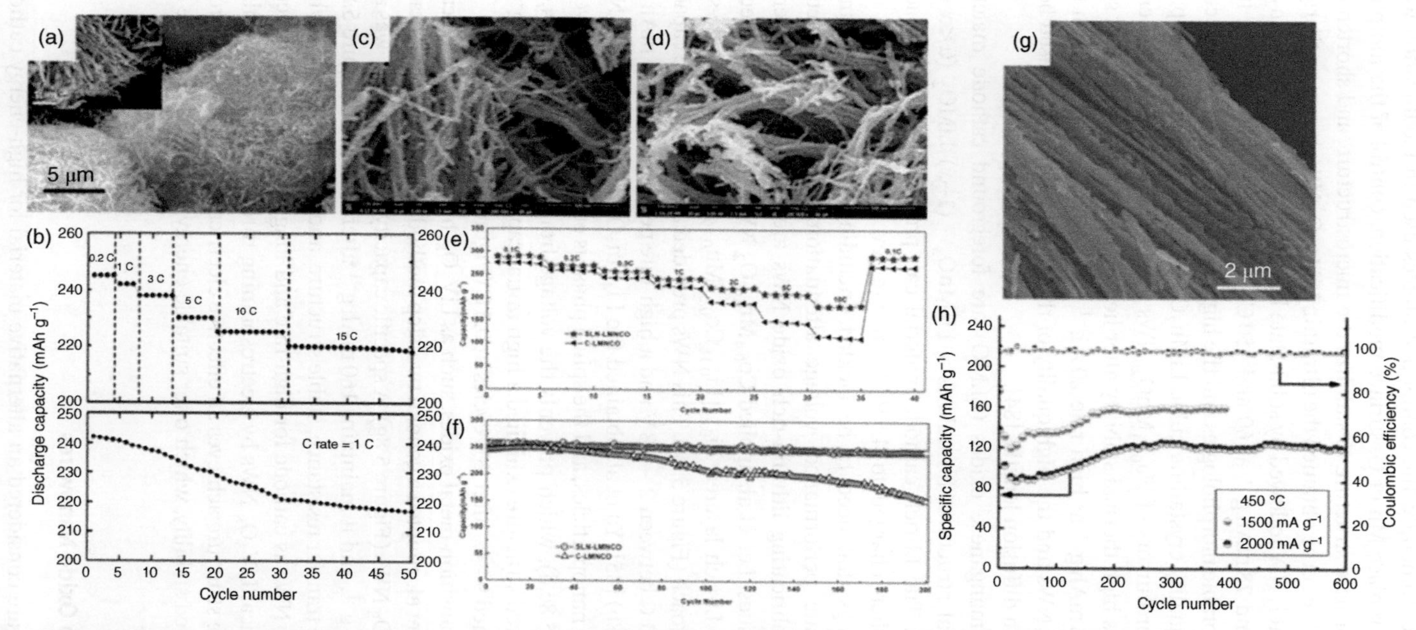

Figure 5.8 (a) SEM image of the $Li_{0.88}[Li_{0.18}Co_{0.33}Mn_{0.49}]O_2$ nanowires discharge capacity and cycle performance of the $Li[Li_{0.08}Co_{0.33}Mn_{0.5}]O_2$ nanowires at 1C rate between 2–4.8 V. (b) Discharge capacity versus cycle number of the $Li[Li_{0.08}Co_{0.33}Mn_{0.5}]O_2$ nanowires in a coin-type half-cell at a 1C rate between 2 and 4.8 V. SEM images of (c) SLN-LMNCO and (d) C-LMNCO. (e) Rate and (f) Cycling performance for Li/SLN-LMNCO and Li/C-LMNCO cells at 0.5 C (g) SEM image of LiV_3O_8 nanowires at 450 °C. (h) Cycling performance of LiV_3O_8 nanowires cathode at 1500 and 2000 mA g^{-1}.

synthesis [89, 90]. However, it has suffered from agglomeration, low electronic conductivity, poor structural stability, and sluggish kinetics, which blocks its application in high energy-density LIBs. Vanadium oxides with nanostructure are designed to ultimately improve its electrochemical performance due to the superiorities of nanostructure, including short ion/electron transport path, large SSA and improved conductivity. Mai et al. constructed hierarchical vanadium oxide nanowires with diameter of 1–10 mm and diameter of 100–200 nm by electrospinning combined with annealing (Figure 5.9a,b) [91]. NWs are composed with several small nanorods. This hierarchical structure is beneficial to greatly reduce self-agglomeration of NWs during the cycle process, thus maintaining effective contact between electrode materials, conductive agents, adhesives, and electrolyte. The composite electrodes exhibit excellent initial capacity of 275 mAh g^{-1} at 2.0–4.0 V and a capacity of 187 mAh g^{-1} remained after 50 cycles at high current density of 10 A g^{-1} (Figure 5.9c), implying a promising candidate for cathode of LIBs. To prevent volume expansion of vanadium oxides-based materials, hollow structure NWs are used to overcome this dilemma. Yan et al. designed NW-template semihollow bicontinuous graphene scroll architecture through "oriented assembly" and "self-scroll" strategies. V_3O_7 NW templated graphene scrolls (VGSs) (Figure 5.9d). for LIBs exhibit an excellent performance with specific capacity of 321 mAh g^{-1} (100 mA g^{-1}) and 87.3% capacity retention after 400 cycles (2 A g^{-1}) (Figure 5.9e). These outstanding electrochemical properties are attributed to the unique structure, which provides continuous electron and ion transfer channels and free space for enduring the volume expansion of NWs during cycling [92]. Niu et al. reported VO_2 hollow microspheres with empty spherical core and radial NWs through an ion modulating approach. It shows higher SSA, and efficiently buffer self-expansion and self-shrinkage during lithiation/delithiation process. As a result, the original capacity of 163 mAh g^{-1} (1 A g^{-1}) is achieved in hollow microspheres and it maintains 80% of initial capacity after 1000 cycles (2 A g^{-1}), showing a high specific capacity and long cycle performance. Assembling the NWs into a functional architecture or constructing compound materials will grant the materials with new properties. In order to tackle the issue that vanadium oxides-based materials have a disposition to agglomerate, Lee et al. embedded ultrathin NWs into the 2D graphene to well maintain the structure of V_2O_5 NWs in the electrochemical process (Figure 5.9f), so the specific capacities of 94.4 mAh g^{-1} were preserved over 100 000 cycles at 10 A g^{-1}(Figure 5.9g) [93]. It mainly originates from the close contact between V_2O_5 NWs and graphene, thus improving electron transport and suppressing the structural degradation. Mai et al. fabricated self-buffered nanoroll VO_2(B) composites composed of nanobelts and NWs by hydrothermal method using surfactants of polyvinyl pyrrolidone and polyethylene glycol (Figure 5.9h). At the current density of 1 A g^{-1}, the capacity retention rate of 82% is obtained after 1000 cycles, and the maximum rate can reach 2000 mA g^{-1} (20 C) (Figure 5.9i,j) [94]. It is mainly due to the nanostructures that it can effectively shorten the Li^+ transfer path, while it plays the roles as buffer intermediates in the cycle process to form buffer voids between nanobelts and NWs. It realizes the buffer of volume swelling stress in the electrochemical process, maintains the stability of the structure, and

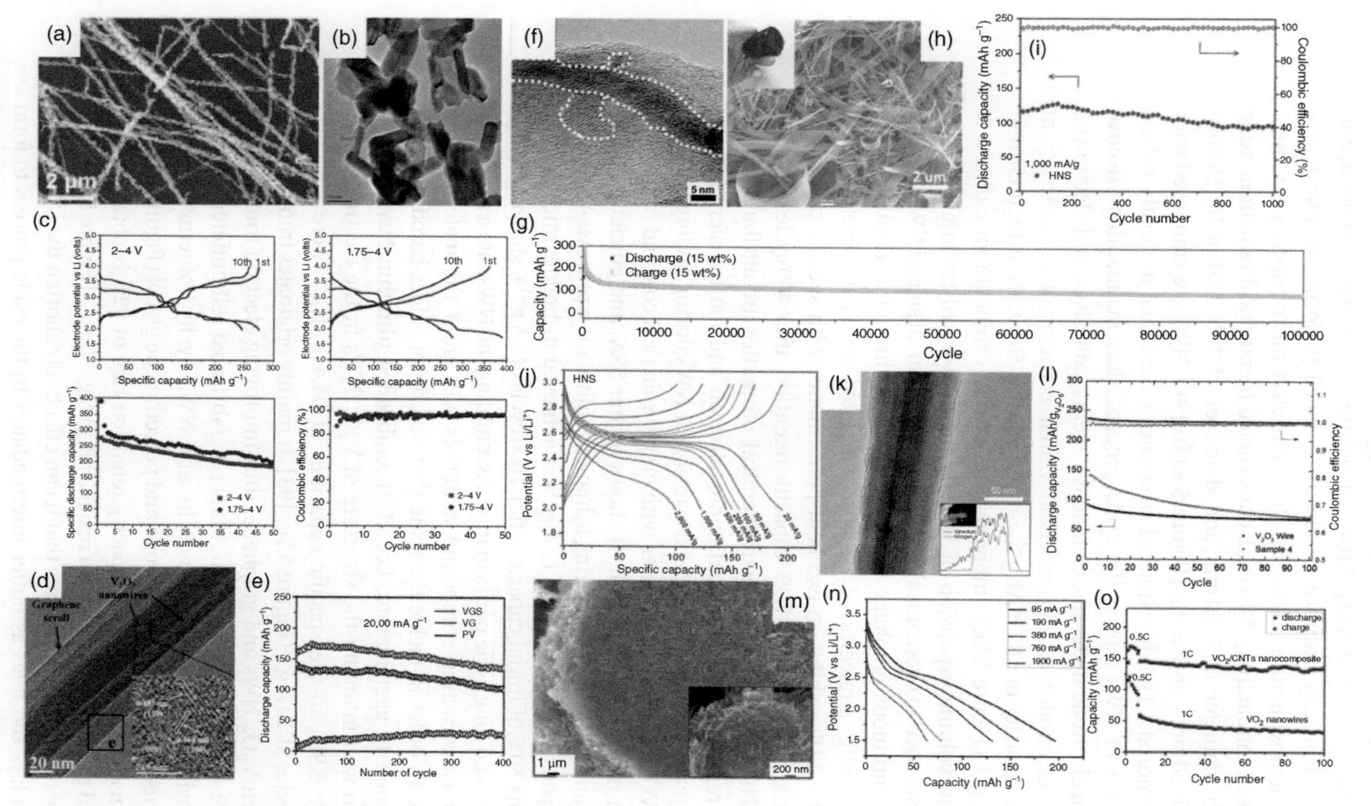

Figure 5.9 (a) SEM and (b) TEM of the ultralong hierarchical vanadium oxide nanowires after annealing. (c) Electrochemical performance of the ultralong hierarchical vanadium oxide nanowires after annealing at voltage profile of 2–4 V and 1.75–4 V. (d) TEM images of VGS (HRTEM image of V_3O_7 nanowire in GSs was inserted). (e) Cycling performance of different samples. (f) HRTEM image of the V_2O_5 NW–graphene composite. (g) Cycling performance at 10 000 mA g^{-1}. (h) SEM images of the (b) hybrid nanostructural VO_2. (i) Cycling performance at 1000 mA g^{-1}. (j) Charge/discharge curves at current density of 20 to 2000 mA g^{-1}. (k) TEM image of V_2O_5-PANI sample 4 with accompanying line scans indicating conformal coating of PANI on V_2O_5 wire bundles. (l) Cycling performance of V_2O_5-PANI sample 4 and V_2O_5 wire at 440 mA g^{-1}. (m) SEM image of VO_2/CNTs nanocomposite film containing 40 wt% of CNTs. (n) Discharge curves of VO_2/CNTs nanocomposite at different current density. (o) Cycling performance of VO_2/CNTs and VO_2 nanowire.

thus finally improves the rate performance and cycle performance. Gittleson et al. gained a core–shell structure, which was V_2O_5 nanowires coated with electroactive polyaniline (PANI) (Figure 5.9k). After correlating the weight percent of PANI polymer in active materials (51% PANI in composite), the actual capacity will display a 57% increase compared with uncoated one at 1 C (Figure 5.9l), which will be useful in the generation of microbattery [95]. Besides, intertwined VO_2 NW/CNT composites and carbon quantum dot-decorated VO_2 on carbon cloth, etc., have been proven to optimize the electrochemical performance. Yan et al. synthesized the VO_2(B) NWs/CNT compounds with stable lithium storage performance (Figure 5.9m). It formed the interpenetrating porous network, and it improved the Li^+ storage capability. Consequently, this composite will stably cycle about 100 times with a capacity retention of 94% (Figure 5.9n,o), which mainly comes from rapid transportation of Li^+ and charge enrichment of carbon networks [96].

5.4.3 Iron Compounds Including Oxides and Phosphates

Iron compounds have been widely studied in the cathode of LIB due to their abundant resources, environmental friendliness, and considerable capacity. Its representatives are oxides and polyanions, where polyanion-based cathodes include $LiFePO_4$, and $Li_3V_2(PO_4)_3$ (LVP).

As for polyanion, it has stable microstructure, high voltage plateau, and satisfactory theoretical capacity. However, its practical applications are impeded by the intrinsic low electrical conductivity and sluggish kinetics. Lim et al. reported the hard templates KIT-6 and SBA-15 to synthesize $LiFePO_4$ NWs (Figure 5.10a,b) and hollow $LiFePO_4$ (Figure 5.10c,d) as cathode materials for LIBs. Electrochemical performance of both the NW and hollow cathodes showed the excellent rate capability even above 10 C rate, showing >89% capacity retention of the initial capacity (Figure 5.10e). What's more, the rate capability of the hollow $LiFePO_4$ cathode at 15 C is about 10 mAh g^{-1} higher than that of the NW cathode, the higher SSA leads to excellent electrochemical performance. However, the NaOH used in this system will react with $LiFePO_4$ to destroy the structure [97]. So, to avoid side effects and improve the faster ion-transport capability, carbon coating on the electrode material is the most effective and convenient method. Drummond et al. reported a new hierarchically porous $LiFePO_4$ electrode materials for LIBs, this cathode forms a meso/macroporous carbon layer structure (Figure 5.10f,g) to increase the electrode/electrolyte interface and improve the rate capability. This macro-porous network structure allows efficient percolation of the electrolyte through the electrode, improving electrolyte access to the mesopores while providing a conductive framework with intimate contact with the $LiFePO_4$. The cathode delivered a reversible capacity of 140 mAh g^{-1} at 0.1 C and 100 mAh g^{-1} at 5 C (Figure 5.10h) [98]. Besides, a carbon-coated single-crystalline $LiFePO_4$ NW with less 100 nm diameter was produced by electrospinning method. This NW structure with diameters less than 100 nm offers better percolation behavior than normal NWs and particles, which creates an efficient conductive network with short Li^+ diffusion pathway and significantly improved electrochemical performance. The

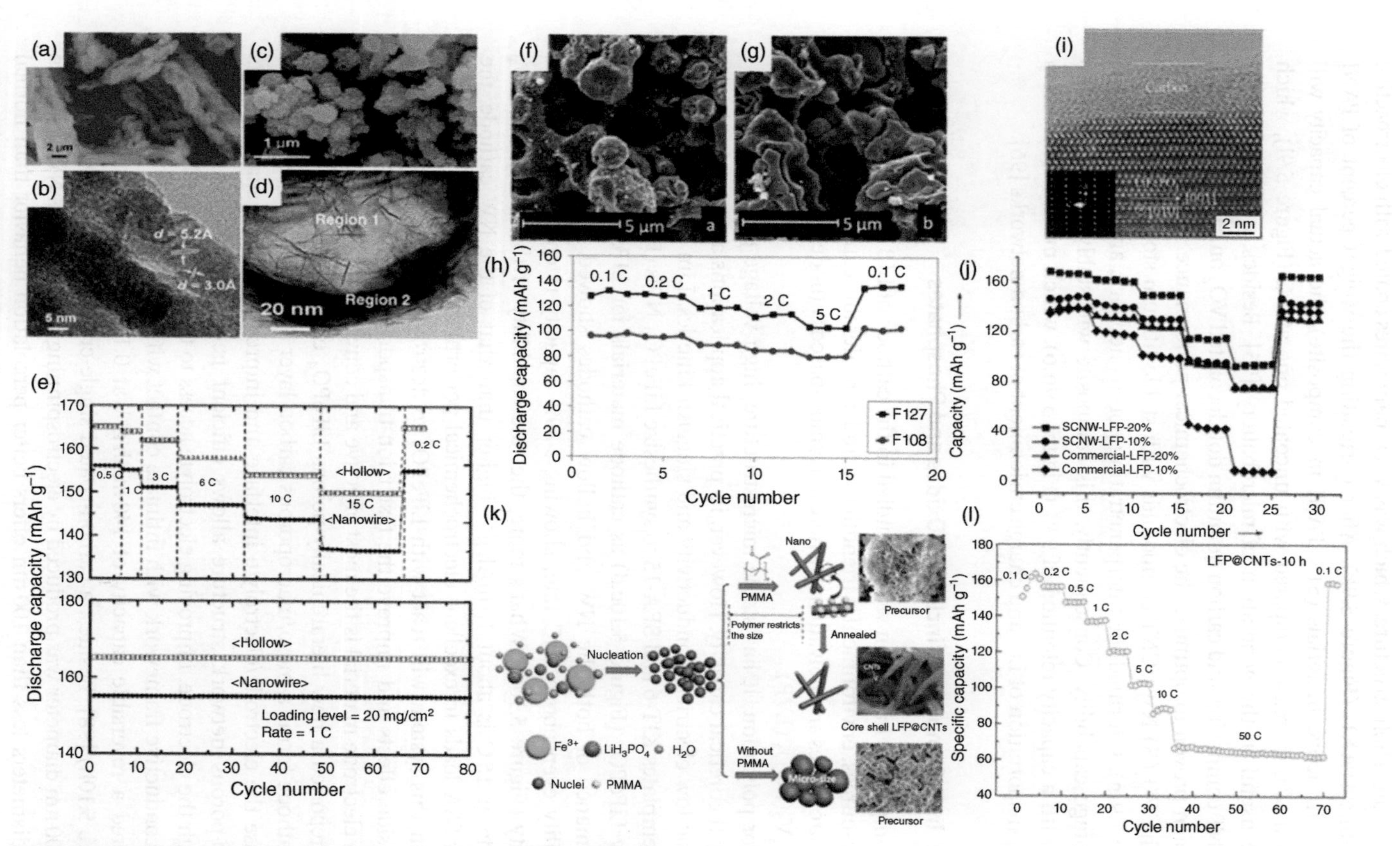

Figure 5.10 (a) SEM image of $LiFePO_4$ nanowires (b) TEM image of $LiFePO_4$ nanowires. (c) SEM image of hollow $LiFePO_4$. (d) TEM image of hollow $LiFePO_4$. (e) Electrochemical performance of both the nanowire and hollow $LiFePO_4$ cathode. SEM images of $LiFePO_4$/carbon monoliths prepared with (f) F127 and (g) F108 triblock copolymer amphiphiles. (h) Rate capability of $LiFePO_4$/carbon monoliths prepared with F127 and F108 triblock copolymer amphiphiles. (i) HRTEM image of $LiFePO_4$ nanowires with SAED pattern inset. (j) Rate capability of SCNW-LFP and commercial $LiFePO_4$ with 10% and 20% additional conductive carbon. (k) Schematics and SEM images of the LFP@CNT nanocomposites. (l) Rate capability of LFP@CNT nanocomposites.

presence of a carbon layer further protects the electrode from direct contact with the electrolyte. The electrochemical performance of single-crystalline $LiFePO_4$ NW delivered a high capacity of 160 mAh g^{-1} at 0.1 C and excellent rate capability of 65 mAh g^{-1} at 50 C (Figure 5.10i,j) [99]. Smaller diameters of $LiFePO_4$ NW were constructed by Sun et al. In situ self-catalyzed core–shell $LiFePO_4$@CNT NWs with a diameter of 20–30 nm were encapsulated into CNTs (Figure 5.10k), and 3D conducting networks of CNTs were obtained from in situ carbonization of a polymer. $LiFePO_4$@CNT NWs deliver a capacity of 160 mAh g^{-1} at 17 mA g^{-1}, and 65 mAh g^{-1} at 8500 mA g^{-1} (Figure 5.10l) [100].

Faster ion and electron transport was achieved in $Li_3V_2(PO_4)_3$ NW cathode. Jing et al. reported a 3D mace-like $Li_3V_2(PO_4)_3$/C NW and nanofiber hybrid membrane, which fabricated by using Ni nanoparticles as a catalyst and a modified electrospinning method (Figure 5.11a,b,c). And this membrane possesses a highly

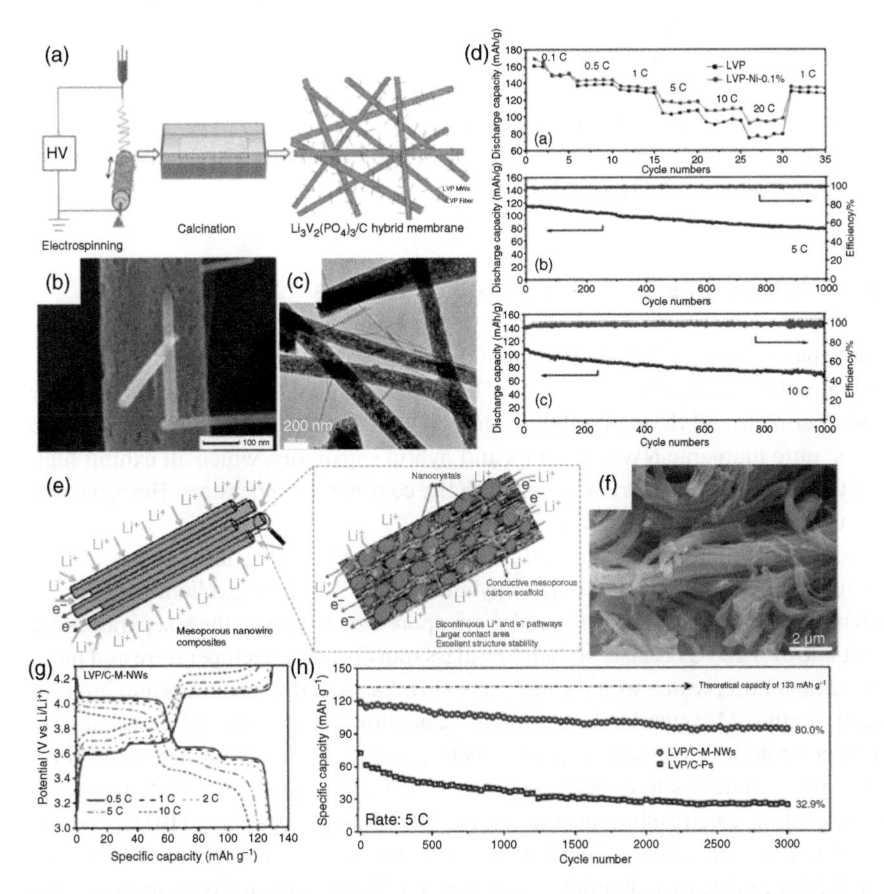

Figure 5.11 (a) Schematic illustration of 3D mace-like structural $Li_3V_2(PO_4)_3$/C hybrid membrane (b) SEM and (c) TEM images of $Li_3V_2(PO_4)_3$/C. (d) Rate performance and cycle performance of $Li_3V_2(PO_4)_3$/C nanowire and nanofiber hybrid membrane at 5 C and 10 C. (e) Schematic illustration of the LVP/C-M-NWs composites with electron/ion transport pathways, larger electrode-electrolyte contact area, and facile strain relaxation during Li^+ extraction/insertion. (f) SEM image of LVP/C-M-NWs. (g) Charge and discharge profiles and (h) Cycle performance of LVP/C-M-NWs.

SSA, making material and electrolyte fully contact to accelerate electrochemical reaction, greatly improve ion and electron transport kinetics. 3D mace-like $Li_3V_2(PO_4)_3/C$ NW exhibits an excellent rate capability in the voltage range of 3.0–4.8 V with discharge capacity of 115.3 mAh g^{-1} and 108.6 mAh g^{-1} at 5 and 10 C, respectively, (Figure 5.11d). This material also exhibits good cyclic stability, with capacity retention of 81.4% after 500 cycles and 78.8% after 1000 cycles. Such superior electrochemical performance is attributed to unique 3D conductive networks and mace-like fiber structure, which favorably improve the reaction kinetics of $Li_3V_2(PO_4)_3$ [101]. Wei et al. discovered a new hierarchical mesoporous NW with bicontinuous electron/ion pathways, which create a larger contact area between electrode and electrolyte (Figure 5.11e,f). This fast ion transport pathway along 1D NW has a low charge transfer resistance. What's more, in the process of Li^+ de/intercalation, robust structure stability of $Li_3V_2(PO_4)_3/C$ mesoporous NW results in a slight volume change, which greatly improves cycling ability. $Li_3V_2(PO_4)_3/C$ mesoporous nanowire exhibits outstanding capacity retention of 80.0% after 3000 cycles at 5 C in the voltage range of 3–4.3 V (Figure 5.11g,h). At a high rate of charge and discharge at 10 C, it can still have 88.0% theoretical capacity retention. Consequently, high-rate capability of polyphosphate NW cathode material makes it possible to apply it in the field of rapid charge and discharge in energy storage applications [102].

5.5 Nanowires-Based Separators in LIBs

The separator is a critical component of the battery and acts as the electrolyte reservoir to isolate the anode and cathode. The separator should exhibit good wettability and electrolyte and thermal stability. The NW-based separators are divided into two types: pure inorganic NW separators and hybrid separators, which all exhibit high safety and heat resistance compared with the commercial separators. Here, we will discuss the NW-based separators in LIBs.

Nowadays, microporous polyolefin membrane is the widely used LIB separator, such as polyethylene (PE) or polypropylene (PP) due to their excellent mechanical strength, stability, processability, and proper pore size. However, these separators are unstable over 120 C because of their low glass transformation temperature and melting temperature, which would cause large shrinkages of the separators and internal short circuits. Meanwhile, the polyolefin separator is nonpolar, and its wettability with polar liquid electrolyte is poor, which would generate high resistance. Therefore, many strategies have been proposed to solve these problems, including inorganic coating, electrospinning, and ceramic membrane preparation [103]. Shaohua Jiang et al. designed hierarchical 3D micro/nano-architecture separator of polyaniline NWs wrapped on polyimide nanofibers (PANI/PI) through electrospinning and in situ polymerization [104]. The separator exhibited a hierarchical 3D structure, as shown in Figure 5.12a. The PANI nanowires coated uniformly on the PI nanofibers, as shown in Figure 5.12b. The separator shows excellent thermal stability of 180 C as shown in Figure 5.12c, there was no obvious shrinkage for the PI and PANI/PI

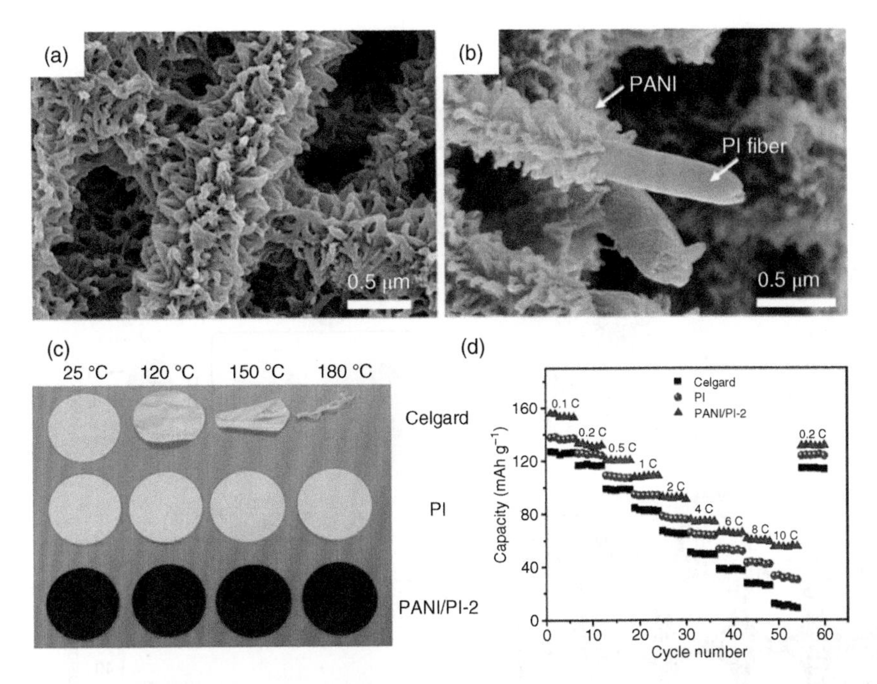

Figure 5.12 SEM images of PANI/PI-2 composite: (a) high magnification; (b) cross-section view; (c) digital photographs of Celgard, PI nanofiber nonwoven, and PANI/PI-2 composite membrane before and after thermal shrinkage at different temperatures (25, 120, 150, and 180 °C) for twohours; (d) rate capabilities of the cells with Celgard membrane, PI nanofiber nonwoven, and PANI/PI composite membrane as cell separators at different C-rate. Source: Ye et al. [104] / with permission of Elsevier.

separators at 180 C. The PANI/PI separator also showed high porosity of 84%, 61.9% liquid electrolyte uptake, high ionic conductivity of 2.33 mS cm^{-1}, which induced enhanced capacity and rate capability (Figure 5.12d).

On the other hand, inorganic ceramic membranes have good wettability with electrolyte and high temperature resistance compared with polyolefin separators. Inorganic NWs are often used as pure inorganic separator or hybrid separator due to their nonflammability and special structure. For example, Shaojin Jia et al. reported an eco-friendly Xonotlite nanowires/Wood pulp (XWP) fiber hybrid separator through low-cost papermaking process [105]. The hybrid separator shows excellent heat resistance. When the separator was heated up to 600 C, there was no obvious shrinkage. The cycle capability of XWP-20% separator was improved because of its high porosity and good electrolyte wettability, and after 100 cycles, the capacity reached 91% of the initial capacity. Therefore, the separator is a promising candidate for low-cost and eco-friendly separator. Yingjie Zhu et al. also fabricated a hybrid separator between hydroxyapatite nanowires and cellulose fibers (HAP/CF) through hybridization, and the schematic illustration of the separator is shown in Figure 5.13a–e [106]. The separator with hierarchical cross-linked network is flexible and porous, shows near-zero contact angle with the liquid electrolyte, and has 253% electrolyte uptake. And there was almost zero shrinkage for the pure

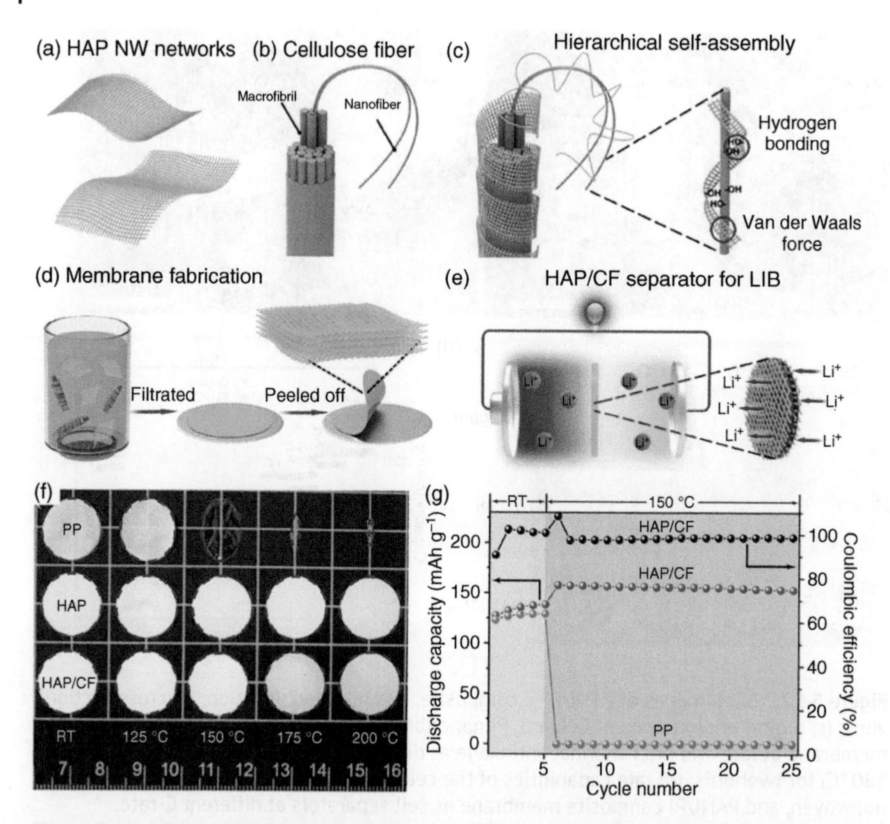

Figure 5.13 Schematic illustration of the new kind of highly flexible and porous separator based on HAP NW networks with excellent thermal stability, fire resistance, and superior electrolyte wettability: (a) HAP NW networks; (b) CFs; (c) hierarchical assembly of HAP NW networks and CFs; (d) suction filtration process for fabricating the layered and highly porous HAP/CF separator; (e) the application of the HAP/CF separator in the LIB; (f) illustration of thermal shrinkages of the separators at different temperatures (top panel: PP separator; middle panel: pure HAP separator; bottom panel: HAP/CF separator); (g) cycling performance of the battery with the HAP/CF separator and the PP separator at 2 C. Source: Li et al. [106] / John Wiley & Sons.

HAP and HAP/CF separators at different temperatures, as shown in Figure 5.13f. The color turned light yellow because of CF's oxidation at elevated temperature, while the shape and diameter of the HAP/CF separator remained the same, which was beneficial for the safety of the battery. The battery equipped with the HAP/CF hybrid separator showed good cycling stability at 150 °C with 98% CE (Figure 5.13g), which indicated good thermal stability of the hybrid separator.

Other strategy for improved separators is pure inorganic separators based on nanowires. For example, Zhendong Huang et al. prepared silica nanowires, extracted from asbestos fibers, for scalable bendable networks as stable separators [107]. The silica nanowires with their 3D porous structure and polar surface ensured good electrolyte wettability, excellent electrolyte uptake of 325% and retention ratio of 63%. Meanwhile, the inorganic separators enhanced the chemical and electrochemical stability at high temperatures and exhibited good thermal stability as high as 700 C. Xinjie Zhang et al. also developed a pure inorganic separator based on aluminum oxide nanowires without any polymer additives or binders through hydrothermal treatment of aluminum-containing precursor [108]. The pore size of the separator was about 100 nm, and the separator was bendable. The separator showed high electrolyte uptake of 190% because of the high porosity and the structure. In addition, the separator shows excellent rate capability and life cycle performance compared with the Celgard 2500. Above all, the inorganic NW separator shows excellent non-flammability, high porosity, and thermal stability at elevated temperatures. The NW-based hybrid separator has special 3D structure, which is beneficial to electrolyte wettability and uptake and enhances battery performance.

5.6 Nanowires-Based Solid-State Electrolytes in LIBs

The above section illustrated the NW-based separators with high safety and other advantages. However, the liquid electrolyte is a must to infiltrate the separators, which could cause safety issues. Liquid electrolyte, composed of salts and organic solvents, is the traditional electrolyte in LIBs. However, the leakage and combustion of the liquid electrolyte cause severe safety problems. Therefore, solid-state electrolyte (SSE) as ions transport medium is proposed to solve the safety problems. Here, we would focus on the NW-based SSEs in LIBs. The SSE is often divided into two types: inorganic and polymer SSE. The inorganic electrolyte exhibited high ionic conductivity and attracted much attention. For example, $Li_7La_3Zr_2O_{12}$ (LLZO) is a promising SSE and the Li^+ ionic conductivity of high temperature cubic phase LLZO (c-LLZO) is about 10^{-4} S cm^{-1} at RT. However, the c-LLZO is unstable at RT. And the LLZO is tend to transform into tetragonal phase (*t*-LLZO) and the ionic conductivity would decrease 2–3 orders of magnitude. Some reports found that the tetragonal structure LLZO could transform to the cubic phase after reducing the particle size through ball milling. Candace K. Chan et al. reported that LLZO NWs via electrospinning synthesis could be stable with cubic structure at RT, and nanostructured LLZO showed higher ionic conductivity, cycling performance, current density, and fracture strength [109].

For the all-solid-state lithium batteries, there are some requirements: first, the solid electrolyte ionic conductivity should be larger than 10^{-4} S cm^{-1}; the interface resistance between electrolyte and electrode should be small, and there should be proper ionic transportation at the electrolyte/electrode interface; third, lithium dendrites should be controlled during charging and discharging processes. Yan-bing He et al. reported a composite electrolyte embedded with $Li_7La_3Zr_2O_{12}$ (LLZO) NWs [110]. LLZO NWs could further improve the electrolyte conductivity because of the formation of the continuous path compared with the LLZO microparticles, and can be favor to the Li ion transportation and form uniform Li metal deposition to suppress the Li dendrite growth. The LLZO NWs reduced the activation energy and improved the mechanical property of PLLN, which improved the stability of Li symmetric cells to 1000 hours without short circuit at 60 C.

Compared with inorganic electrolyte, polymer electrolyte is stable, flexible, stretchable, low-cost, and have low interfacial resistance, which attracts more and more attention. Unfortunately, the conductivity of the polymer electrolyte is only 10^{-6} S cm^{-1} at RT. So many methods were reported to enhance the conductivity of the polymer electrolyte, such as the addition of inorganic nanomaterials, plasticizer, cross-linking, and block copolymerization. The majority of the ceramic fillers are 0D particles. And there are junctions or interface between the nanofillers and the polymer matrix, which could enhance the ionic conductivity. However, the length of ionic transport pathway is really limited. Meanwhile, the tiny particles are prone to agglomeration and could cause the opposite effect of conductivity or mechanical strength. Shi Wang et al. fabricated a composite polymer electrolyte (CPE) composed of polymer electrolyte, ceramic NWs, and nanoparticles [111]. As known from the reports, ceramic nanofillers could reduce the crystallinity of polymers and facilitate the motion of segments. Meanwhile, the strong Lewis acid–base interaction between the surface chemical groups of the nanofillers and electrolyte ion species can increase the dissociation of lithium salts and also stabilize the anions. And NWs have high length-diameter and could form much longer continuous ionic transport layers compared with nanoparticle fillers. The polymer electrolyte is PE oxide-polypropylene oxide-polyethylene oxide (PEO–PPO–PEO) because the copolymer structure could reduce the crystallinity and glass transition temperature of PEO. Meanwhile, the ceramic film could build long-range interfacial ion channels and act as a mechanical support for the polymer electrolyte. The combination of the nanowires and nanoparticles could induce more continuous ion channels at the interface between the polymer and ceramic fillers and higher ionic conductivity. On the other hand, the extra SiO_2 nanoparticles could scavenge trace impurities. Also, Candace K. Chan et al. reported $Li_7La_3Zr_2O_{12}$ (LLZO) NWs and nanoparticles as the ceramic fillers and improved the ionic conductivity of the polyacrylonitrile (PAN)-$LiClO_4$-based electrolyte at room temperature to 1.31×10^{-4} S cm^{-1} by three orders of magnitude [112]. They found that LLZO NWs changed the local condition of the PAN polymer and Li^+ diffused preferentially at the LLZO/polymer interface. In addition, Yi Zhang et al. fabricated different shapes of $BaTiO_3$ nanofillers (nanowires, nanocubes, and nanospheres) based on PEO solid polymer electrolyte [113]. They found that there is interaction between $BaTiO_3$ nanofillers and PEO chains and lithium ions, and caused the weakness of complexation between O atoms and lithium ions. And the electrochemical stable window increased to 4.7 V versus Li/Li^+ for $BaTiO_3$ nanospheres and the SPE with $BaTiO_3$ spheres has the highest ionic conductivity of 1.8×10^{-5} S cm^{-1} at 25 °C because of the highest SSA of $BaTiO_3$ nanospheres.

The advantage of the polymer-based electrolytes is that they are flexible and have an intact interface with the electrodes. However, their mechanical properties and ionic conductivities at RT were poor. It is an effective way to disperse ceramic particles into polymer matrix for enhancement of the ionic conductivity and electrochemical stability, because the addition of the ceramic filler could hinder the polymer crystallization and form highly conductive interface layers. Generally, the ceramic fillers consist of inactive fillers (Al_2O_3 and SiO_2) and active fillers

(Li_3N and $Li_{1.3}Al_{0.3}Ti_{1.7}(PO_4)_3$) for lithium ion conduction. For example, Jianqi Sun et al. prepared $Li_{6.75}La_3Zr_{1.75}Nb_{0.25}O_{12}$ (LLZN) NWs with high crystallinity and high aspect ratio through a electrospinning and calcination process, and fabricated composite solid polymer electrolytes (CSPEs) with poly(methyl methacrylate) (PMMA) matrix, LLZN NWs fillers and $LiClO_4$ salt [114]. The 3D network structure interlaced by LLZN NWs creates more amorphous regions, and the interaction between surface of NWs and polymer chains provide the unique conductive routes for lithium ions. And the ionic conductivity is sharply increased from 5.98×10^{-7} of filler-free solid electrolyte to $2.20 \times 10^{-5}\ S\ cm^{-1}$ of CPSE at RT. Wei Liu et al. prepared $Li_{0.33}La_{0.557}TiO_3$ NWs through electrospinning method and dispersed the NWs into PAN-$LiClO_4$ polymer, gained $2.4 \times 10^{-4}\ S\ cm^{-1}$ ionic conductivity at RT [115]. The synthesis procedure, morphologies, and ionic conductivity are shown in Figure 5.14. They found that the NW fillers have high aspect ratio and form continuous ionic transport pathways over longer distances, which could reduce the junction cross significantly and enhance the lithium ionic conductivity. Lin Zhu et al. fabricated PEO, poly (propylene carbonate) (PPC), lithium bis(fluorosulfonyl)imide (LiTFSI), and $Li_{0.33}La_{0.557}TiO_3$ NWs (LLTO-NWs) solid polymer composite electrolyte (SPCE) [116]. The addition of PPC could reduce the crystallinity of PEO. And the addition of LLTO-NWs could further reduce the PEO crystallinity and form a lot of ion transport channels at the interface between fillers and polymers. The LLTO-NWs were fabricated by electrospinning and have a large SSA and wide electrochemical window. And the novel SPCE with 8 wt% LLTO-NWs had the highest ionic conductivities of $5.66 \times 10^{-5}\ S\ cm^{-1}$ at 25 °C and $4.72 \times 10^{-4}\ S\ cm^{-1}$ at 60 °C, which could favor the SPCE's development.

Except lithium ionic conductors, Al_2O_3, TiO_2, oxygen ionic conductors are also often used as ceramic fillers. The ionic conductivity based on PEO or PAN solid polymer electrolytes is about $10^{-7}\ S\ cm^{-1}$ at RT. To improve the ionic conductivity of the polymer electrolyte, ceramic fillers such as SiO_2, Al_2O_3, and TiO_2 on were added to interact with the polymer matrix. For polymer electrolytes, Li^+ transportation is related to the polymer segmental motion. And the introduction of the ceramic fillers could hinder the polymer crystallization and was beneficial to fast Li^+ transportation. On the other hand, the Lewis-acid sites on the surfaces of the ceramic fillers could attract the salt anions and set more positively charged Li^+ free and enhanced Li^+ conductivity. Meanwhile, the NWs have more continuous ion-conducted pathway across long distance and percolation network, while nanoparticles are isolated. Therefore, Wei Liu et al. added oxide-ion-conducted NWs into polymer electrolyte to improve the ionic conductivity [117]. They found that the more oxygen vacancies on the surface, the more free Li^+ was released and the higher the ionic conductivity. And 7 mol% Y_2O_3 stabilized ZrO_2 NWs had the highest conductivity of $1.07 \times 10^{-5}\ S\ cm^{-1}$ at 30 °C. In addition, there are competitive interactions between ceramic fillers and polymers, which could favor ionic conductivity, and improve the electrochemical stability window and thermal stability. The addition of the ceramic fillers could reduce the polymer reorganization tendency and the polymer crystallization, which is beneficial to the enhancement of the lithium-ion conductivity. Peiqi Lun et al. dispersed oxygen-ion conducting Sm-doped CeO_2 (SDC) NWs into

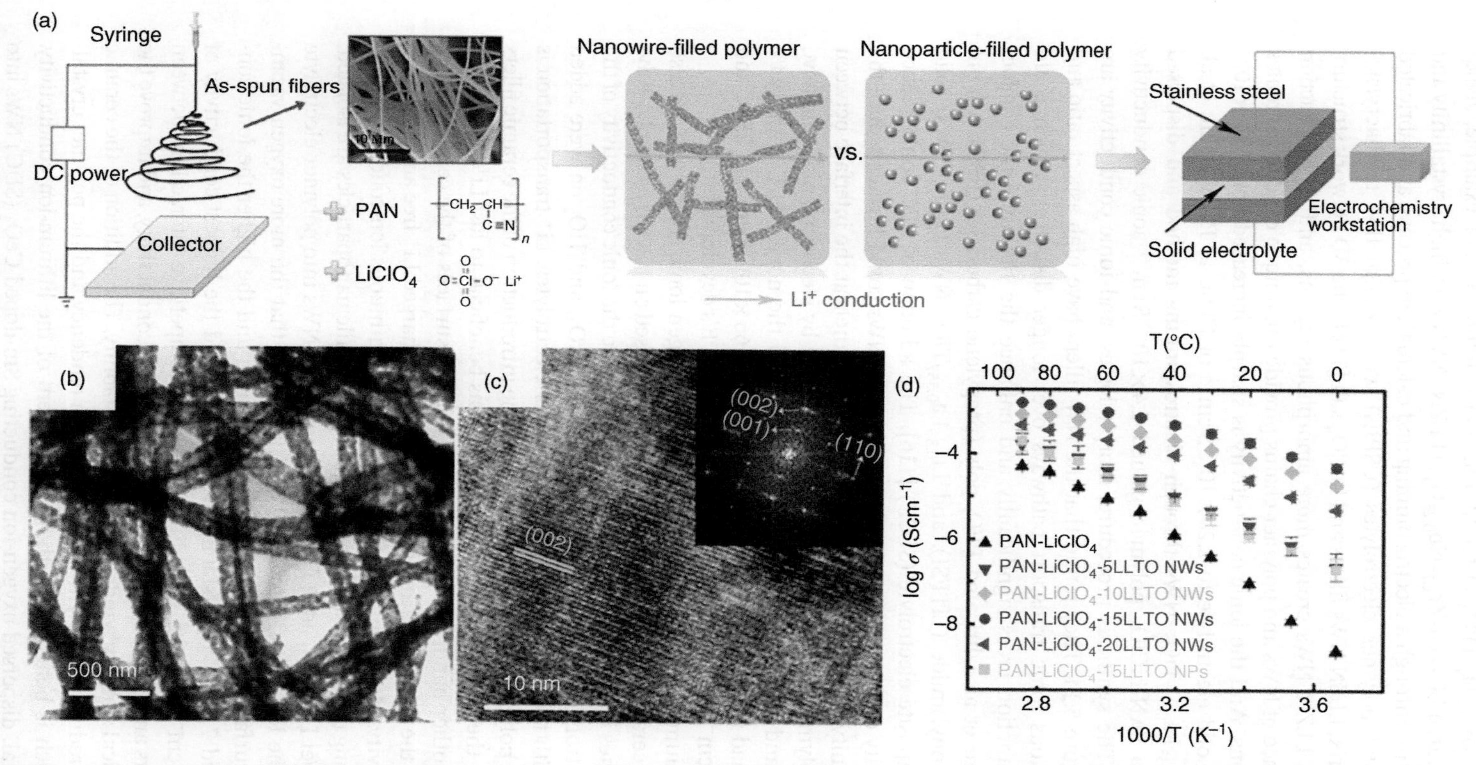

Figure 5.14 (a) Schematic illustration for the synthesis of ceramic nanowire-filled polymer-based composite electrolytes, together with the comparison of possible lithium-ion conduction pathway in nanowire-filled and nanoparticle-filled composite electrolytes, and illustration of the electrode configuration for the AC impedance spectroscope measurement; (b) TEM image of LLTO nanowires; (c) HRTEM image of an individual LLTO nanowire, with inset FFT pattern; (d) Arrhenius plots of the composite electrolytes with carious LLTO nanowire concentrations, together with the date for LLTO nanoparticle-filled PAN-$LiClO_4$ electrolyte. Source: Liu et al. [115] / American Chemical Society.

the polyvinylidene fluoride matrix to form CPEs, and gained an ionic conductivity of 9.09×10^{-5} S cm^{-1} at 30 °C with a lithium-ion transfer number of ~0.40 [118]. And the enhancement of the lithium-ion conductivity is attributed to the Lewis acid-base interaction resulting from the formation of positively charged oxygen vacancies. The 1D nanostructures of the SDC NWs could provide better pathways for lithium-ion transfer and migration. Their results indicate that employing an oxygen-ion conductor could be a facile and promising way to enhance the ionic conductivity of the polymer electrolytes.

Al_2O_3, TiO_2 are also proper ceramic fillers. For example, Ming-Chou Chen et al. fabricated a new type organic–inorganic hybrid electrolytes based on the triblock copolymer with lithium trifluoromethanesulfonate and Al_2O_3 NWs, and the maximum ionic conductivity was 9.8×10^{-5} S cm^{-1} [119]. There were interactions among salt, Al_2O_3 NWs, and polymer, which caused the formation of the hybrid structure. Yu Wang et al. fabricated a core–shell protein@TiO_2 NWs CPE via protein-functionalization on electrospun TiO_2 NWs [120]. The protein can considerably adsorb anions, promote the dissociation of lithium salts, and enhance the conduction of Li^+ at the polymer-nanofiller interfaces. Compared with protein-TiO_2 nanoparticles, protein-TiO_2 NWs have more continuous networked structure than the 1D NWs, resulting in enhanced ionic conductivity and mechanical properties of the CPEs.

Some other inactive fillers are also reported. For example, Xinyong Tao el al. fabricated $Mg_2B_2O_5$ NWs and added them to the PEO-based SSEs [121]. The fabrication of the $Mg_2B_2O_5$ nanowires was low cost and facile. The nanowires were lightweight, flame-retardant, and high-tenacity. They found that there was interaction between $Mg_2B_2O_5$ nanowires and $-SO_2-$ in the $TFSI^-$ anion, and the interaction could enhance lithium ion release and transportation, which improved the ionic conductivity. And the addition of $Mg_2B_2O_5$ improved the mechanical properties and the flame-retardant properties of the polymer electrolyte. In addition, PEO electrolyte is easy to oxidize above 4 V versus Li/Li^+. Therefore, it is critical to develop electrolyte paired with 4 V $Li(Ni_{1/3}Mn_{1/3}Co_{1/3})O_2$ (NMC) cathodes and enhance the ionic conductivity. Poly(vinylidene fluoride) (PVDF) is nonflammability, easy processing, and the ionic conductivity is about 10^{-4} S cm^{-1}. To address this problem, Yuan Yang et al. randomly dispersed palygorskite ($(Mg, Al)_2Si_4O_{10}(OH)$) NWs into PVDF electrolyte, and gained flexible PVDF/palygorskite composite electrolyte with enhanced ionic conductivity and mechanical properties [122]. The synthesis process and the photos of the membrane are shown in Figure 5.15a–d. Palygorskite has an open structure and improve the PVDF membrane's stiffness and toughness. The lithium dendrite is suppressed due to the mechanical robustness, which is beneficial to the cycling stability as shown in Figure 5.15e,f. The PVDF/palygorskite composite membrane meets the requirements of high energy density and safety for solid-state electrolyte.

In summary, NWs with high length anddiameter play an important role in the SSEs. For example, LLZO nanowires could be stable with cubic structure. On the other hand, lithium ionic conductors, Al_2O_3, SiO_2, oxygen ionic conductors, and

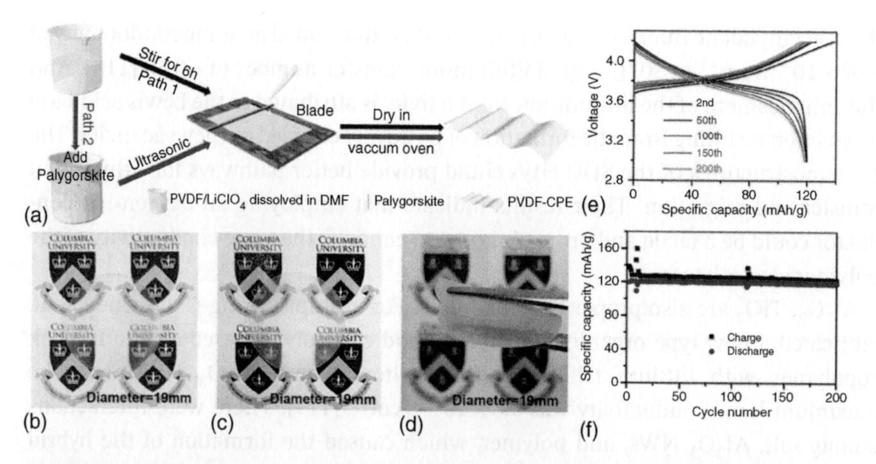

Figure 5.15 (a) A schematic diagram of the synthesis of PVDF-based polymer electrolytes and PVDF/palygorskite nanowires/CPE; (b) PVDF polymer electrolyte membrane and (c) PVDF/palygorskite CPE membrane dried at 60 °C in vacuum chamber for 24 hours; (d) bent PVDF/palygorskite CPE showing excellent flexibility; (e) voltage profile and (f) cycling performance of NMC|PVDF/5 wt% palygorskite CPE|Li cell (1 C = 150 mA g^{-1}). Source: Yao et al. [122] / American Chemical Society.

inactive inorganic NWs with long distance and percolation network have been used as ceramic fillers to improve the ionic conductivity of the polymer electrolyte.

5.7 Nanowires-Based Electrodes for Flexible LIBs

With the rapid development of flexible and wearable electronic products, such as electronic paper, smart clothes, electronic skin, flexible display, bendable smart phones, and implantable medical devices, there is an urgent need for bendable, foldable, and stretchable electronic devices in the market. Correspondingly, the power supply to these electronic devices should be flexible while maintaining certain electrochemical performances. A LIB with high energy density, power density, and cycle life is one of the most ideal power supply options for electronic devices. However, because the fabrication process currently used is based on the traditional mud manufacturing method, the traditional lithium ion battery is usually rigid and bulky, the performance of which will decline sharply after repeated folding and bending, and even lead to safety risks. Nanomaterials are ideal materials for preparing flexible electrodes for energy storage devices, because they have at least one dimension in the nanoscale range and often have better mechanical and electrical properties than bulk materials. Cheng Huiming et al. [30] have summarized the application of carbon nanotubes and graphene materials in foldable, stretchable and compressible flexible electrodes, and discussed the gain effect of various structures on electrode flexibility, as shown in Figure 5.16. Although nanomaterials possess many advantages as flexible electrode materials, the study of flexible, stretchable energy storage devices is still in the early stages

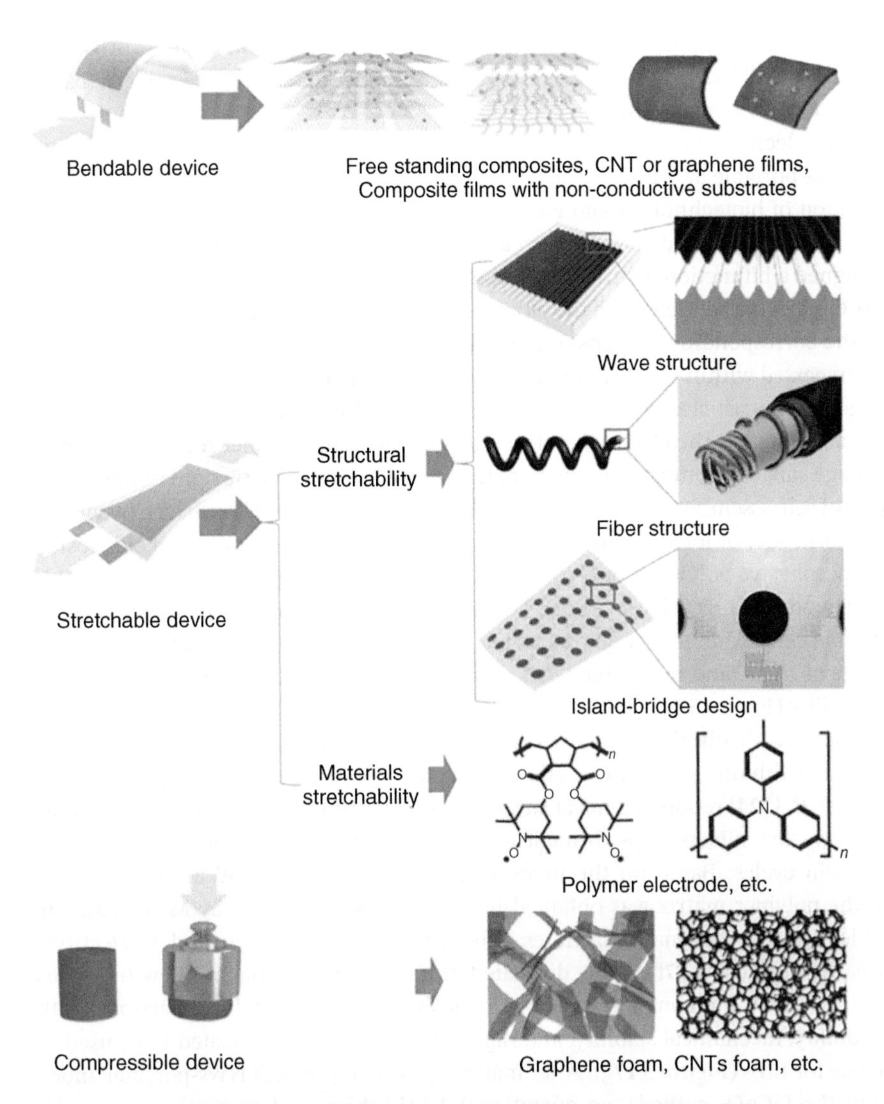

Figure 5.16 Classification of flexible electrodes and the applications of CNTs and graphene. Source: Wen et al. [30] / John Wiley & Sons.

of laboratory research. They still face the following three key challenges in the preparation of practical flexible lithium ion battery devices: (i) large-scale production of flexible electrodes; (ii) stable electrochemical performance during dynamic bending; (iii) high energy and power densities of devices. NWs materials tend to exhibit fast charge transfer and excellent mechanical flexibility due to their radial nanoscale size and ultrahigh aspect ratio. Additionally, the NW structure can make the active material grow closely on the collector or cross-link each other to form a network structure so that it does not fall off during the bending process of the flexible device. Thus, the suitable NW-based electrode will provide a marvelous opportunity for practical production of flexible wearable LIBs. In recent years, the unique advantages of NWs have also been well developed in flexible LIBs. The following will focus on some representative examples and summaries of NW-based electrodes for flexible LIBs.

The selection and assembly of NW materials are central issues in the development of portable and flexible LIBs. Nam and co-workers [123] pioneered the combination of biotechnology and energy storage technology, using viruses to synthesize Au-Co_3O_4 hybrid NWs at room temperature as anode materials for LIBs. They designed a bifunctional virus template that simultaneously expressed both Au- and Co_3O_4-specific peptides with the virus coat (Figure 5.17a). After incubating the virus in the corresponding suspension, hybrid NWs of 5-nm Au nanoparticles spatially interspersed within the Co_3O_4 wires were obtain (Figure 5.17b,c). The introduction of gold nanoparticles cannot only improve the conductivity of the whole nanowire, but also catalyze the electrochemical reaction, enabling a higher reversible lithium storage capacity compared to the pure Co_3O_4 NWs. Moreover, by right of the principles of self-assembly and bio-template, the hybrid Au-Co_3O_4 NWs driven by viruses can form very ordered two-dimension liquid crystalline layers on top of electrostatically cohesive polymer films, constructing a transparent and free-standing flexible film anode of LIBs (Figure 5.17d). The assembled monolayer of Au-CO_3O_4 NWs/Li flexible cells exhibited a high reversible capacity (94% of its theoretical capacity) at a rate of 1.12 C and maintained 65% at the rate of 5.19 C, demonstrating a high rate capability (Figure 5.17e) [123].

Silicon with ultrahigh theoretical lithium storage capacity is one of the most potential lithium anode materials, which has always been a hot topic material. Vald et al. [124] reported a novel strategy to obtain lithium-ion battery components from wasted silicon chips through tunable and repeatable chemical etching permeation cycles. Based on the predesigned parameters, a SiNW array embedded in the polymer matrix was obtained from recycle silicon wafer by MACE, among which the polymer matrix operates as both Li-ion gel-electrolyte and an electrode separator (Figure 5.17f). Then, deposited uniformly around the SiNWs with porous and electric- interconnected copper nanoshells, a free-standing electrode with flexibility, mechanical stability, and high conductivity was fabricated to be used as anode for LIBs (Figure 5.17g). After matching the composite Si NWs-polymer anode with the $LiCoO_2$ cathode, an operational 3.4 V lithium-polymer silicon nanowire mechanically flexible battery was demonstrated, and the initial discharge capacity of this battery achieved ~3500 mAh g^{-1} at a C/20 cycling rate (Figure 5.17h) [124].

Besides using the above up-down method to prepare flexible SiNWs electrodes, the down-up synthesis method is also a good choice to prepare excellent nanowire electrodes. Chockla and co-workers [125] proposed a paper-like nonwoven fabric flexible film electrode, consisting of entangled silicon NWs. By the supercritical-fluid-liquid–solid process, the SiNWs were obtained with diameter of 10 to 50 nm and average length over 100 μm, coated with a thin chemisorbed polymer nanoshell (Figure 5.18a,b). After annealing in a reducing atmosphere at a proper temperature, the polymer coating was converted into an electrically conductive carbon layer on the SiNW surface, which can significantly improve the electrical conductivity and mechanical stability of the Si nanowires (Figure 5.18c).

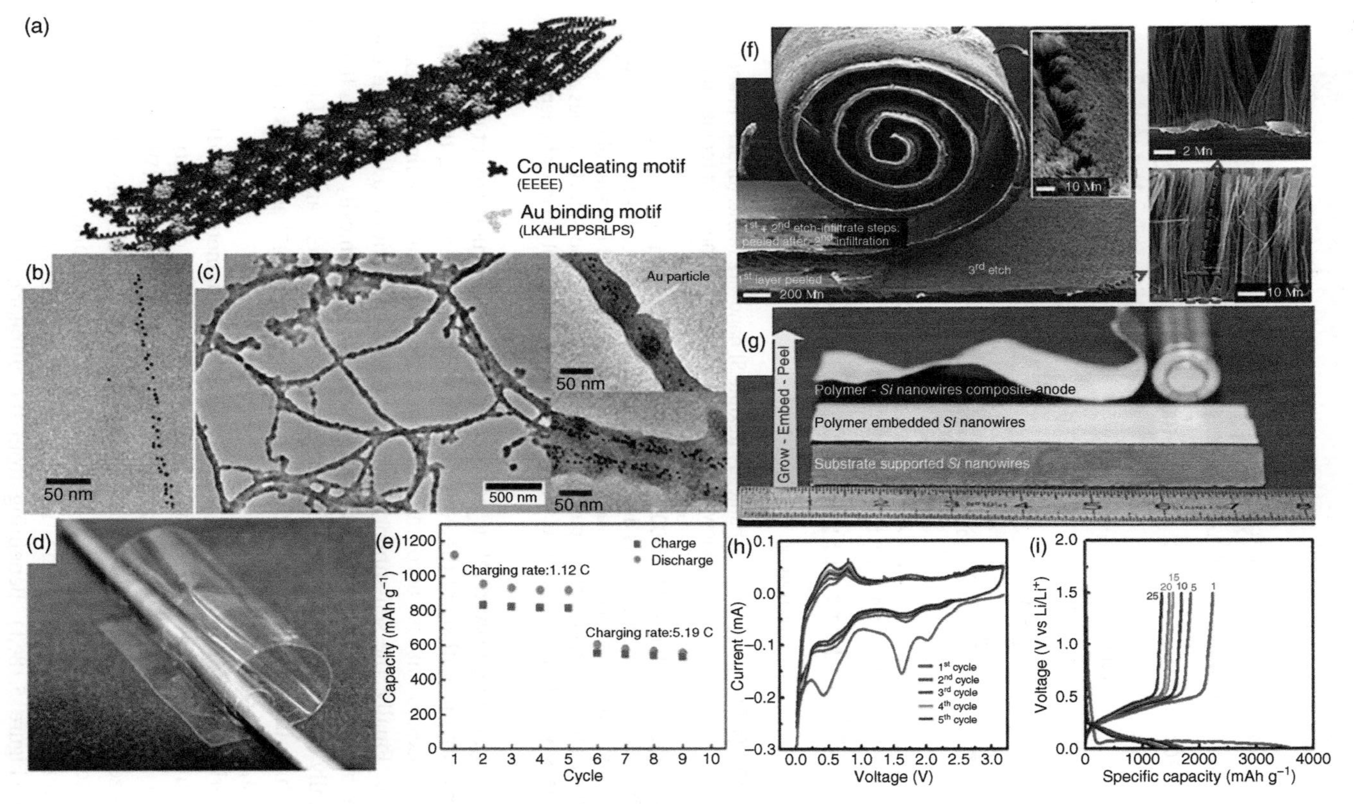

Figure 5.17 (a) Visualization of the genetically engineered M13 bacteriophage viruses. P8 proteins containing a gold-binding motif (yellow) were doped by the phagemid method in E4 clones, which can grow Co_3O_4. (b) TEM images of the assembled gold nanoparticles on the virus. (c) TEM image of hybrid nanowires of Au nanoparticles/Co_3O_4. (d) Digital camera image of a flexible and transparent free-standing film of $(LPEI/PAA)_{100.5}$ on which Co_3O_4 viral nanowires are assembled into nanostructured monolayer with dimensions of 10 cm by 4 cm. (e) Specific capacity of hybrid Au-Co_3O_4 nanowires [6]. (f) Three consecutive etch-infiltrate-peeling processing loops have been performed on the same Si chip and using the same MACE mask. (g) Ensemble view of the large-scale LIPOSIL processing flow. Commercial battery-scale fabrication is achieved. (h) Cyclic voltammetry curves of the polymer-wrapped silicon nanowire anodes. (i) Galvanostatic discharge/charge profiles of the Si – polymer composite cycled at a rate of C/20 between 1.5 and 0.02 V versus Li/Li^+. Source: Vlad et al. [124] / PNAS.

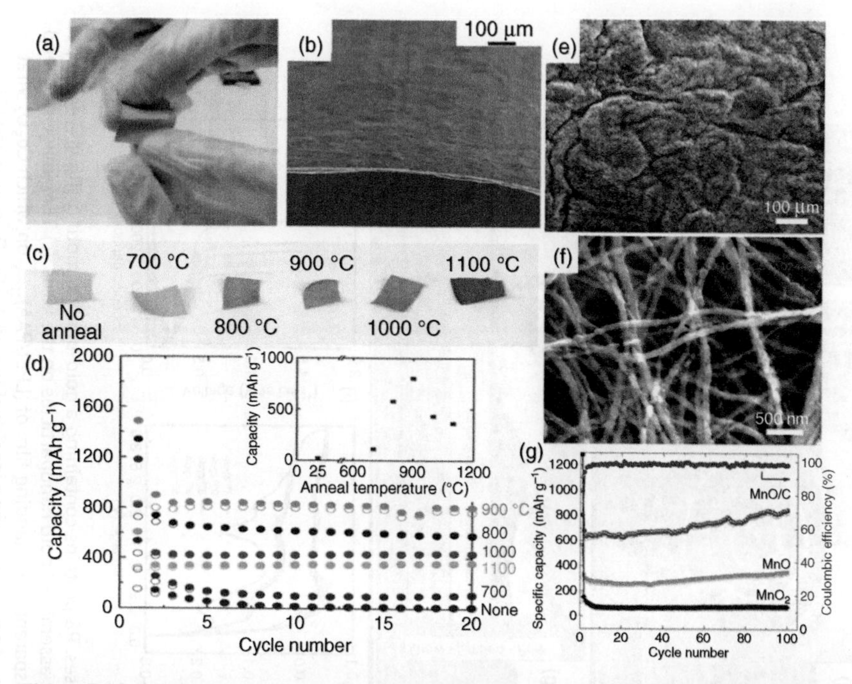

Figure 5.18 (a) Photograph of mechanically flexible Si nanowire fabric. (b) SEM images of the fabric. (c) Photographs of Si nanowire fabrics annealed at the indicated temperatures under a reducing atmosphere [8]; (d) Galvanostatic cycling tests of Si nanowire fabric before and after annealing at 700, 800, 900, 1000, and 1100 °C at a cycle rate of C/20 (C = 3579 mAh g^{-1}). The inset shows discharge capacities after 20 cycles for batteries with nanowire fabric anodes annealed at various temperatures. (e, f) SEM images of MnO/C at different magnifications. (g) Cycling performance of MnO_2, MnO, and MnO/C anodes at a current density of 100 mA g^{-1}. Source: Wang et al. [126] / The Royal Society of Chemistry.

Additionally, the nonwoven fabric has 90% void volume, which cannot only boost ion transport and withstand the volume expansion of silicon during the repeated cycling processes but also accommodate repeated bending and folding as flexible electrode. Finally, the self-supporting fabric electrode, without adding the conductive carbon and binder matched, was used as a mechanically flexible, highly electrically conductive anode material in a lithium ion battery anode of LIBs and present high lithium storage capacity of more than 800 mA g^{-1} as well as excellent cycling stability (Figure 5.18d) [125].

Similarly, the ultralong MnO/C coaxial NWs with diameters of ~100 nm and lengths up to hundreds of micrometers were synthesized, constructing the free-standing flexible membrane anode for LIBs [126]. The MnO NWs were first synthesized by a facile hydrothermal method. Then, the flexible membrane of composite MnO/C coaxial NWs with carbon coated was fabricated through an in situ novel interfacial polymerization technique and the following carbonization process (Figure 5.18e,f). Benefiting from the ultralong NW structure and the network framework with outstanding electric conduction provided by carbon layer, the MnO/C membrane electrode not only has good mechanical flexibility but also presents superior electrochemical performance. When acting as a flexible freestanding anode for LIBs, the composite MnO/C NW electrode exhibits a decent reversible lithium ion storage capacity of 832 mA g^{-1} at the current density of 100 mA g^{-1} after 100 cycles, as well as superior cycling stability with a specific capacity of 480 mAh g^{-1} being retained after 600 cycles at a high current density of 1000 mA g^{-1}(Figure 5.18g) [126].

Different from the aforementioned synthesis method of preparing raw materials directly into NWs under certain conditions, Kawamori et al. [127] reported an flexible metal sheet called metal nanowire nonwoven cloth (MNNC) as a kind of universal substrate for loading proper LIB active material. Consisting of intertwined metal NWs 100 nm in diameter with several dozen micrometers of length, the MNNC possesses both a flexibility and mechanically robustness of the metal and a large SSA of the NWs (Figure 5.19a). When an active material was deposited on the MNNC, a flexible membrane electrode with any conductive additives, binders, and current collectors was obtained for LIBs, which is significant for increasing the gravimetric specific capacity (Figure 5.19b,c). Besides, there are many merits for this electrode design, including high SSA for the efficient ion/electron transports, and abundant

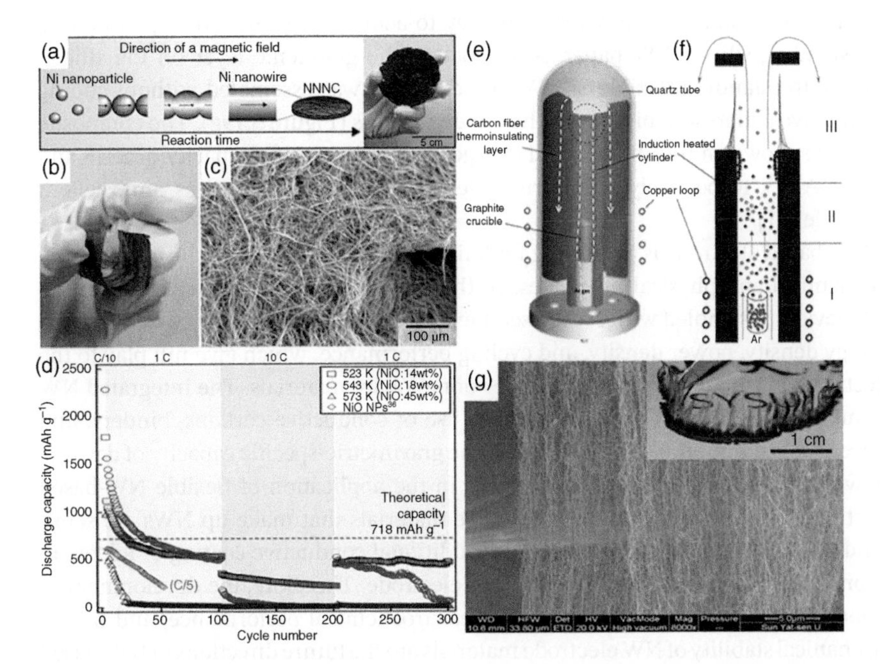

Figure 5.19 (a) Schematic illustration of the formation process of the NNNC synthesized by electroless deposition under a magnetic field. (b) Photographic and (c) SEM images of the sheet-like NNNC. (d) Cycle performance of the NNNCs annealed at various temperatures. The charge–discharge experiments were conducted at room temperature between 0.02 and 2.50 V at a rate of C/10, 1 C, and 10 C [128]. (e) Schematic diagrams of the vertical high-frequency induction furnace. The direction of gas flow is marked by the yellow dashed lines. The location where Si nanowires grow is marked by a red circle. (f) Growth mechanism of Si nanowires and FTS-SiNWsP under the action of carrier gas flow. (g) SEM image of the flexible, transparent, and self-standing Si nanowire paper. The inset shows a digital image of originally synthesized FTS-SiNWsP with excellent transmission. Source: Pang et al. [31] / American Chemical Society.

interspaces between the neighboring nanowires accommodating the critical volume change during lithiation/delithiation process. A nickel NW nonwoven cloth covered with NiO active materials was prepared to demonstrate its operability as a prototype flexible anode for LIBs. The fabricated NiO-covered nickel nanowire nonwoven cloth (NNNC) electrode present good mechanical flexibility for twisting and tying, along with superior cycling stability, 500 mAh g^{-1} at 1 C after 300 cycles, compared to that of a NiO nanoparticle-based electrode (Figure 5.19d) [127].

By virtue of the congenital advantage of being compatible with mature integrated circuit technology for bulk silicon, a flexible electrode composed of nanosilicon materials will be more suitable for advanced flexible energy storage devices than one composed of other materials. Pang and co-workers [31] have studied paper-like freestanding silicon nanowires with macroscopic brittleness and transparency characteristics as anode for LIBs. They fabricated the transparent and freestanding flexible paper-like silicon NW membrane by gas flow-directed assembly of a unique interlocking alignment of the Si nanowires, among which the silicon nanowires with an average diameter of about 10 nm interconnected together forming a network structure (Figure 5.19e,f). The ultralarge straining ability provided by the nanoscale silicon nanowires is the key to achieving flexible, transparent, and self-standing silicon NW paper. After coating the grapheme layer on the silicon NWs, a freestanding flexible Si NW film electrode was assembled without adding conductive adders and binders as the anode for LIBs (Figure 5.19g). The composite Si anode shows an ultrahigh initial charge/discharge specific capacity of 5336 and 2699 mA g^{-1}, respectively at a current density of 1 A g^{-1}, as well as an excellent cycle life [31].

NW-based electrode is promising for building flexible LIBs due to its unique electrochemical and physical properties. In the present reported works, those flexible LIB devices assembled with NW-based membrane electrodes exhibited outstanding energy density, power density, and cycling performance, which give full play to the electrochemical and physical properties of nanowire materials. The integrated NW membrane electrode is exempt from the use of conductive carbons, binders, and current collectors, efficiently enhancing the gravimetric-specific capacity of devices. However, there are still some challenges in the application of flexible NW-based electrode. For example, most of the active materials that make up NWs are semiconductors or insulators, which require additional conductive coating or adders to improve the electric conduction of whole electrode. Therefore, the development of higher-level NW loading process, better electrochemical performance, and higher mechanical stability of NW electrode materials are the future directions of NW-based flexible electrode materials.

5.8 Summary and Outlook

In this chapter, we discussed NWs as anodes, cathodes, separators, and SSEs in LIBs. NWs with nano-sized diameters are good for ion diffusion in the radical direction, which is beneficial to enhance the diffusion dynamics of carriers and improve the rate capability. And NW-based electrodes could release the stress properly, and adjust the volume changes during charging and discharging process, improving the structural stability. In addition, nanowires exhibit good bending and robust mechanical properties due to their high aspect ratio and are suitable to construct flexible devices. In addition, NWs are also used as separators and ceramic fillers for SSEs, they can interweave with other materials, which is beneficial to liquid electrolyte retention and ionic conductivity. On the other hand, challenges still exist for NWS in LIBs. For NW electrode-based intercalation reaction, the intercalation leads to volume expansion and fracture of the NWs, which can be improved by pre-intercalation and topological substitution. For NW electrodes based on conversion reactions, the properties are improved by constructing core–shell NW coatings and designing hierarchical heterogeneous nanowire structures. For an alloy NW anode, the active materials are easy to destroy or even fall off, the SEI is easy to form, and the alloy NWs are easy to aggregate. The effective solutions are to combine the NWs with other materials, or reduce the size of the alloy nanowires. Meanwhile, the mass loading of NW electrodes is relatively low, and the effective strategies are to construct 3D structured electrodes, employ lightweight current collectors, and so on.

References

1 Goodenough, J.B. (2012). Rechargeable batteries: challenges old and new. *Journal of Solid State Electrochemistry* 16: 2019–2029.
2 Goodenough, J.B. and Park, K.-S. (2013). The Li-ion rechargeable battery: a perspective. *Journal of the American Chemical Society* 135: 1167–1176.
3 Tarascon, J.M. and Armand, M. (2001). Issues and challenges facing rechargeable lithium batteries. *Nature* 414: 359–367.
4 Whittingham, M.S. (1976). Electrical energy storage and intercalation chemistry. *Science Advances* 192: 1126–1127.
5 Goodenough, J.B. (2018). How we made the Li-ion rechargeable battery. *Nature Electronics* 1: 204–204.
6 Mizushima, K., Jones, P.C., Wiseman, P.J. et al. (1980). Li_xCoO_2 $(0<x<-1)$: a new cathode material for batteries of high energy density. *Materials Research Bulletin* 15: 783–789.
7 Goodenough, J.B. (2007). Cathode materials: a personal perspective. *Journal of Power Sources* 174: 996–1000.
8 Placke, T., Kloepsch, R., Dühnen, S. et al. (2017). Lithium ion, lithium metal, and alternative rechargeable battery technologies: the odyssey for high energy density. *Journal of Solid State Electrochemistry* 21: 1939–1964.
9 Manthiram, A. (2017). An outlook on lithium ion battery technology. *ACS Central Science* 3: 1063–1069.
10 Etacheri, V., Marom, R., Elazari, R. et al. (2011). Challenges in the development of advanced Li-ion batteries: a review. *Energy & Environmental Science* 4 (9): 3243–3250. https://doi.org/10.1039/c1ee01598b.

11 Goodenough, J.B. (2013). Evolution of strategies for modern rechargeable batteries. *Accounts of Chemical Research* 46: 1053–1061.

12 Goodenough, J.B. and Kim, Y. (2010). Challenges for rechargeable Li batteries. *Chemistry of Materials* 22: 587–603.

13 You, Y., Celio, H., Li, J. et al. (2018). Modified high-nickel cathodes with stable surface chemistry against ambient air for lithium-ion batteries. *Angewandte Chemie International Edition* 57: 6480–6485.

14 Blomgren, G.E. (2016). The development and future of lithium ion batteries. *Journal of the Electrochemical Society* 164: A5019–A5025.

15 Lim, C.H., Jung, Y.H., Yeom, S.J. et al. (2017). Encapsulation of lithium vanadium phosphate in reduced graphene oxide for a lithium-ion battery cathode with stable elevated temperature performance. *Electrochimica Acta* 253: 208–217.

16 Zhang, Y.-C., You, Y., Xin, S. et al. (2016). Rice husk-derived hierarchical silicon/nitrogen-doped carbon/carbon nanotube spheres as low-cost and high-capacity anodes for lithium-ion batteries. *Nano Energy* 25: 120–127.

17 Yu, Z.-L., Xin, S., You, Y. et al. (2016). Ion-catalyzed synthesis of microporous hard carbon embedded with expanded nanographite for enhanced lithium/sodium storage. *Journal of the American Chemical Society* 138: 14915–14922.

18 Shan, T.-T., Xin, S., You, Y. et al. (2016). Combining nitrogen-doped graphene sheets and MoS_2: a unique film–foam–film structure for enhanced lithium storage. *Angewandte Chemie International Edition* 55: 12783–12788.

19 Zhou, W., Zhu, Y., Grundish, N. et al. (2018). Polymer lithium-garnet interphase for an all-solid-state rechargeable battery. *Nano Energy* 53: 926–931.

20 Kamali-Heidari, E., Kamyabi-Gol, A., Heydarzadehsohi, M. et al. (2018). Electrode materials for lithium ion batteries: a review. *Journal of Ultrafine Grained and Nanostructured Materials* 51: 1–12.

21 Ji, L., Lin, Z., Alcoutlabi, M. et al. (2011). Recent developments in nanostructured anode materials for rechargeable lithium-ion batteries. *Energy & Environmental Science* 4: 2682–2699.

22 Zhang, X.-Q., Chen, X., Cheng, X.-B. et al. Highly stable lithium metal batteries enabled by regulating the Li^+ solvation in nonaqueous electrolyte. *Angewandte Chemie International Edition* .

23 Nitta, N. and Yushin, G. (2014). High-capacity anode materials for lithium-ion batteries: choice of elements and structures for active particles. *Particle & Particle Systems Characterization* 31: 317–336.

24 Wu, X.-L., Guo, J.-Z., and Guo, Y.-G. (2019). Traditional nanostructures and nanomaterials in batteries. In: *Nanostructures and Nanomaterials for Batteries: Principles and Applications* (ed. Y.-G. Guo), 313–357. Singapore: Springer Singapore.

25 Liang, Y., Zhao, C.-Z., Yuan, H. et al. (2019). A review of rechargeable batteries for portable electronic devices. *InfoMat* 1: 6–32.

26 Walter, M., Kovalenko, M.V., and Kravchyk, K.V. (2020). Challenges and benefits of post-lithium-ion batteries. *New Journal of Chemistry* 44: 1677–1683.

27 Choi, J.W. and Aurbach, D. (2016). Promise and reality of post-lithium-ion batteries with high energy densities. *Nature Reviews Materials* 1: 16013.

28 Zhou, G., Xu, L., Hu, G. et al. (2019). Nanowires for electrochemical energy storage. *Chemical Reviews* 119: 11042–11109.

29 Liu, C., Li, F., Ma, L.P., and Cheng, H.M. (2010). Advanced materials for energy storage. *Advanced Materials* 22 (8): E28–E62.

30 Wen, L., Li, F., and Cheng, H.M. (2016). Carbon nanotubes and graphene for flexible electrochemical energy storage: from materials to devices. *Advanced Materials* 28 (22): 4306–4337.

31 Pang, C., Cui, H., Yang, G., and Wang, C. (2013). Flexible transparent and free-standing silicon nanowires paper. *Nano Letters* 13 (10): 4708–4714.

32 Mai, L., Tian, X., Xu, X. et al. (2014). Nanowire electrodes for electrochemical energy storage devices. *Chemical Reviews* 114 (23): 11828–11862.

33 Luo, W., Calas, A., Tang, C. et al. (2016). Ultralong Sb_2Se_3 nanowire-based free-standing membrane anode for lithium/sodium ion batteries. *ACS Applied Materials & Interfaces* 8 (51): 35219–35226.

34 Xu, J., Dou, Y., Wei, Z. et al. (2017). Recent progress in graphite intercalation compounds for rechargeable metal (Li, Na, K, Al)-ion batteries. *Advanced Science (Weinh)* 4: 1700146.

35 Teki, R., Datta, M.K., Krishnan, R. et al. (2009). Nanostructured silicon anodes for lithium ion rechargeable batteries. *Small* 5: 2236–2242.

36 Chan, C.K., Ruffo, R., Hong, S.S. et al. (2009). Surface chemistry and morphology of the solid electrolyte interphase on silicon nanowire lithium-ion battery anodes. *Journal of Power Sources* 189: 1132–1140.

37 Yu, S., Hong Ng, V.M., Wang, F. et al. (2018). Synthesis and application of iron-based nanomaterials as anodes of lithium-ion batteries and supercapacitors. *Journal of Materials Chemistry A* 6: 9332–9367.

38 Su, X., Wu, Q., Li, J. et al. (2014). Silicon-based nanomaterials for lithium-ion batteries: a review. *Advanced Energy Materials* 4: 1300882.

39 Namdari, P., Daraee, H., and Eatemadi, A. (2016). Recent advances in silicon nanowire biosensors: synthesis methods, properties, and applications. *Nanoscale Research Letters* 11: 406.

40 Peng, K.-Q., Wang, X., Li, L. et al. (2013). Silicon nanowires for advanced energy conversion and storage. *Nano Today* 8: 75–97.

41 Schmidt, V., Wittemann, J.V., Senz, S. et al. (2009). Silicon nanowires: a review on aspects of their growth and their electrical properties. *Advanced Materials* 21: 2681–2702.

42 Ge, M., Rong, J., Fang, X. et al. (2012). Porous doped silicon nanowires for lithium ion battery anode with long cycle life. *Nano Letters* 12: 2318–2323.

43 Hwang, C., Lee, K., Um, H.-D. et al. (2017). Conductive and porous silicon nanowire anodes for lithium ion batteries. *Journal of the Electrochemical Society* 164: A1564–A1568.

44 Chen, X., Bi, Q., Sajjad, M. et al. (2018). One-dimensional porous silicon nanowires with large surface area for fast charge(−)discharge lithium-ion batteries. *Nanomaterials (Basel)* 8: 285.

45 Liu, X.H., Zhang, L.Q., Zhong, L. et al. (2011). Ultrafast electrochemical lithiation of individual Si nanowire anodes. *Nano Letters* 11: 2251–2258.

46 Huang, R., Fan, X., Shen, W. et al. (2009). Carbon-coated silicon nanowire array films for high-performance lithium-ion battery anodes. *Applied Physics Letters* 95: 133119.

47 Wang, B., Li, X., Zhang, X. et al. (2013). Adaptable silicon-carbon nanocables sandwiched between reduced graphene oxide sheets as lithium ion battery anodes. *ACS Nano* 7: 1437–1445.

48 Sandu, G., Coulombier, M., Kumar, V. et al. (2018). Kinked silicon nanowires-enabled interweaving electrode configuration for lithium-ion batteries. *Scientific Reports* 8: 9794.

49 Chen, H., Xiao, Y., Wang, L. et al. (2011). Silicon nanowires coated with copper layer as anode materials for lithium-ion batteries. *Journal of Power Sources* 196: 6657–6662.

50 Memarzadeh, E.L., Kalisvaart, W.P., Kohandehghan, A. et al. (2012). Silicon nanowire core aluminum shell coaxial nanocomposites for lithium ion battery anodes grown with and without a TiN interlayer. *Journal of Materials Chemistry* 22: 6655.

51 Kohandehghan, A., Cui, K., Kupsta, M. et al. (2014). Nanometer-scale Sn coatings improve the performance of silicon nanowire LIB anodes. *Journal of Materials Chemistry A* 2: 11261.

52 Sadeghipari, M., Mashayekhi, A., and Mohajerzadeh, S. (2018). Novel approach for improving the performance of Si-based anodes in lithium-ion batteries. *Nanotechnology* 29: 055403.

53 Song, H., Wang, H.X., Lin, Z. et al. (2016). Hierarchical nano-branched c-Si/SnO_2 nanowires for high areal capacity and stable lithium-ion battery. *Nano Energy* 19: 511–521.

54 Yao, Y., Liu, N., McDowell, M.T. et al. (2012). Improving the cycling stability of silicon nanowire anodes with conducting polymer coatings. *Energy & Environmental Science* 5: 7927.

55 Wang, J., Wang, H., Zhang, B. et al. (2015). A stable flexible silicon nanowire array as anode for high-performance lithium-ion batteries. *Electrochimica Acta* 176: 321–326.

56 Chen, Y., Liu, L., Xiong, J. et al. (2015). Porous Si nanowires from cheap metallurgical silicon stabilized by a surface oxide layer for lithium ion batteries. *Advanced Functional Materials* 25: 6701–6709.

57 Liang, W., Yang, H., Fan, F. et al. (2013). Tough germanium nanoparticles under electrochemical cycling. *ACS Nano* 7: 3427–3433.

58 Liu, X.H., Huang, S., Picraux, S.T. et al. (2011). Reversible nanopore formation in Ge nanowires during lithiation-delithiation cycling: an in situ transmission electron microscopy study. *Nano Letters* 11: 3991–3997.

59 Chan, C.K., Zhang, X.F., and Cui, Y. (2008). High capacity Li ion battery anodes using Ge nanowires. *Nano Letters* 8: 307–309.

60 Yuan, F.W., Yang, H.J., and Tuan, H.Y. (2012). Alkanethiol-passivated Ge nanowires as high-performance anode materials for lithium-ion batteries: the role of chemical surface functionalization. *ACS Nano* 6: 9932–9942.

61 Kennedy, T., Bezuidenhout, M., Palaniappan, K. et al. (2015). Nanowire heterostructures comprising germanium stems and silicon branches as high-capacity Li-ion anodes with tunable rate capability. *ACS Nano* 9: 7456–7465.

62 Wang, J., Raistrick, I.D., and Huggins, R.A. (1984). Behavior of some binary lithium alloys as negative electrodes in organic solvent based electrolytes. *Electrochemical Society Extended Abstracts* 84: 184–185.

63 Guo, J., Yang, Z., and Archer, L.A. (2013). Aerosol assisted synthesis of hierarchical tin–carbon composites and their application as lithium battery anode materials. *Journal of Materials Chemistry A* 1: 8710.

64 Zhong, Y., Li, X., Zhang, Y. et al. (2015). Nanostructued core-shell Sn nanowires @ CNTs with controllable thickness of CNT shells for lithium ion battery. *Applied Surface Science* 332: 192–197.

65 Yuan, C., Wu, H.B., Xie, Y. et al. (2014). Mixed transition-metal oxides: design, synthesis, and energy-related applications. *Angewandte Chemie (International Ed. in English)* 53: 1488–1504.

66 Kumar, V., Kim, J.H., and Sunkara, M.K. (2008). Gas phase production of metal oxide nanowires and nanoparticles using a microwave plasma reactor. *The Journal of Physical Chemistry C* 112: 17750–17754.

67 Li, Y., Tan, B., and Wu, Y. (2008). Mesoporous Co_3O_4 nanowire arrays for lithium ion batteries with high capacity and rate capability. *Nano Letters* 8: 265–270.

68 Du, N., Xu, Y., Zhang, H. et al. (2011). Porous $ZnCo_2O_4$ nanowires synthesis via sacrificial templates: high-performance anode materials of Li-ion batteries. *Inorganic Chemistry* 50: 3320–3324.

69 Cheng, F., Tao, Z., Liang, J. et al. (2008). Template-directed materials for rechargeable lithium-ion. *Chemistry of Materials* 20: 667–681.

70 Liao, J.Y., Higgins, D., Lui, G. et al. (2013). Multifunctional TiO_2-C/MnO_2 core-double-shell nanowire arrays as high-performance 3D electrodes for lithium ion batteries. *Nano Letters* 13: 5467–5473.

71 Yao, X., Guo, G., Zhao, Y. et al. (2016). Synergistic effect of mesoporous Co_3O_4 nanowires confined by N-doped graphene aerogel for enhanced lithium storage. *Small* 12: 3849–3860.

72 Yang, L., Li, X., Ouyang, Y. et al. (2016). Hierarchical MoO_2/Mo_2C/C hybrid nanowires as high-rate and long-life anodes for lithium-ion batteries. *ACS Applied Materials & Interfaces* 8: 19987–19993.

73 Bogart, T.D., Oka, D., Lu, X. et al. (2014). Lithium ion battery peformance of silicon nanowires with carbon skin. *ACS Nano* 8: 915–922.

74 Choi, S., Kim, J., Hwang, D.Y. et al. (2016). Generalized redox-responsive assembly of carbon-sheathed metallic and semiconducting nanowire heterostructures. *Nano Letters* 16: 1179–1185.

75 Hu, J., Sun, C.-F., Gillette, E. et al. (2016). Dual-template ordered mesoporous carbon/Fe_2O_3 nanowires as lithium-ion battery anodes. *Nanoscale* 8: 12958–12969.

76 Chen, W., Lu, L., Maloney, S. et al. (2015). Coaxial Zn_2GeO_4@carbon nanowires directly grown on cu foils as high-performance anodes for lithium ion batteries. *Physical Chemistry Chemical Physics* 17: 5109–5114.

77 Lyu, Y., Wu, X., Wang, K. et al. (2021). An overview on the advances of $LiCoO_2$ cathodes for lithium-ion batteries. *Advanced Energy Materials* 11: 2000982.

78 Qian, J., Liu, L., Yang, J. et al. (2018). Electrochemical surface passivation of $LiCoO_2$ particles at ultrahigh voltage and its applications in lithium-based batteries. *Nature Communications* 9: 4918.

79 Liu, Q., Su, X., Lei, D. et al. (2018). Approaching the capacity limit of lithium cobalt oxide in lithium ion batteries via lanthanum and aluminium doping. *Nature Energy* 3: 936–943.

80 Yoon, M., Dong, Y., Yoo, Y. et al. (2020). Unveiling nickel chemistry in stabilizing high-voltage cobalt-rich cathodes for lithium-ion batteries. *Advanced Functional Materials* 30: 1907903.

81 Xiao, X., Yang, L., Zhao, H. et al. (2012). Facile synthesis of $LiCoO_2$ nanowires with high electrochemical performance. *Nano Research* 5: 27–32.

82 Xia, H., Wan, Y., Assenmacher, W. et al. (2014). Facile synthesis of chain-like $LiCoO_2$ nanowire arrays as three-dimensional cathode for microbatteries. *NPG Asia Materials* 6: e126–e126.

83 Lee, H.-W., Muralidharan, P., Ruffo, R. et al. (2010). Ultrathin spinel $LiMn_2O_4$ nanowires as high power cathode materials for Li-ion batteries. *Nano Letters* 10: 3852–3856.

84 Hosono, E., Kudo, T., Honma, I. et al. (2009). Synthesis of single crystalline spinel $LiMn_2O_4$ nanowires for a lithium ion battery with high power density. *Nano Letters* 9: 1045–1051.

85 Lee, Y., Kim, M.G., and Cho, J. (2008). Layered $Li_{0.88}[Li_{0.18}Co_{0.33}Mn_{0.49}]O_2$ nanowires for fast and high capacity Li-ion storage material. *Nano Letters* 8: 957–961.

86 Yu, R., Zhang, X., Liu, T. et al. (2017). Spinel/layered heterostructured lithium-rich oxide nanowires as cathode material for high-energy lithium-ion batteries. *ACS Applied Materials & Interfaces* 9: 41210–41223.

87 Xu, X., Luo, Y.-Z., Mai, L.-Q. et al. (2012). Topotactically synthesized ultralong LiV_3O_8 nanowire cathode materials for high-rate and long-life rechargeable lithium batteries. *NPG Asia Materials* 4: e20–e20.

88 Ren, W., Zheng, Z., Luo, Y. et al. (2015). An electrospun hierarchical LiV_3O_8 nanowire-in-network for high-rate and long-life lithium batteries. *Journal of Materials Chemistry A* 3: 19850–19856.

89 Whittingham, M.S. (2004). Lithium batteries and cathode materials. *Chemical Reviews* 104: 4271–4301.

90 Ellis, B.L., Lee, K.T., and Nazar, L.F. (2010). Positive electrode materials for Li-ion and Li-batteries. *Chemistry of Materials* 22: 691–714.

91 Mai, L., Xu, L., Han, C. et al. (2010). Electrospun ultralong hierarchical vanadium oxide nanowires with high performance for lithium ion batteries. *Nano Letters* 10: 4750–4755.

92 Yan, M., Wang, F., Han, C. et al. (2013). Nanowire templated semihollow bicontinuous graphene scrolls: designed construction, mechanism, and enhanced energy storage performance. *Journal of the American Chemical Society* 135: 18176–18182.

93 Lee, J.W., Lim, S.Y., Jeong, H.M. et al. (2012). Extremely stable cycling of ultra-thin V_2O_5 nanowire–graphene electrodes for lithium rechargeable battery cathodes. *Energy & Environmental Science* 5: 9889–9894.

94 Mai, L., Wei, Q., An, Q. et al. (2013). Nanoscroll buffered hybrid nanostructural VO_2 (B) cathodes for high-rate and long-life lithium storage. *Advanced Materials* 25: 2969–2973.

95 Gittleson, F.S., Hwang, J., Sekol, R.C. et al. (2013). Polymer coating of vanadium oxide nanowires to improve cathodic capacity in lithium batteries. *Journal of Materials Chemistry A* 1: 7979–7984.

96 Yan, C., Chen, Z., Peng, Y. et al. (2012). Stable lithium-ion cathodes from nanocomposites of VO_2 nanowires and CNTs. *Nanotechnology* 23: 475701.

97 Lim, S., Yoon, C.S., and Cho, J. (2008). Synthesis of nanowire and hollow $LiFePO_4$ cathodes for high-performance lithium batteries. *Chemistry of Materials* 20: 4560–4564.

98 Doherty, C.M., Caruso, R.A., Smarsly, B.M. et al. (2009). Hierarchically porous monolithic $LiFePO_4$/carbon composite electrode materials for high power lithium ion batteries. *Chemistry of Materials* 21: 5300–5306.

99 Zhu, C., Yu, Y., Gu, L. et al. (2011). Electrospinning of highly electroactive carbon-coated single-crystalline $LiFePO_4$ nanowires. *Angewandte Chemie International Edition* 50: 6278–6282.

100 Yang, J., Wang, J., Tang, Y. et al. (2013). In situ self-catalyzed formation of core–shell $LiFePO_4$@CNT nanowires for high rate performance lithium-ion batteries. *Journal of Materials Chemistry A* 1: 7306–7311.

101 Jing, M.-x., Pi, Z.-C., Zhai, H.-A. et al. (2016). Three-dimensional $Li_3V_2(PO_4)_3$/C nanowire and nanofiber hybrid membrane as a self-standing, binder-free cathode for lithium ion batteries. *RSC Advances* 6: 71574–71580.

102 Wei, Q., An, Q., Chen, D. et al. (2014). One-pot synthesized bicontinuous hierarchical $Li_3V_2(PO_4)_3$/C mesoporous nanowires for high-rate and ultralong-life lithium-ion batteries. *Nano Letters* 14: 1042–1048.

103 La Monaca, A., Paolella, A., Guerfi, A. et al. (2019). Electrospun ceramic nanofibers as 1D solid electrolytes for lithium batteries. *Electrochemistry Communications* 104: 106483.

104 Ye, W., Zhu, J., Liao, X. et al. (2015). Hierarchical three-dimensional micro/nano-architecture of polyaniline nanowires wrapped-on polyimide nanofibers for high performance lithium-ion battery separators. *Journal of Power Sources* 299: 417–424.

105 Jia, S., Yang, S., Zhang, M. et al. (2020). Eco-friendly xonotlite nanowires/wood pulp fibers ceramic hybrid separators through a simple papermaking process for lithium ion battery. *Journal of Membrane Science* 597: 117725.

106 Li, H., Wu, D., Wu, J. et al. (2017). Flexible, high-wettability and fire-resistant separators based on hydroxyapatite nanowires for advanced lithium-ion batteries. *Advanced Materials* 29: 1703548.

107 Du, Q.C., Yang, M.T., Yang, J.K. et al. (2019). Bendable network built with ultralong silica nanowires as a stable separator for high-safety and high-power lithium-metal batteries. *ACS Applied Materials & Interfaces* 11: 34895–34903.

108 He, M., Zhang, X., Jiang, K. et al. (2015). Pure inorganic separator for lithium ion batteries. *ACS Applied Materials & Interfaces* 7: 738–742.

109 Yang, T., Gordon, Z.D., Li, Y. et al. (2015). Nanostructured garnet-type solid electrolytes for lithium batteries: electrospinning synthesis of $Li_7La_3Zr_2O_{12}$ nanowires and particle size-dependent phase transformation. *The Journal of Physical Chemistry C* 119: 14947–14953.

110 Wan, Z., Lei, D., Yang, W. et al. (2019). Low resistance-integrated all-solid-state battery achieved by $Li_7La_3Zr_2O_{12}$ nanowire upgrading polyethylene oxide (PEO) composite electrolyte and PEO cathode binder. *Advanced Functional Materials* 29: 1805301.

111 Wang, S., Zhang, L., Li, J. et al. (2019). A nanowire-nanoparticle double composite polymer electrolyte for high performance ambient temperature solid-state lithium batteries. *Electrochimica Acta* 320: 134560.

112 Ting Yang, J.Z., Cheng, Q., Yan Yan, H., and Chan, C.K. (2017). Composite polymer electrolytes with $Li_7La_3Zr_2O_{12}$ garnet type nanowires as ceramic fillers mechanism of conductivity enhancement and role of doping and morphology. *ACS Applied Materials & Interfaces* 9: 21773–21780.

113 Zhang, Y., Wang, X., Feng, W. et al. (2018). Effects of the shapes of $BaTiO_3$ nanofillers on PEO-based electrolytes for all-solid-state lithium-ion batteries. *Ionics* 25: 1471–1480.

114 Sun, J., Li, Y., Zhang, Q. et al. (2019). A highly ionic conductive poly(methyl methacrylate) composite electrolyte with garnet-typed $Li_{6.75}La_3Zr_{1.75}Nb_{0.25}O_{12}$ nanowires. *Chemical Engineering Journal* 375: 121922.

115 Liu, W., Liu, N., Sun, J. et al. (2015). Ionic conductivity enhancement of polymer electrolytes with ceramic nanowire fillers. *Nano Letters* 15: 2740–2745.

116 Zhu, L., Zhu, P., Yao, S. et al. (2019). High-performance solid PEO/PPC/LLTO-nanowires polymer composite electrolyte for solid-state lithium battery. *International Journal of Energy Research* 43: 4854–4866.

117 Liu, W., Lin, D., Sun, J. et al. (2016). Improved lithium ionic conductivity in composite polymer electrolytes with oxide-ion conducting nanowires. *ACS Nano* 10: 11407–11413.

118 Lun, P., Liu, P., Lin, H. et al. (2019). Ionic conductivity promotion of polymer membranes with oxygen-ion conducting nanowires for rechargeable lithium batteries. *Journal of Membrane Science* 580: 92–100.

119 Chu, Y.-Y., Liu, Z.-Y., Saikia, D. et al. (2015). Effect of Al_2O_3 nanowires on the electrochemical properties of di-ureasil-based organic–inorganic hybrid electrolytes. *Ionics* 21: 2523–2534.

120 Fu, X., Wang, Y., Fan, X. et al. (2018). Core-shell hybrid nanowires with protein enabling fast ion conduction for high-performance composite polymer electrolytes. *Small* 14: e1803564.

121 Sheng, O., Jin, C., Luo, J. et al. (2018). $Mg_2B_2O_5$ nanowire enabled multifunctional solid-state electrolytes with high ionic conductivity, excellent mechanical properties, and flame-retardant performance. *Nano Letters* 18: 3104–3112.

122 Yao, P., Zhu, B., Zhai, H. et al. (2018). PVDF/Palygorskite nanowire composite electrolyte for 4 V rechargeable lithium batteries with high energy density. *Nano Letters* 18: 6113–6120.

123 Nam, K.T., Kim, D.W., Yoo, P.J. et al. (2006). Virus-enabled synthesis and assembly of nanowires for lithium ion battery electrodes. *Science* 312 (5775): 885–888.

124 Vlad, A., Reddy, A.L., Ajayan, A. et al. (2012). Roll up nanowire battery from silicon chips. *Proceedings of the National Academy of Sciences of the United States of America* 109 (38): 15168–15173.

125 Chockla, A.M., Harris, J.T., Akhavan, V.A. et al. (2011). Silicon nanowire fabric as a lithium ion battery electrode material. *Journal of the American Chemical Society* 133 (51): 20914–20921.

126 Wang, J.-G., Zhang, C., Jin, D. et al. (2015). Synthesis of ultralong MnO/C coaxial nanowires as freestanding anodes for high-performance lithium ion batteries. *Journal of Materials Chemistry A* 3 (26): 13699–13705.

127 Kawamori, M., Asai, T., Shirai, Y. et al. (2014). Three-dimensional nanoelectrode by metal nanowire nonwoven clothes. *Nano Letters* 14 (4): 1932–1937.

128 Dunn, B., Kamath, H., and Tarascon, J.-M. (2011). Electrical energy storage for the grid: a battery of choices. *Science* 334: 928–935.

6

Nanowires for Sodium-ion Batteries

6.1 Advantages and Challenges of Sodium-ion Batteries

6.1.1 Development of Sodium-ion Batteries

As early as the 1980s, sodium-ion batteries (SIBs) and lithium-ion batteries (LIBs) were proposed almost at the same time. Since the successful commercialization of LIBs in the 1990s, they have dominated today's markets for portable and large energy storage devices. However, the research reports on SIBs are very limited, and only a few papers and patents have been published in the 1990s. The main reasons can be listed as follow: (i) The research on lithium-ion intercalation materials has just started in this period, and a large number of researchers have locked in the research direction on LIBs; (ii) Limited by the research conditions (such as the purity of the electrolyte, the sealing of the glove box, and the purity of argon gas), it is difficult to accurately evaluate the performance of the electrode material in the half-cell using the active metal sodium as the anode; (iii) The graphite for commercial LIBs exhibits no sodium storage ability in carbonate electrolytes, resulting in the lack of suitable anode materials for the research of SIB [1–4]. In fact, before the successful commercialization of LIBs, some companies in the United States and Japan carried out the research and development of sodium-ion full batteries using P2-Na_xCoO_2 and Na-Pb alloy as the cathode and anode materials, respectively. Although the cycle life of the sodium-ion battery can reach 300 cycles, the average discharge voltage is lower than 3 V, which has no advantage compared with the C||$LiCoO_2$ battery (3.7 V) reported at that time and failed to attract the attention of researchers [5].

In 2000, Stevens and Dahn developed a hard carbon material by pyrolysis of glucose, which delivered a high capacity of 300 mAh g^{-1} when applied as the SIBs anode [6]. Until now, hard carbon materials were still considered the most promising anode materials for SIBs. With further research on cathode materials such as polyanionic compounds, layered oxides, and Prussian white, SIBs have been greatly developed in recent years. Due to the low lithium resources in the crust and increased prices of lithium raw materials, low-cost SIBs are considered promising candidates for energy storage devices. Because of the larger radius of Na^+ (0.102 nm) than Li^+ (0.076 nm), the research on SIBs can learn from LIBs, but

Nanowire Energy Storage Devices: Synthesis, Characterization, and Applications, First Edition.
Edited by Liqiang Mai.

it cannot be completely transplanted. Especially the electrode materials for Na^+ intercalation generally exhibit sluggish diffusion kinetics, which severely affects the performance of SIBs. Therefore, developing high-performance electrode materials through structural regulation and morphology design is a great challenge for further practical application of SIBs [2].

6.1.2 Characteristic of Sodium-ion Batteries

6.1.2.1 The Working Principle of Sodium-ion Battery

The working principle of SIB is "rocking chair," which means that Na^+ ions shuttle between cathode and anode materials to achieve reversible charging/discharging. The cathode and anode are reservoirs of sodium and work for electron storage/release. The separator, which is placed between two electrodes, is to avoid short circuits in the SIBs. The electrolyte ensures the transport of Na^+ ions in the process of charging and discharging. While charging, Na^+ ions are released from the positive electrode and move to the negative electrode through the electrolyte. Due to the potential difference, electrons flow from the positive electrode to the negative electrode through the external circuit to maintain the electrical neutrality of the electrode. This process is accompanied by an increase in the battery voltage, which is used to store energy. While discharging, Na^+ ions and electrons are released from the negative electrode to the positive electrode by an opposite process. This process is accompanied by a decrease in the battery voltage, which will release energy [7] (Figure 6.1).

6.1.2.2 Advantages of Sodium-ion Batteries

(i) Abundant material resources, wide distribution, and lower price.
(ii) The battery structure of SIB is similar to that of LIB, so they can use the same production equipment [9].
(iii) Sodium ions won't form aluminum alloy, so aluminum foil can be used as a negative current collector to further reduce the cost and weight.
(iv) Bipolar electrodes can be designed for solid batteries. Coating the positive and negative electrode materials on the two sides of the aluminum foil and periodically stacking these electrodes together to achieve higher voltage in a single battery, so that other inactive materials can be saved to improve the volume energy density [10, 11];
(v) The half-cell potential of SIB is higher than that of LIB, which means that SIBs can use electrolytes with lower decomposition potential and have more choices of electrolytes.
(vi) The solvation energy of sodium-ion is lower than that of lithium-ion, and the SIBs' interfacial ion diffusion ability is better.
(vii) According to the current low and high temperature test results, SIBs have better performance compared with LIBs.

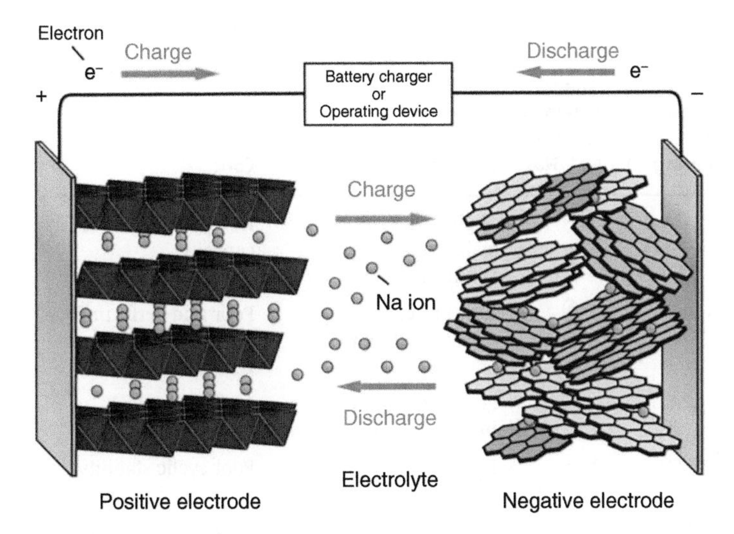

Figure 6.1 The working principle of SIB. Source: Yabuuchi et al. [8] / with permission of American Chemical Society.

(viii) Better safety performance. The internal resistance of SIBs is slightly higher than that of LIBs, resulting in less heat and lower temperature rise in short circuits and other safety tests [12].

However, compared with Li^+ ions, the larger and heavier Na^+ ions often lead to sluggish reaction kinetics in SIBs. In the process of sodiation, the SIBs tend to cause large volume change or even irreversible structure failure of electrode materials, deteriorating the cyclic stability of batteries. Therefore, developing appropriate materials that are capable of accommodating fast and stable Na^+ ion (de)insertion has become a research hotspot to push forward the industrial deployment of SIBs [13].

6.1.3 Key Materials for Sodium-ion Batteries

SIB is mainly composed of four parts: positive electrode, negative electrode, separator, and electrolyte. At present, Prussian blue analogue, layered oxides and polyanionic compounds are widely studied as cathode materials for SIBs. The electrolyte is a medium that is responsible for the transport of Na^+ ions between positive and negative electrodes. The electrolyte of SIBs is composed of electrolyte and organic solvent at a certain concentration. The commonly used electrolyte is sodium salt, and the solvents are classified into ether and ester electrolytes. The separator works by separating the positive and negative electrodes, allowing Na^+ ions to pass through and blocking the transport electrons at the same time, preventing the battery from short-circuiting [14].

Table 6.1 Electrochemical-related features of some typical cathodes for SIBs.

Material	Specific capacity ($mAh\,g^{-1}$)	Average potential (V versus Na^+/Na)	Pros.	Cons.
$Na_{0.44}MnO_2$	121[b)]	2.8[a)]	High structure stability, simple synthesis	Low actual specific capacity
$NaFePO_4$	154[b)]	2.7[a)]	Low price, high structure stability	Poor rate performance and low working voltage
$Na_3V_2(PO_4)_3$	117.6[b)]	3.4[a)]	good cycle stability, high rate performance	Low theoretical specific capacity
$NaMnO_2$	243[b)]	3.7[a)]	High theoretical capacity, low price	Poor cyclic stability
$Na_3MnTi(PO_4)_3$	110	3.8[a)]	High structure stability and thermal stability	Low specific capacity, poor electronic conductivity

Data obtained from Refs. [15–17].
a) Midpoint voltage.
b) Theoretical value.

6.1.3.1 Cathode

The cathode has appropriate channels and spaces for Na^+ ions transport and storage. It is the key factor affecting the voltage, energy density, power density, safety, and cycle life of SIBs. Therefore, the research and development of cathode materials have become essential issue in the research of SIBs. At present, the research mainly focuses on layered oxides, polyanionic materials, polymer materials, Prussian blue analogues, etc. These materials develop very well in the field of LIBs. Due to the small Li^+ ion radius, the effect of Li^+ insertion/deinsertion on the material structure is small. But in SIBs, Na^+ ions insertion/deinsertion will lead to larger material structure distortion, and then the cycle of the battery's performance is poorer. Here we summarize the electrochemical-related features of some typical cathodes in Table 6.1.

6.1.3.2 Anode

The anode materials mainly provide sites for ions storage and low-potential redox pairs for SIBs. The available anode materials mainly include carbon-based materials, alloy materials, and metal compounds. Here we summarize some typical anode materials in Table 6.2.

SIBs, which are relatively low-cost compared to current mainstream LIBs, are expected to be used on a large scale in the future. However, due to the influence of the properties of positive and negative electrode materials suitable for Na^+ ion insertion/deinsertion, the SIBs development and application are relatively slow, especially for the anode materials with excellent performance and practical application.

Table 6.2 Some typical anode materials.

Material	Classification	Pros.	Cons.
Carbon-based	Graphene, hard carbon, soft carbon, etc.	Low sodium embedded platform, high capacity, long cycle life, and simple preparation	Low structure stability, poor rate performance, forming sodium dendrites, and safety hazards
Alloy	Multivariate composite	Good conductivity, high specific capacity, and high safety	The large volume change will easily lead to electrode powder and poor cycle performance
metallic compound	Metal oxide	High theoretical capacity, good safety, stable voltage platform, cheap, and easy to get	Poor cycling stability, large volume expansion, and poor cycle performance
	Metal sulfide	High structural stability, thermodynamic stability, mechanical stability, and good electrical conductivity	
	Metal phosphide	High theoretical specific capacity, stable cycle reversibility, and high safety	

Data obtained from Refs. [18–20].

As the core component of SIB, the sodium storage behavior of anode material is different from that of LIB, so it needs to be modified according to the material type and charging and discharging mechanism [21].

6.1.3.3 Electrolyte

The composition of SIB electrolyte is similar to that of LIB electrolyte, and the experience of LIB electrolyte is mainly used for reference. The performance requirements are mainly focused on the following five aspects [22]:

(i) The conductivity of sodium electrolyte requires a high ionic conductivity within the normal temperature range.
(ii) Water content: the water content of the electrolyte is required to be controlled below 20 PPM.
(iii) Stability: the electrolyte does not react with the components inside the battery.
(iv) Safety: requiring the electrolyte to have strong thermal stability and not be easy to burn.
(v) Low cost: it is hoped that the price of electrolytes will be low [23].

6.1.3.4 Separator

The separator plays an important role in isolating positive and negative electrodes, preventing batteries from short-circuiting, storing electrolytes, and allowing sodium ions to travel quickly. The function of the separator is to isolate the cathode and anode, store the electrolyte, and allow Na^+ transport between positive and negative electrodes. Therefore, the separator must be able to stably exist in the battery electrolyte. At the same time, the separator and electrolyte should have good affinity and porosity to ensure the liquid absorption rate, good mechanical performance, and thermal stability are also needed. Other characteristics of the separator are also important, such as aperture size, thickness, and cost. It must be pointed out that, compared with the LIBs separator, due to the larger ion radius of sodium ions, the aperture of the SIBs separator should be larger than that of the LIBs, which can be conducive to the transport of Na^+ ions. Here we summarize the separator specifications for SIBs [24].

(i) Chemical and electrochemical stabilities: the separator must be able to maintain its chemical and structural integrity in the electrolyte, so it is required to neither dissolve in the electrolyte nor react with the electrolyte [25, 26].
(ii) Wettability: wettability refers to the affinity between the electrolyte and the separator, and the ability of the electrolyte to spread over the separator. Good wettability is also the premise of the electrochemical performance of the battery, because the transport of Na^+ relies on the electrolyte. The Na^+ ions can't be freely transported if the separator isn't wetted by the electrolyte. Therefore, the ion conductivity of the battery will be very poor [27].
(iii) Pore size: Theoretically speaking, a large through-hole is conducive to the transport of ions, but the pore size on the separator can't be too large or it may cause the shuttle of positive and negative material powders. The uniformity of pore size is also important. In the process of battery charging and discharging, if the pore size is too different, the current density will be uneven, which makes it easy to cause the growth of sodium dendrites [28].
(iv) Porosity: Porosity refers to the percentage of the volume of the hole in the separator to the volume of the whole separator. It is critical for ion transport and electrolyte storage, and an appropriate increase in porosity is beneficial to the power characteristics of the battery [29, 30].
(v) Mechanical property: The mechanical strength of the separator is related to its composition, pore size, porosity, and thickness. It includes tensile strength and puncture strength. The battery needs to be wound in the production process, so a certain tensile strength is necessary. At the same time, a stronger puncture can prevent the sodium dendrite from piercing the separator and causing a battery short circuit.
(vi) Thermal stability: The thermal stability of the separator refers to its ability to maintain its structure and morphology when affected by the heat source. When extreme conditions occur, the battery temperature may rise sharply. If the separator has poor heat resistance, it will shrink and cause short-circuiting, which is hazardous.
(vii) Cost: Cost is an important consideration for the SIB separator. In LIBs, the separator accounts for 15–20% of the whole battery cost. It is believed that in the future, the SIB separator's proportion in commerce will not be low. Thus, the development of low-cost, high-safety, large-scale production of SIB separators is necessary [31].

6.1.4 Challenges for Sodium-ion Batteries

In the late 1970s, the research on SIBs and LIBs was almost carried out simultaneously, but LIBs were commercialized as early as the 1990s, and SIBs are still in the research stage [32]. Lithium and sodium are located in the same master and have similar chemical properties in some aspects. However, due to the large gap between ion radius, research on SIBs can learn from the research experience of LIBs but cannot be completely copied [33]. Li has a low abundance in the crust, and the price of Li raw materials has continued to rise, making it difficult to meet the needs of large-scale energy storage devices in recent years. Sodium elements are abundant in the crust, and sodium salt can be obtained directly from seawater [34]. As a result, research on SIBs in recent years has also become more extensive and deeper. Although sodium is the second light and the second small alkaline metal that is second only to lithium in terms of atomic quality and atomic radius in the element cycle table, the differences between the two in physical chemistry will inevitably cause the corresponding electrode materials [35]. The ion radius of Na^+ (0.102 nm) is 34% larger than that of Li^+ (0.076 nm), so sodium ions are slightly difficult to embed/out of the same material. The comparison between the two is shown in Table 6.3. The heavier sodium ion quality and a larger sodium ion radius make the weight and volume energy density of SIBs comparable to LIBs. The large ion radius of a large sodium ion will also cause differences in electrode materials in ion transportation, structure evolution, and interface nature [36].

At present, the research on the positive electrode materials of SIBs is mainly concentrated on layer-shaped transition metal oxides, polyanionic compounds, and Prussian blue materials [38]. To get a large scale in energy storage in the future, it is necessary to consider the cost of the pupae, cobalt, and the expensive metal that they use. The Prussian blue object has the advantages of low cost and long life, but it is still not satisfactory in terms of capacity and stability. The P2-type transition metal oxide (TMO) shows high capacity and good cycle performance, but its characteristics caused by sodium defects hinder the assembly of the entire battery [39]. The O3-type has the advantages of high energy density due to its high working voltage, which is easy to match and realize industrialized applications, but the problem of moisture in materials still needs to be paid attention to. For negative

Table 6.3 Sodium versus lithium characteristics. Source: Zhang et al. [37] / John Wiley & Sons / CC BY 4.0.

Category	Lithium	Sodium
Cation radius (Å)	0.76	1.06
Atomic weight	6.9 g mol^{-1}	23 g mol^{-1}
E (vs Li/Li^+)	0	0.3 V
Cost, carbonates	$5000/ton	$150/ton
Capacity (mAh g^{-1}), met	3829	1165

materials, graphite with excellent lithium storage capacity in LIBs does not have sodium storage capacity due to thermodynamic reasons [40].

Layer-shaped transition metal oxides are divided into single metal-shaped oxides and multimetal layer oxides. Single metal oxides include $NaNiO_2$, $NaFeO_2$, $NaCrO_2$, $NaCoO_2$, $NaMnO_2$, etc. [41]. Among these materials, due to its high theoretical capacity and lower synthetic costs, $NaMnO_2$ is an excellent SIB positive electrode material. Layer-shaped oxides have a periodic layer structure, simple preparation method, high capacity, and high voltage, which is the main positive material of SIBs. In addition, the oxidation reaction of the lattice oxygen can further improve the energy density of such materials [42]. However, most of the layers of materials are easily absorbed or react with air, affecting the stability and electrochemical properties of the material, so it cannot be stored in the air for a long time. The crystal structure of the tunnel-type oxide has a unique "S" channel with good multiplier performance, and exhibits high stability in air and water [43]. However, its charging ratio is lower in the first cycle, so the actual use of the ratio is smaller. During the cycle, the unstable obstacles of the unstable changes in the cycle and the phase instability of the sodium dehydration structure are the biggest obstacle. The theoretical capacity is far from the actual capacity of these layered electrode materials [44].

Most of the polyanionic compounds have an open three-dimensional (3D) skeleton, good rate performance, and better cycle performance. However, the conductivity of this type of compound is generally poor. To improve its electrons and ionic conductivity, carbon coverage and doping are often needed, but they will also cause its volume energy density to decrease [45]. The unique polyanion structural unit in the polyanion cathode material is tightly connected by a strong covalent bond, isolating the polyanion group from the valence electrons of the transition metal ions. Although the isolated electronic structure of transition metal ions makes these materials have high operating voltage, it also leads to their low electronic conductivity, which greatly limits their charge and discharge performance at high rates and brings some difficulties to practical applications [46].

Therefore, the preparation and modification of such materials mainly focus on improving their electronic conductivity of materials. The modification methods mainly include nanocrystallization and carbon coating. Nanomaterials can increase the solid (active particles)-liquid (electrolyte) contact area and shorten the sodium ion diffusion path [47]. Carbon coating helps to improve the surface electronic conductivity of the material and the electrical contact between particles. At the same time, the presence of a carbon coating layer can also inhibit the growth of particles, so the nanomaterial can improve the agglomeration of particles.

Prussian blue material is a new type of cathode material with great potential that has developed in recent years. It has an open three-dimensional channel, and Na^+ can migrate rapidly in the channel, so it has good structural stability and rate performance [48]. However, Prussian blue compounds have the disadvantages of difficult removal of crystal water, dissolution of transition metal ions, low capacity utilization, poor rate capability, and cycle instability [49]. The main reason may be related to the vacancies and lattice water molecules in the Prussian blue structure, which will seriously affect the electrochemical sodium storage performance of Prussian blue and destroy the lattice integrity. Sodium ions are easy to cause lattice distortion and even structural collapse when they are released and embedded, resulting in a serious attenuation of cycle performance [50].

At present, SIBs are in a state of parallel development of a variety of material systems, and the processing performance of some positive and negative material systems needs to be further improved. The negative amorphous carbon (AC) materials also have the problems of low Coulombic efficiency in the first week and an unclear sodium storage mechanism. In addition, the research and development of electrolyte systems that match the positive and negative electrode materials are also insufficient [37]. Although most of the nonactive substances (current collectors, binders, conductive agents, separators, enclosures, etc.) of SIBs can be learned from the mature industrial chain of LIBs, the core of the positive and negative materials and electrolyte active materials, such as large-scale supply channels, are still missing, so their source stability cannot be guaranteed, thus affecting the production process and product quality stability.

Thus, to achieve continuous improvements in the power and energy densities of sodium-ion storage devices, the key issues are the promotion of the performance of electrode materials and the design of suitable electrode nanoarchitectures [51]. Among various nanostructures, 1D nanowire structures become promising candidates for the SIB electrodes owing to the advantages of easy electrolyte access to the active material, short ionic diffusion length, and the ability to accommodate volume changes during the electrochemical reaction process [52].

6.2 Nanowires as Cathodes in Sodium-ion Batteries

6.2.1 Layered Oxide Nanowires

Layered oxides are one of the most extensively studied cathode materials for SIBs, which exhibit high specific capacity and facile 2D Na^+ diffusion path. The research on layered oxides for sodium systems can be traced back to the 1980s and has made great progress as the electrode materials for sodium insertion hosts in recent years. The general formula of sodium-containing layered oxides can be written as Na_xMeO_2, where Me refers to transition metals. Layered Na_xMeO_2 materials are commonly composed of stacked MeO_2 and NaO_2 slabs along the c-axis direction. The edge-sharing MeO_6 octahedra build up the structural frame. According to the sodium ions accommodated sites (octahedral or prismatic) and oxygen stacking sequence (ABCABC, ABBA) Layered Na_xMeO_2 can be categorized into two main groups: O3 type or P2 type (Figure 6.2).

Generally, Layered Na_xMeO_2 materials are synthesized by a solid-state reaction or sol–gel reaction; the products are mainly composed of micrometer-sized particles or flakes. For the common synthetic methods, the obtained precursors usually have to be sintered at high temperature for more than ten hours, and many well-designed morphologies are difficult to maintain. Compared to bulk particles, nanowires generally exhibit better electrochemical performance due to their shortened ion

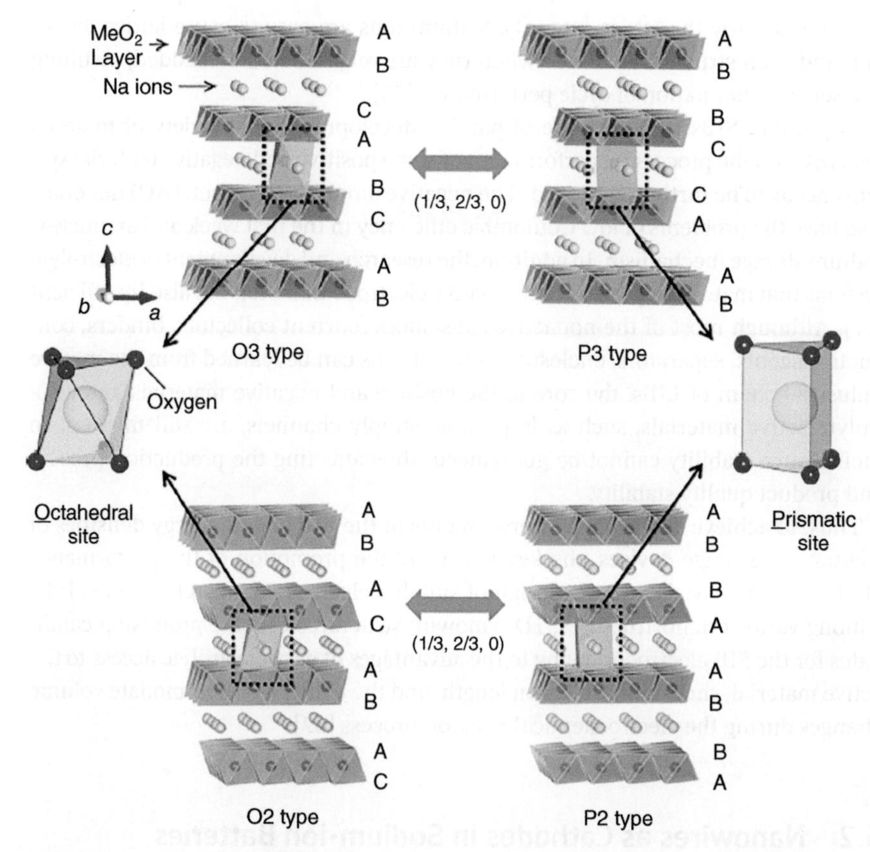

Figure 6.2 Classification of Na–Me–O-layered materials with sheets of edge-sharing MeO_6 octahedra. Source: Yabuuchi et al. [8] / with permission of American Chemical Society.

diffusion distance and enhanced structural stability. In addition, the design of nanowires enables electrochemical characterization and the construction of flexible devices. Therefore, designing nanowire-structured Layered Na_xMeO_2 is a promising direction to promote the development of SIBs.

In previous reports, most of the layered Na_xMeO_2 nanowires were constructed by the electrospinning method. As displayed in Figure 6.3a,b, Liang et al. [53] fabricated ultralong layered $NaCrO_2$ nanowires by electrospinning synthetic methods. The resultant 1D $NaCrO_2$ nanoarchitecture is endowed with oriented and shortened electronic/ionic transport and remarkable structural tolerance to stress change over sodiation–de-sodiation processes. These structural superiorities enable the $NaCrO_2$ nanowires cathode to exhibit prominent Na^+ storage performance. Liu et al. [54] reported novel porous P2- $Na_{2/3}Ni_{1/3}Mn_{2/3}O_2$ nanowires (fibers) assembled by nanoparticles through electrospinning for SIBs, the hierarchical engineering structure can effectively boost the reaction kinetics and stabilize the structure (Figure 6.3c,d). Additionally, Shen et al. [55] constructed a pearl necklace-like hierarchical P2-$Na_{0.76}Cu_{0.22}Fe_{0.30}Mn_{0.48}O_2$ nanowires (fibers) material. Benefiting from the nanostructured design, this cathode exhibited high Na^+ coefficients

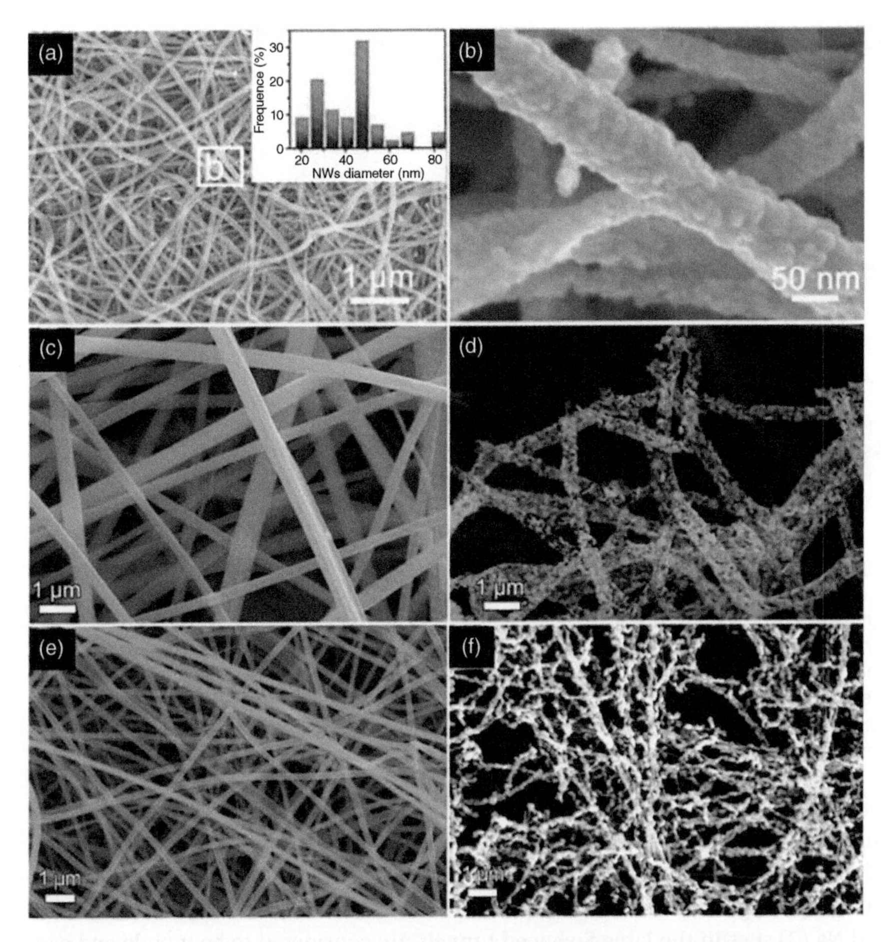

Figure 6.3 The research of layered oxide nanowires. (a,b) SEM images of $NaCrO_2$. Source: Liang et al. [53] / with permission of American Chemical Society. (c,d) the electrospun precursor, and the resultant $Na_{2/3}Ni_{1/3}Mn_{2/3}O_2$ nanowires. Source: Liu et al. [54] / with permission of John Wiley & Sons, Inc. (e,f) the electrospun precursor and the resultant $Na_{0.76}Cu_{0.22}Fe_{0.30}Mn_{0.48}O_2$ nanowires. Source: [55] /with permission of John Wiley & Sons, Inc.

and a low ionic migration energy barrier (Figure 6.3e,f). He et al. [56] developed NaV_3O_8 nanowires as the cathode for SIBs, which exhibited good cycling stability and charge–discharge plateaus during the Na-ion insertion/extraction processes. Therefore, nanowire-structured engineering for layered oxides will significantly improve the electrochemical performance.

6.2.2 Tunnel-type Oxide Nanowires

Tunnel-type manganese oxide has also received extensive attention due to its internal, double ion channels and superior structural stability to tolerate some stress during the Na^+ extraction/insertion process. $Na_{0.44}MnO_2$ is a typical tunnel-type

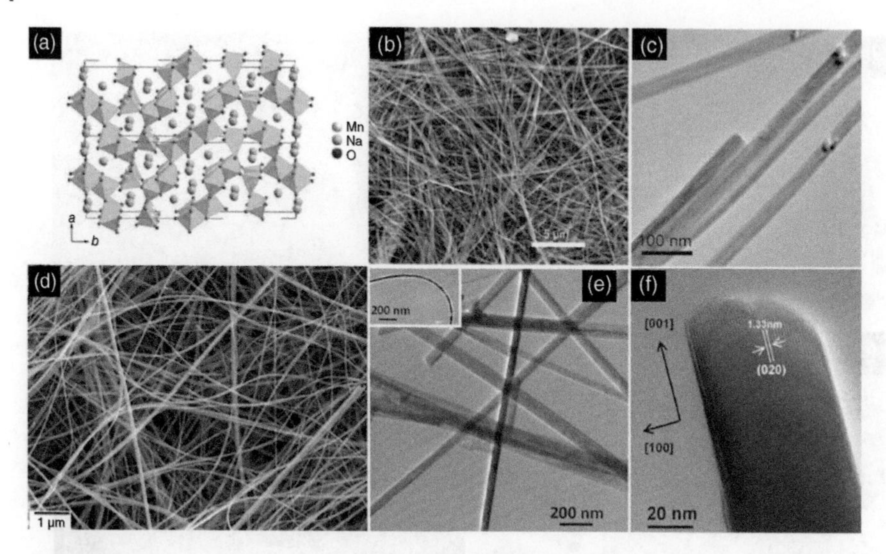

Figure 6.4 The research of tunnel-type $Na_{0.44}MnO_2$ nanowires. (a) Schematic illustrations of the crystal structures of $Na_{0.44}MnO_2$. (b,c) SEM and TEM images of the $Na_{0.44}MnO_2$ nanowires synthesized by a stress-induced splitting mechanism for the conversion of birnessite nanosheets. Source: Li et al. [57] / with permission of Springer Nature. (d) SEM image of the $Na_{0.44}MnO_2$ nanowires synthesized by a low-temperature hydrothermal method. Source: Hosono et al. [58] / with permission of Elsevier. (e,f) TEM images of the $Na_4Mn_9O_{18}$ ($Na_{0.44}MnO_2$) nanowires synthesized by a polymer-pyrolysis method. Source: Cao et al. [59] / with permission of John Wiley & Sons, Inc.

manganese oxide for the cathode of SIBs. The framework structure of $Na_{0.44}MnO_2$ consists of four MnO_6 octahedrals and one MnO_5 square-pyramidal (Figure 6.4a), which are arranged to form two types of tunnels: large S-shaped tunnels and smaller pentagon tunnels. There are three Na sites in this structure, and the Na (2) and Na (3) sites in the large S-shaped tunnels are considered to be mobile and can potentially be reversibly extracted.

Different from the layered manganese oxides, tunnel-type $Na_{0.44}MnO_2$ can be synthesized by many methods. For example, Li et al. [57] synthesized the $Na_{0.44}MnO_2$ nanowires by a stress-induced splitting mechanism that transforms birnessite nanosheets, the nanowires feature an average length of over 10 μm and a diameter smaller than 100 nm (Figure 6.4b,c). Hosono et al. [58] reported a $Na_{0.44}MnO_2$ nanowire synthesized by a low-temperature hydrothermal method that exhibits a high electronic conductivity for good fast charge–discharge properties (Figure 6.4d). Cao et al. [59] fabricated a single crystalline $Na_4Mn_9O_{18}$ ($Na_{0.44}MnO_2$) nanowires by a polymer-pyrolysis method. The $Na_4Mn_9O_{18}$ nanowires delivered a high reversible capacity with excellent capacity retention for sodium storage (Figure 6.4e,f).

6.2.3 Polyanionic Compound Nanowires

Polyanionic compounds are divided into phosphate, sulfate, pyrophosphate, mixed phosphate, and so on. The structure of the polyanion cathode material with good stability, high security, high theoretical capacity, and environmentally friendly performance, in the field of SIB cathode material, has been widely studied, but the polyanion cathode material also has some problems, such as poor conductivity, which can lead to high overpotential and poor rate performance. A common modification method is to change the material morphology and combine the nanowires with the polyanionic cathode material, so as to improve the specific surface area of the cathode material and shorten the transmission distance of ion transport [60]. Among polyanionic cathode materials, the most common one is sodium vanadium phosphate (NVP), which has attracted much attention due to its stable structure, high operating voltage, and good thermal stability. However, the low electronic conductivity of this material limits its electrochemical performance.

Jiang et al. designed a carbon-coated NVP embedded in one-dimensional porous carbon nanowires and synthesized it by hydrothermal method (NVP@C-CNW) (Figure 6.5a–k). As a cathode material for SIB, NVP@C-CNW shows excellent performance (Figure 6.5l–o). After 1000 cycles at 1 C, NVP@C-CNW still maintains a capacity of 96.6 mAh g^{-1}, equivalent to 82.6% of the theoretical capacity (Figure 6.5m). In the same case, NVP@C only has a capacity of 41.5 mAh g^{-1} after 200 cycles at 1 C. The capacity of pure NVP is reduced to 19.1 mAh g^{-1} after 50 cycles at 1 C. NVP@C-CNW has a good rate performance at 0.5–60 C current (Figure 6.5n). The excellent properties of NVP@C-CNW thanks to a wiener rice noodle structure material. First, the porous structure of a wiener rice noodle shortened the sodium ion diffusion path, and in the process of charging and discharging cycles, it provided enough space for volume expansion of active material. Second, the large surface area of the porous structure of a wiener rice noodle made it easier for electrolyte materials, Increase the diffusion rate of sodium ions. Third, the embedded structure can inhibit the growth of particles during heating and improve the conductivity of the material [61]. In addition, Wang et al. synthesized a bundle of carbon-coated potassium vanadium phosphate ($K_3V_2(PO_4)_3/C$) nanowires by an organic acid-assisted method for application in SIBs and showed excellent electrochemical performance [62]. $K_3V_2(PO_4)_3/C$ exhibits an initial discharge capacity of 119 mAh g^{-1} at a current density of 100 mAh g^{-1}. After 100 cycles of this current, the capacity retention rate is up to 99.4%.

Peng et al. prepared Ruthenium oxide (RuO_2)-coated sodium vanadium fluorophosphate nanowires ($Na_3V_2O_2(PO_4)_2F$) via microemulsion mediated hydrothermal synthesis [63]. RuO_2-coated $Na_3V_2O_2(PO_4)_2F$ core–shell nanowires have a diameter of about 50 nm. Their unique 1D morphology, efficient electrochemical coupling, and 3D conductive network provide excellent electrochemical performance. The total mass of $Na_3V_2O_2(PO_4)_2F$ nanowires coated with RuO_2 coating of 6 nm thickness and the initial discharge and charge-specific capacities are 134 and 126 mAh g^{-1}, respectively. The initial Coulomb efficiency was 94.0%, with a capacity of 95 mAh g^{-1} remaining after 1000 cycles. Compared with $Na_3V_2O_2(PO_4)_2F$ without Ruthenium oxide coating, the multiplication performance of $Na_3V_2O_2(PO_4)_2F$ coated with RuO_2 is significantly improved.

Because of its unique structure, mixed phosphate can also be used as an excellent cathode material for SIBs. Chen et al. synthesized $Na_3Mn_{2-x}Fe_x$ (P_2O_7) (PO_4) nanowires (NMFP) [64]. When NMFP was applied to the positive electrode of SIB,

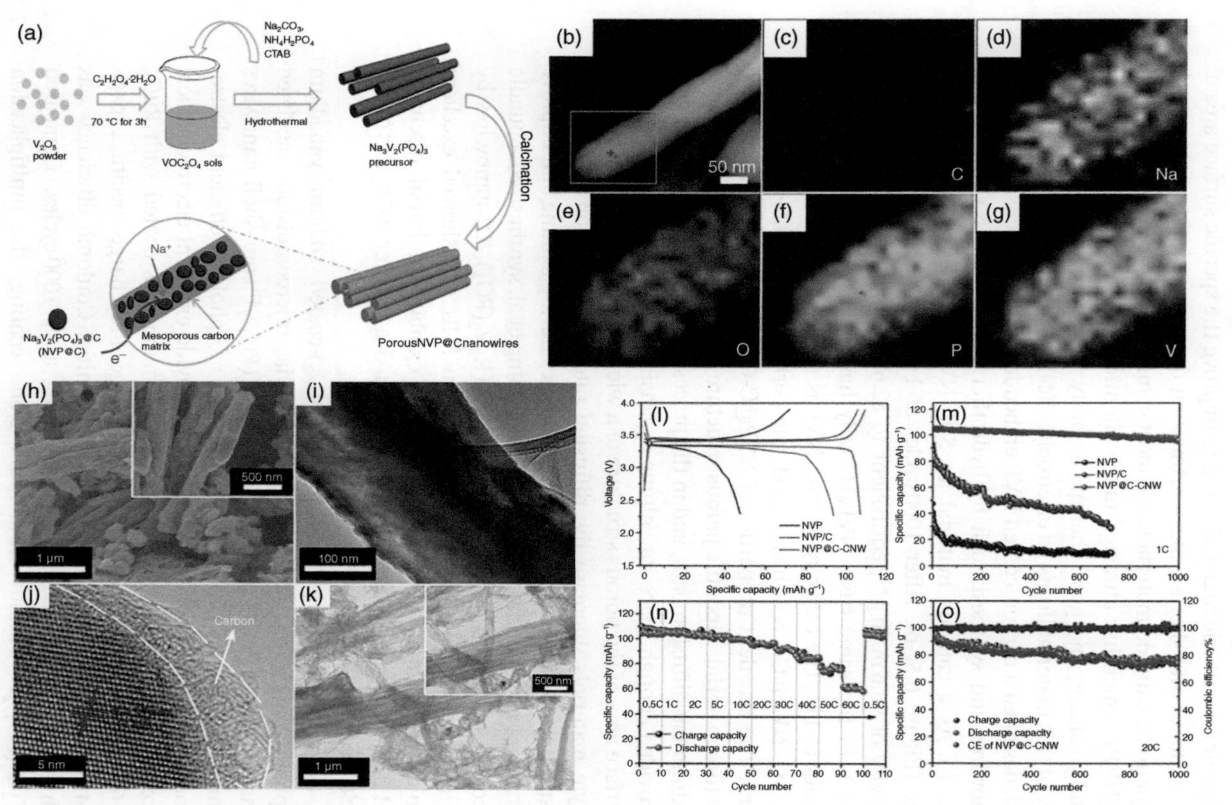

Figure 6.5 (a) NVP@C-CNW synthesis method. (b) Scanning transmission electron microscopy (STEM) image of NVP@C-CNW and (c–g) corresponding element mapping, showing that C elements are evenly distributed on the surface of NVP@C-CNW. (h) SEM image of NVP@C-CNW. (i) TEM image of NVP@C-CNW. (j) HRTEM image of NVP@C-CNW. (k) TEM image of carbon net after corrosion of some NVP particles NVP@C-CNW. The red arrow in (j) indicates the spacing of 0.437 nm corresponding to the (110) plane. (l) Charge and discharge curves of NVP, NVP/C, and NVP@C-CNW in the first cycle at 1 C. (m) Cycling performance of all samples at 1C for 1000 cycles. (n) NVP@C-CNW Rate performance at different current densities (0.5 to 60 C). (o) Specific capacity of NVP@C-CNW for 1000 cycles at 20 C and Coulomb efficiency of NVP@C-CNW for 1000 cycles at 20 C. Source: Jiang et.al. [61] / with permission of John Wiley & Sons, Inc.

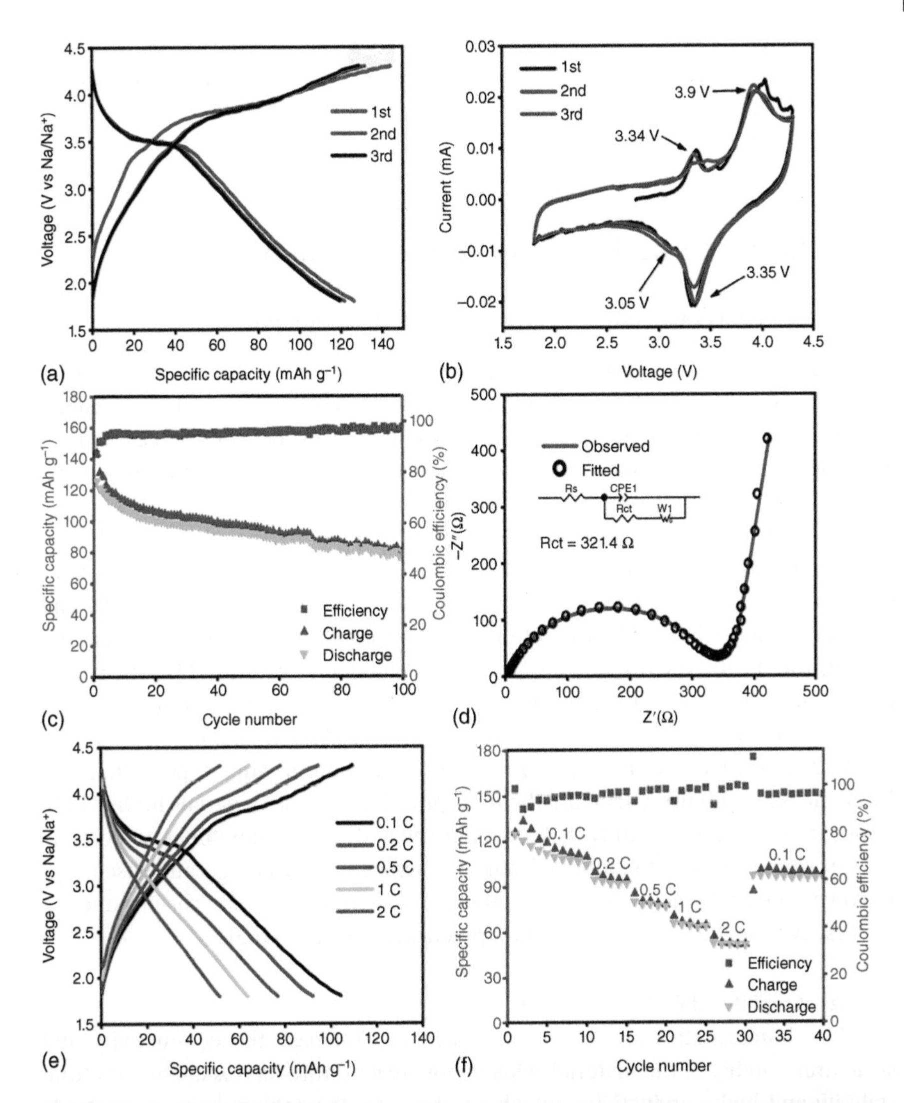

Figure 6.6 (a) The charge–discharge plots of the first, second and third turns of NMFP-0.4. (b) The CV plots of NMFP-0.4. (c) The long-cycle plots of NMFP-0.4. (d) The Nyquist plots of NMFP-0.4 applied in SIB. (e) The multiplier curves and (f) rate performance of NMFP-0.4. Source: Chen et al. [64] / with permission of Elsevier.

$X = 0.4$, it showed the best electrochemical activity, and the initial charge–discharge capacities were 143.8 and 126 mAh g^{-1} (Figure 6.6a,b). The NMFP-0.4 electrode material still has 80.95 mAh g^{-1} capacity after 100 cycles at 0.1 C (Figure 6.6c). The rate performance of NMFP-0.4 nanowire is shown in the figure, and the material exhibits good rate performance. When the current density is restored to 0.1 C, the discharge specific capacity is also restored to 97.1 mAh g^{-1}, which is 93.4% of the initial capacity (Figure 6.6e–f). The synthesized nanowire structure increases the surface area, which also makes the material more closely integrated with the collecting flow and reduces the diffusion resistance of Na^+. All these indicate that polyanionic nanowire materials have a very high research potential in SIBs.

6.3 Nanowires as Anodes in Sodium-ion Batteries

6.3.1 Carbonaceous Materials and Polyanionic Compounds

In recent years, more and more researchers are focusing on carbonaceous materials due to their large interlayer distance and unordered structure, which facilitate Na-ion insertion and extraction [36]. As the main body of sodium storage, the negative electrode of SIB realizes the insertion and extraction of sodium ions during charging and discharging.

At present, the reported anode materials for SIBs mainly include carbon-based, titanium-based, organic, alloy, and other anode materials. The research on carbon-based materials mainly focuses on graphite carbon materials, AC materials, and nanocarbon materials [65]. Graphite is the most commonly used anode material in LIBs due to its low discharge platform and moderate capacity [32]. However, due to the large radius of sodium ions and the small spacing of graphite layers, it is not suitable to de-intercalate sodium ions. AC-based anode materials with high disorder have high sodium storage capacity, low sodium storage potential, and excellent cycle stability, which are the most promising anode materials for SIB [47]. Nanocarbon materials mainly include graphene and carbon nanotubes, which rely on surface adsorption to store sodium and achieve rapid charge and discharge. However, the problems of low Coulomb efficiency and poor cycle performance in the first cycle make it difficult to obtain practical application [65].

6.3.1.1 Graphitized Carbon Materials

The 1D characteristics of carbon nanowires can increase the conductivity and structural stability of the material. This nanowire structure has excellent structural stability and high conductivity, which can improve the performance of electrode materials.

Carbonaceous materials in the form of nanowires will be more conducive to playing to their respective advantages. For example, the sodium storage performance of hollow carbon nanowires (HCNWs) prepared through pyrolization of the polyaniline precursor was studied by Cao et al. [66] (Figure 6.7a–e). When serving as an anode for SIBs, this material demonstrated superior electrochemical performance. At a current density of 500 mA g^{-1} (2 C), a high reversible capacity of 149 mAh g^{-1} could be obtained (Figure 6.7g). It also delivered a capacity retention of 82.2% after 400 cycles at a current density of 50 mA g^{-1} (Figure 6.7f), showing outstanding cycling stability. The authors attributed this excellent electrochemical performance to the following two points: (i) the hollow structure endows this material with a high specific surface, which provides much more active sites as well as contact area between the electrode and electrolyte, and (ii) the large interlayer spacing (0.37 nm, Figure 6.7d) between the graphitic sheets, leading to effective insertion and transfer of Na ions during the charge–discharge process.

However, there are few reports on the synthesis of graphitized carbon materials with nanowire morphology, but it is still very promising to synthesize graphitized carbon materials with nanowire structure according to the existing nanowire synthesis strategy [67]. It is still of great scientific value to further improve the performance of graphitized carbon materials as cathode materials for new SIBs by using the unique structural advantages of nanowires.

6.3.1.2 Amorphous Carbon Materials

In the field of carbon materials, AC materials are usually divided into easily graphitized carbon and difficult graphitized carbon. Easily graphitized carbon, also known as soft carbon, usually refers to the carbon material that can be graphitized above 2800 °C, and the disordered structure is easily eliminated. Hard carbon, also known as hard carbon, usually refers to carbon that is difficult to completely graphitize above 2800 °C, and its disordered structure is difficult to eliminate at high temperatures. The main difference between the two AC materials is the arrangement of the carbon layers that make them [68].

HCNWs are synthesized by the pyrolysis of hollow polyaniline nanowire precursors. This nanowire structure has excellent structural stability and high conductivity, which can improve the performance of electrode materials [69]. When used as a negative electrode material for SIBs, it has a reversible specific capacity of 251 $mAh\,g^{-1}$ at a current density of 50 $mA\,g^{-1}$, and excellent stability over 400 cycles. At a high current density of 500 $mA\,g^{-1}$, it still has a reversible specific capacity of 149 $mAh\,g^{-1}$. As early in 1993, the sodium-ion storage performance of hard carbon was studied, and a reversible specific capacity of nearly 85 $mAh\,g^{-1}$ was obtained [70]. Subsequently, when glucose-derived hard carbon was used as the anode material for SIBs, the first discharge specific capacity was about 300 $mAh\,g^{-1}$. A stable specific capacity of 176 $mAh\,g^{-1}$ was obtained after 600 cycles at a current density of 200 $mA\,g^{-1}$ from cellulose-derived carbon nanofibers [71] (Figure 6.8a,b).

Usually, the construction of open nanopores inside the material is an effective method to improve the electrochemical performance of the material. However, experiments have shown that increasing the specific surface area of hard carbon leads to lower reversible capacity, while the porous carbon anode leads to lower Coulombic efficiency, which may be due to the higher specific surface area. It is easier to form more solid-electrolyte-interface (SEI) films [72]. Therefore, reasonably reducing the specific surface area of hard carbon materials becomes an effective means to improve their initial Coulombic efficiency [73].

6.3.1.3 Carbon Nanomaterials

A novel 1D nanohybrid comprised of conductive graphitic carbon (GC)-coated hollow $FeSe_2$ nanospheres decorating reduced graphene oxide (rGO) nanofiber (hollow nanosphere $FeSe_2$@GC–rGO) was designed as an efficient anode material

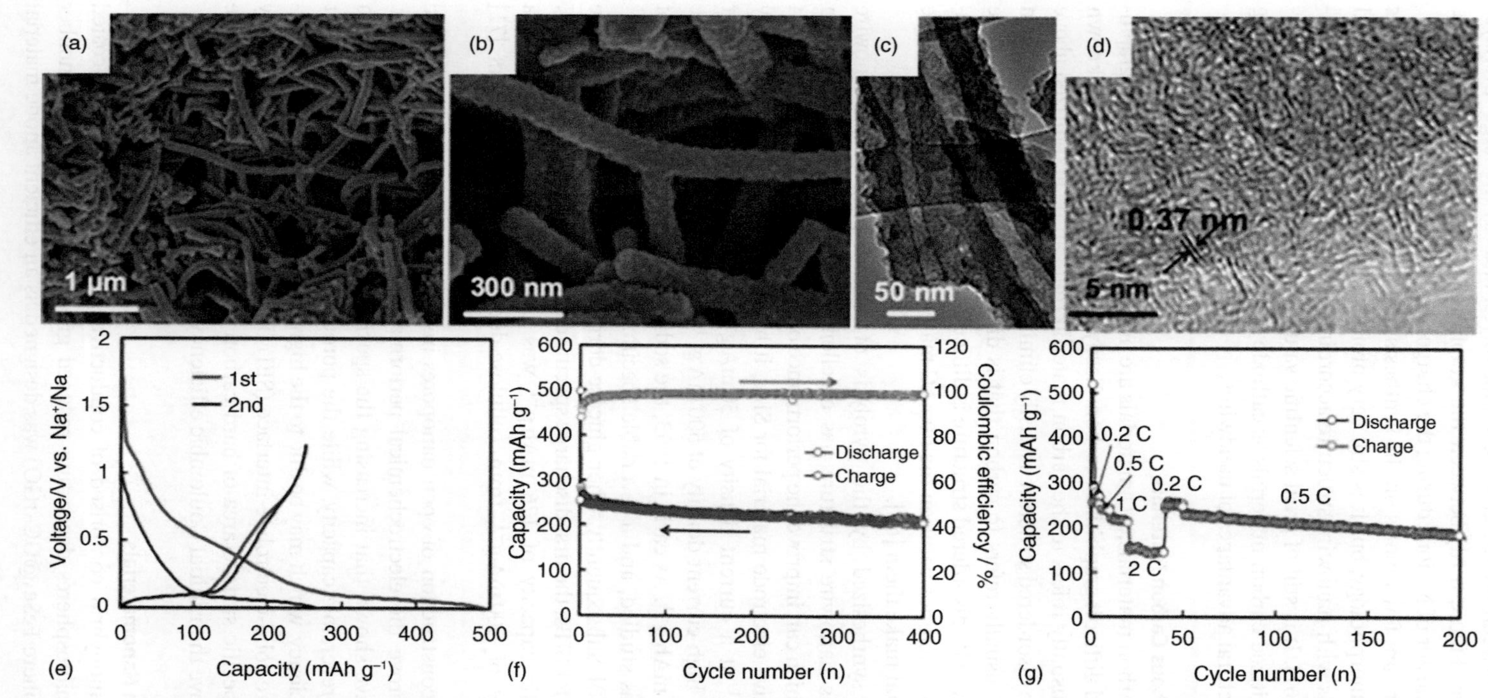

Figure 6.7 (a, b) SEM images of the HCNWs. (c) TEM image of the HCNWs. (d) High-resolution TEM image of the HCNWs. (e) First two charge–discharge profiles of the HCNW electrode between 0 and 1.2 V at a current density of 50 mA g^{-1} (0.2 C). (f) Cycle performance of the HCNW electrode at a current density of 50 mA g^{-1} (0.2 C). (g) Discharge capacity of the HCNW electrode as a function of charge–discharge cycles at different charge–discharge current densities of 50 (0.2 C), 125 (0.5 C), 250 (1 C), and 500 (2 C) mAh g^{-1}, respectively. Source: Cao et al. [66] / with permission of American Chemical Society.

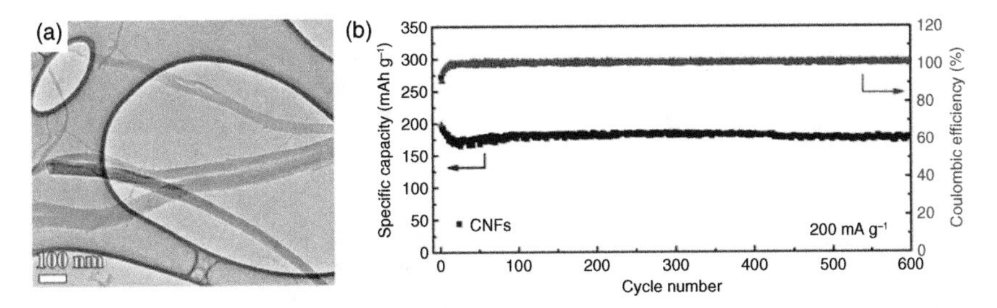

Figure 6.8 SEM and cycle performance of cellulose-derived carbon nanofibers (a) SEM images; (b) Cycle performance diagram. Source: Mai et al. [71] / with permission of American Chemical Society.

for SIBs and synthesized by introducing the nanoscale Kirkendall effect into the electrospinning method [74]. The discharge capacities of the bare $FeSe_2$ nanofibers, nanorod $FeSe_2$-rGO AC hybrid nanofibers (Figure 6.9a,b), and hollow nanosphere $FeSe_2$@GC-rGO hybrid nanofibers at a current density of $1\,A\,g^{-1}$ for the 150th cycle are 63, 302, and $412\,mAh\,g^{-1}$, respectively, and their corresponding capacity retentions measured from the 2nd cycle were 11, 73, and 82%, respectively (Figure 6.9c). The hollow nanosphere $FeSe_2$@GC-rGO hybrid nanofibers delivered a high discharge capacity of $352\,mA\,h\,g^{-1}$ even at an extremely high current density of $10\,A\,g^{-1}$ (Figure 6.9d). The enhanced electrochemical properties of the hollow nanosphere $FeSe_2$@GC-rGO composite nanofibers arose from the synergetic effects of the $FeSe_2$ hollow morphology and highly conductive rGO matrix.

6.3.2 Polyanionic Compounds

Polyanionic compounds, described as $Na_xM_yN_z$ (among them, M represents metallic elements, such as Mn, Fe, Co, Ni, Ti, V, etc., and N represents $(PO_4)^{3-}$, $(SO_4)^{4-}$, $(SiO_4)^{4-}$, etc.), have more robust crystal structures and show relatively higher Na^+ conductivity. As a typical example of polyanionic compounds, $Na_3V_2(PO_4)_3$ has received extensive attention in both anodes and cathodes for SIBs owing to its large interstitial spaces for quick diffusion of the Na ion and multiple valence states of the V element. The $Na_3V_2(PO_4)_3$/C composite can display remarkable rate performance when applied in SIBs. Zhu et al. [75] successfully prepared $Na_3V_2(PO_4)_3$/C composite nanofibers via coaxial electrospinning followed by calcination (Figure 6.10a). When applied as an anode for SIBs, the $Na_3V_2(PO_4)_3$/C nanofibers demonstrate a high reversible capacity of $189.9\,mAh\,g^{-1}$ at a current density of $23.6\,mA\,g^{-1}$ during 0.01–3 V (vs Na^+/Na). This anode still displays 105.1 and $80.4\,mAh\,g^{-1}$ at the second cycle, even at high rates of 590 and $1180\,mA\,g^{-1}$ (Figure 6.10b). Meanwhile, capacity retention is 83 and 77% after 200 and 500 cycles, respectively, showing outstanding Na^+ storage performance (Figure 6.10c). In addition, NASION-type $NaTi_2(PO_4)_3$ is also studied as an anode material for SIBs. The specific capacity corresponding to the reversible insertion/extraction of two sodium ions is $133\,mAh\,g^{-1}$, the voltage platform is about 2.1 V, and the cycle stability is good. However, due to the high potential and the low voltage of the full battery matched with the positive

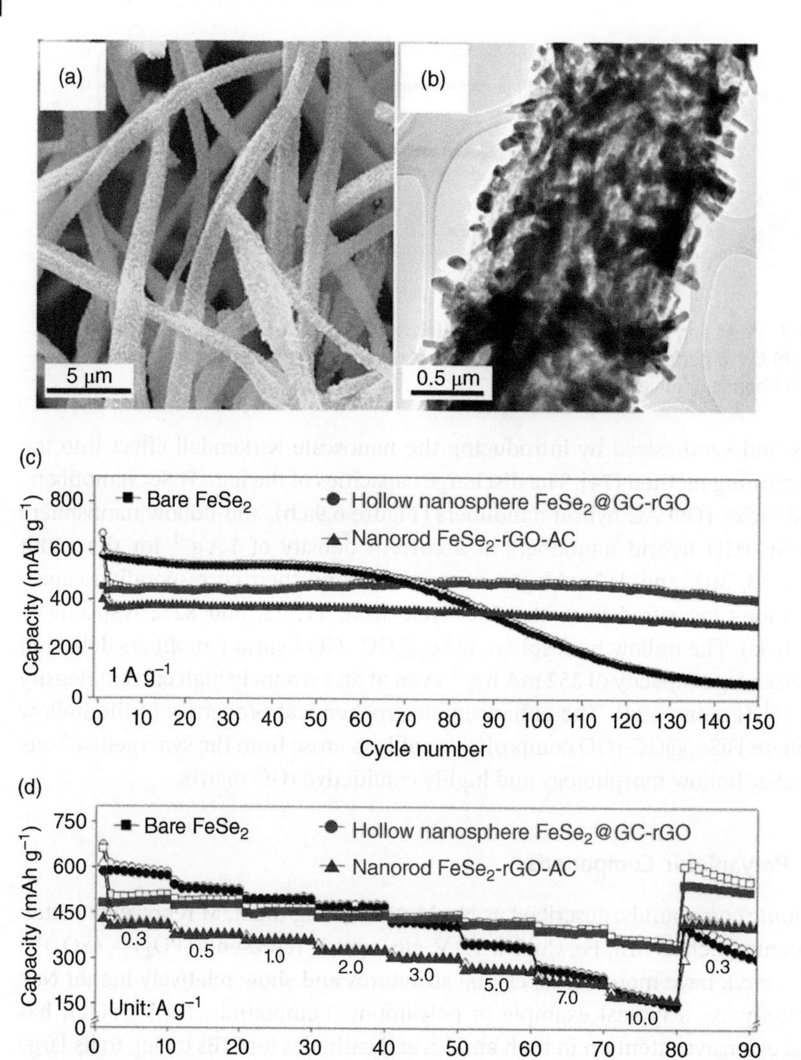

Figure 6.9 Morphologies, selected area electron diffraction (SAED) pattern, and elemental mapping images of the $FeSe_2$@GC−rGO hybrid nanofibers: (a) SEM images, (b) TEM images, (c) cycling performances at a current density of 1.0 A g^{-1}, and (d) rate performances. Source: Cho et al. [74] / with permission of Springer Nature.

electrode, it is less competitive in the organic electrolyte system. However, this material has certain application potential in aqueous SIBs with low voltage [76]. Hu et al. designed and synthesized a free-standing $Na_4MnV(PO_4)_3$@C (NMVP@C) nanofiber, which can be directly used as a SIB cathode. The NMVP@C nanofiber delivers a high capacity of 97.8 mAh g^{-1} at 0.2 C with a remarkable capacity retention of 80% after 800 cycles at 5 C. The in situ X-ray diffraction results demonstrate that the $V^{3+/4+}$ redox reaction occurs through a solid-solution reaction. In contrast, a two-phase $Mn^{2+/3+}$ redox reaction is identified, and both are highly reversible. When coupled with pre-Na^+ Sb@C anode, the NMVP@C//Sb@C full cell exhibits

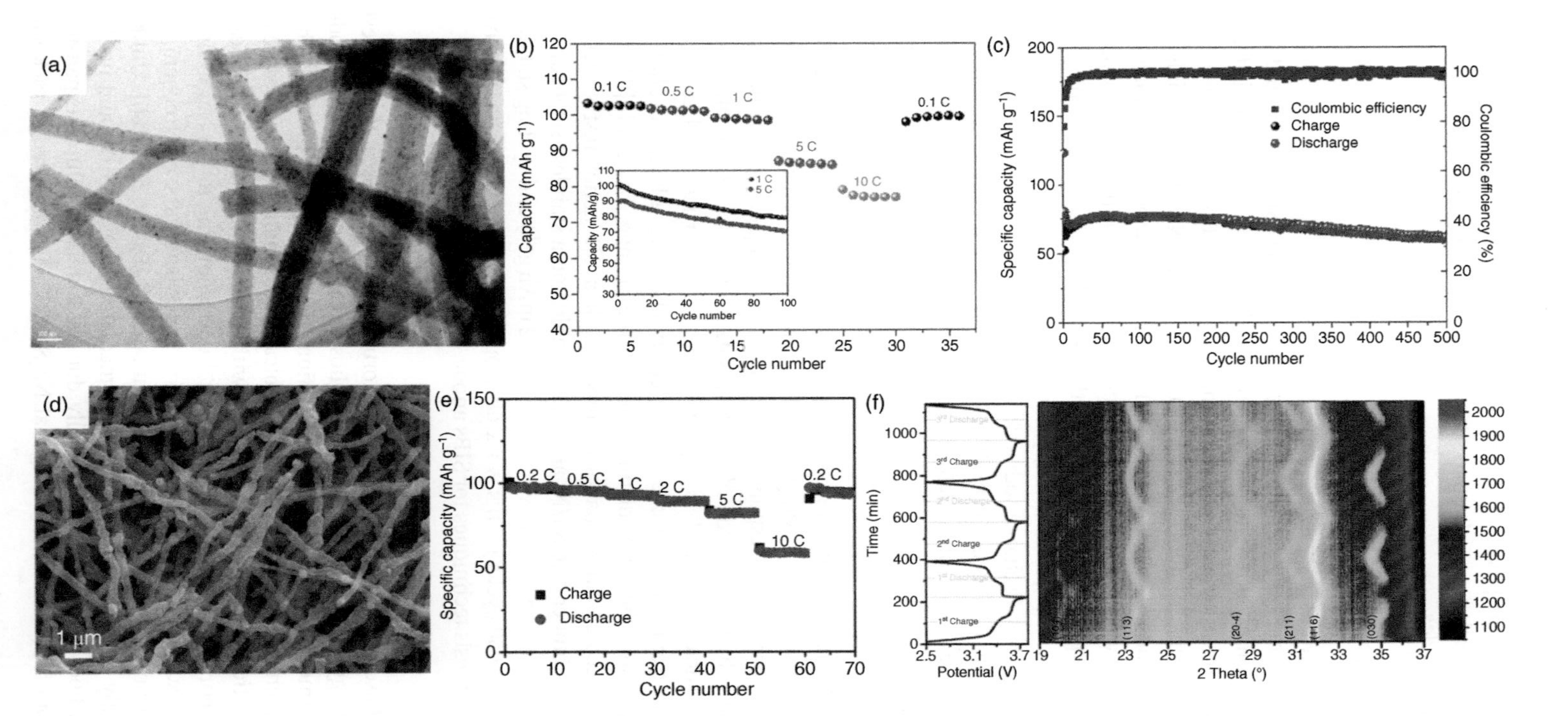

Figure 6.10 (a) TEM images of NF-NVP/C. (b) Rate performance of NF-NVP/C (Inset shows the cycle performance at 1 and 5 C). (c) Cycle performance of NF-NVP/C at 10 C. Source: Zhu et al. [75] / with permission of Springer Nature. (d) The SEM image, (e) rate performance, and (f) in situ XRD patterns of NMVP@C nanofiber. Source: Hu et al. [77] / with permission of John Wiley & Sons, Inc.

two reduction/oxidation peaks (3.35/2.95 and 3.05/2.56 V). The initial discharge capacity of this full cell can reach up to 98.2 mAh g^{-1} with an impressive retention of 87.5% after 100 cycles.

6.3.3 Metals and Metal Oxides

6.3.3.1 Metal Nanowires

Among pure elements, metals (Sn, Bi), metal-like metals (Si, Ge, As, and Sb), and polyatomic nonmetallic compounds (P) have been extensively studied due to their high theoretical capacities (Sn: 847 mAh g^{-1}, P: 2596 mAh g^{-1}) with alloy mechanism. Benefiting from high capacity and appropriate Na^+/Na potentials, the alloy mechanism anodes are suitable for SIBs [78–83]. However, these materials undergo significant volume changes during cycling that result in the loss of active substances and sharp capacity decay due to the larger Na^+ ionic radius compared to Li^+. For instance, Sn experiences a volume expansion of 420%, and Sb expands by 393% during the full discharge sodiation phase ($Na_{15}Sn_{14}$ [79] and Na_3Sb [80]). Combining carefully designed electrode microstructures with buffering matrices is a common strategy to accommodate volume expansion and improve the cycling stability of alloy anodes.

Sn Nanowires

Among these alloy anodes, the Sn anode can undergo electrochemical reactions with sodium to form $Na_{0.6}Sn$, $Na_{1.2}Sn$, Na_5Sn_2, and finally, $Na_{15}Sn_4$ as the fully discharged alloy formula, corresponding to a theoretical capacity of 846 mAh g^{-1}. Meanwhile, the volume expansion of $Na_{15}Sn_4$ is 420% based on the in situ results [79]. The Sn particles also tend to aggregate into large particles and then pulverize to isolate from the electrode during electrochemical alloying/dealloying. In order to overcome the effect of this high volume change, Liu et al. [84] fabricated a carbon-coated Sn/Ni nanorod anode with a 3D nanoforest structure (Figure 6.11). This binder-free Sn/Ni anode, assembled on a stainless steel current collector, exhibits supreme capacity utilization (722 mAh g^{-1}) and cycling stability (450 mAh g^{-1} after 450 cycles). This work develops a new strategy for inexpensive and versatile synthesis techniques for SIBs anode.

Sb Nanowires

Among alloy-based anodes, though phosphorus anode possesses the highest gravimetric energy density (2596 mAh g^{-1}), phosphorus is nonconductive anode with huge volume expansion (491%) and a final Na_3P discharge product. This severe volume change limits its electrochemical performance and widespread practical application. Sb is another SIB anode with an alloying reaction, which can show a high theoretical capacity of 660 mAh g^{-1} corresponding to the Na_3Sb final product when discharged. Monconduit et al. first investigated the electrochemical performance of pure micrometric Sb particle anode in LIBs and SIBs [80]. Unlike Sb in LIB, mostly amorphous of Na, the intermediate phases in SIB could not be precisely identified. This might be partially due to decreased volume expansion upon going from Sb (181.1 Å) to hexagonal Na_3Sb (237 Å) compared to rock salt

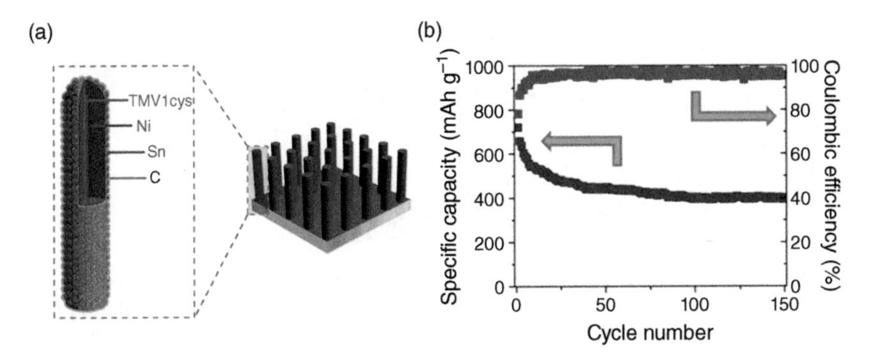

Figure 6.11 Structural diagram and cyclic performance diagram of carbon-coated Ni/Sn nanorods. (a) Structural diagram and (b) cycle performance diagram. Source: Liu et al. [84] / with permission of American Chemical Society.

Li_3Sb (283.8 Å). Afterwards, Baggetto et al. first rationalized the amorphization of Na_xSb phases by the long-ranged strain propagation due to Na-vacancy compared to Li-Sb [85]. Based on the above reports, Sb is one of the most attractive anode materials for SIBs with the following equation [86].

$$\text{Alloying reaction}: \ \mathrm{Sb} + x\mathrm{Na}^+ + xe^{-1} \rightarrow \mathrm{Na}_x\mathrm{Sb} \quad (x = 1\text{–}3) \tag{6.1}$$

The influence of size and electrode morphology on sodium ion storage performance has been demonstrated. Liang et al. fabricated highly ordered arrays of Sb nanorods using a large-scale method, which exhibited uniform spacing intervals of 190 nm (Figure 6.12a) [87]. This Sb nanorod array exhibited a high capacity of 620 mAh g^{-1} at the 100th cycle, along with superior rate capability (557.7 mAh g^{-1} at 20 A g^{-1}). When coupled with P2-$Na_{2/3}Ni_{1/3}Mn_{2/3}O_2$ (NNMO) cathode, NNMO//Sb full cell delivered a superior capacity retention of ~94% before 110 cycles, maintaining a capacity of around 620 mAh g^{-1}. Notably, a nanoporous pure Sb anode, synthesized by a morphology-controlled chemical dealloying method, ensures high accessibility to sodium ions and structural solid integrity even without conductive modifications [90]. 1D nanomaterials facilitate electron transport along their long axis but also maintain radial confinement effects, offering promising prospects for SIB anodes in various applications [91]. Zhu et al. reported an Sb/C free-standing electrode with ~30 nm Sb nanoparticles uniformly encapsulated in interconnecting 1D 400 nm carbon fibers (SbNP@C) by electrospinning method (Figure 6.12b). This SbNP@C delivered an initial capacity of 422 mAh g^{-1} and retained 350 mAh g^{-1} after 300 cycles at 100 mAh g^{-1}. Similar work involving electrospun Sb/C as a SIB anode was concurrently accomplished by Cao et al., followed by the establishment of various other 1D composite anodes such as Sb/CNT composites and coaxial nanotubes composed of Sb@C (Figure 6.12c) [89, 92, 93].

6.3.3.2 Transition Metal Oxide Nanowires

Compared to traditional intercalation reactions, transition metal oxides (TMOs) based on conversion reactions exhibit excellent lithium storage properties and are promising anode materials for SIBs. The concept of conversion reaction was

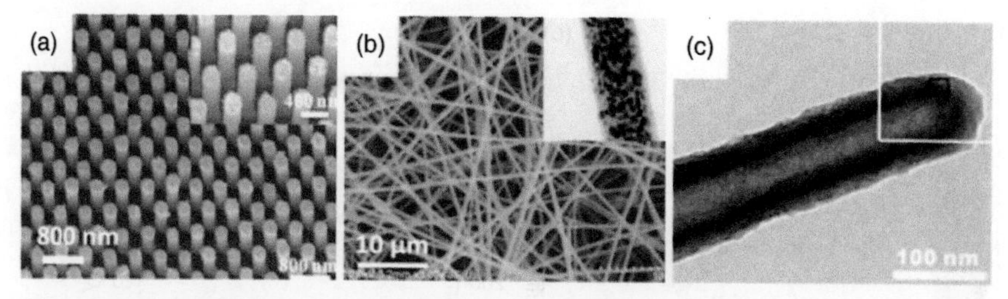

Figure 6.12 Representative morphologies (SEM or TEM images) of several Sb-based hybrid nanostructured electrodes: (a) Ordered Sb nanorod array. Source: Liang et al. [87] / with permission of Royal Society of Chemistry; (b) Electrospun Sb/C Fibers. Source: Zhu et al. [88] / with permission of American Chemical Society; (c) Sb@C coaxial nanotubes. Source: Liu et al. [89] / with permission of Royal Society of Chemistry.

first introduced by Tarascon and coworkers, who demonstrated the reversible electrochemical lithiation of TMOs. TMO anodes can undergo a conversion reaction with sodium to form transition metal nanoclusters and Na_2O matrices, offering low-cost and high gravimetric specific capacities. Unlike in LIBs, the conversion reactions in SIBs may involve more intermediate phases, such as Na_2O, Na_2O_2, and NaO_2, providing richer electrochemical reactivity and potentially different reaction pathways. Among various TMO materials with conversion reactions investigated so far, iron oxide (Fe_2O_3, Fe_3O_4), cobalt oxide (Co_3O_4, CoO), manganese oxide (MnO_x), copper oxide (CuO, Cu_2O), ruthenium oxide (RuO_2), chromium oxide (Cr_2O_3), molybdenum oxide (MoO_x), and nickel oxide (NiO) have attracted significant attention due to their ease of handling and high capacities. However, the large volume change and structural damage associated with the cycling-induced expansion/contraction caused by the large radius of Na^+ ions result in poor cycling stability and rate performance for TMO-based conversion reactions. To address these challenges, the LIB conversion reaction materials mainly focused on the following aspects: (i) nanostructure engineering with favorable morphologies, (ii) hybridization with conductive carbon (CNT, graphene, etc.), and (iii) heteroatoms doping. Importantly, the morphology and particle size significantly influence the sodium ion storage performance. In detail, the transport efficiency of e^-/Na^+ depends on both the morphology and particle size of the electrode material. The reduction in particle size not only shortens the e^- path and enhances Na^+ diffusion but also increases the contact area between the electrode and electrolyte, thereby improving the utilization rate of the active material. However, nanostructured materials with high specific surface areas tend to induce side reactions on the electrode surface. Additionally, hollow/porous structures can effectively mitigate volume expansion during cycling, but their low tapping density sacrifices the volume-specific capacity of these materials. Hybridization with conductive carbon can enhance the electrical conductivity of the electrode, facilitating fast electrochemical kinetics and effectively maintaining morphology while restraining the pulverization of electrode materials during cycling. Although these strategies have improved sodium storage performance in conversion reactions to some extent (Figure 6.13) [94], cyclic

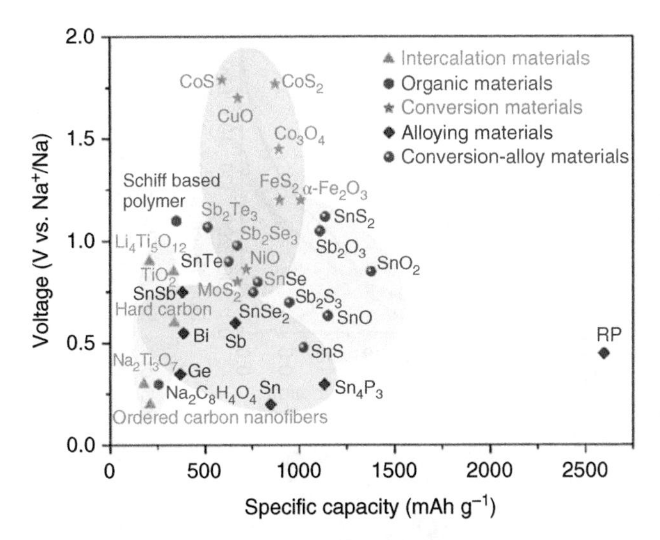

Figure 6.13 Theoretical specific capacity and operating voltage of anode materials for SIBs. Source: Qiao et al. [94] / with permission of American Chemical Society.

stability and rate capacity remain unsatisfactory. In future research, emphasis should be placed on developing high-capacity TMO anodes for long-term cycling.

MnO_2 Nanowires

Manganese oxides (MnO_x) have also been extensively investigated for their applications in LIBs, SIBs, KIBs, and supercapacitors due to their low cost and environmental friendliness. However, similar to other conversion-based materials, MnO_x suffers from poor cycling stability and rate capability caused by its inherent low conductivity and large volume effect despite its high theoretical capacity (756 mAh g^{-1}) (Eqs. (6.2)–(6.4)). Currently, efforts towards improving the sodium ion storage performance of MnO_x mainly focus on nanostructure design, hybrid conductive materials integration, and porous structure engineering.

$$\text{First discharge process}: Na^+ + MnO_2 + e^- \rightarrow NaMnO_2 \tag{6.2}$$

$$NaMnO_2 + Na + e^- \rightarrow Na_2O + MnO \tag{6.3}$$

$$\text{Subsequent cycles}: \quad Na_2O + MnO \rightarrow NaMnO_{2+} Na + e^- \tag{6.4}$$

Porous MnO@C nanorods fabricated by annealing Mn-based metal-organic frameworks precursor. The unique porous structure of the MnO@C nanorods (~200 nm in diameter and 3–10 nm in length) contained embedded MnO nanoparticles (3–5 nm) within a carbon matrix, which significantly enhanced electrical conductivity and mitigated volume changes during charge/discharge processes. After 100 cycles at 50 mA g^{-1}, the obtained MnO@C nanorods exhibited a high reversible specific capacity of 260 mAh g^{-1}, demonstrating superior long-term cycling performance with a capacity retention of 140 mAh g^{-1} at 2 A g^{-1} (Figure 6.14a–c) [95]. Wang et al. synthesized MnO_2 nanowires coated with reduced graphene oxide (MnO_2@rGO) using a simple hydrothermal method.

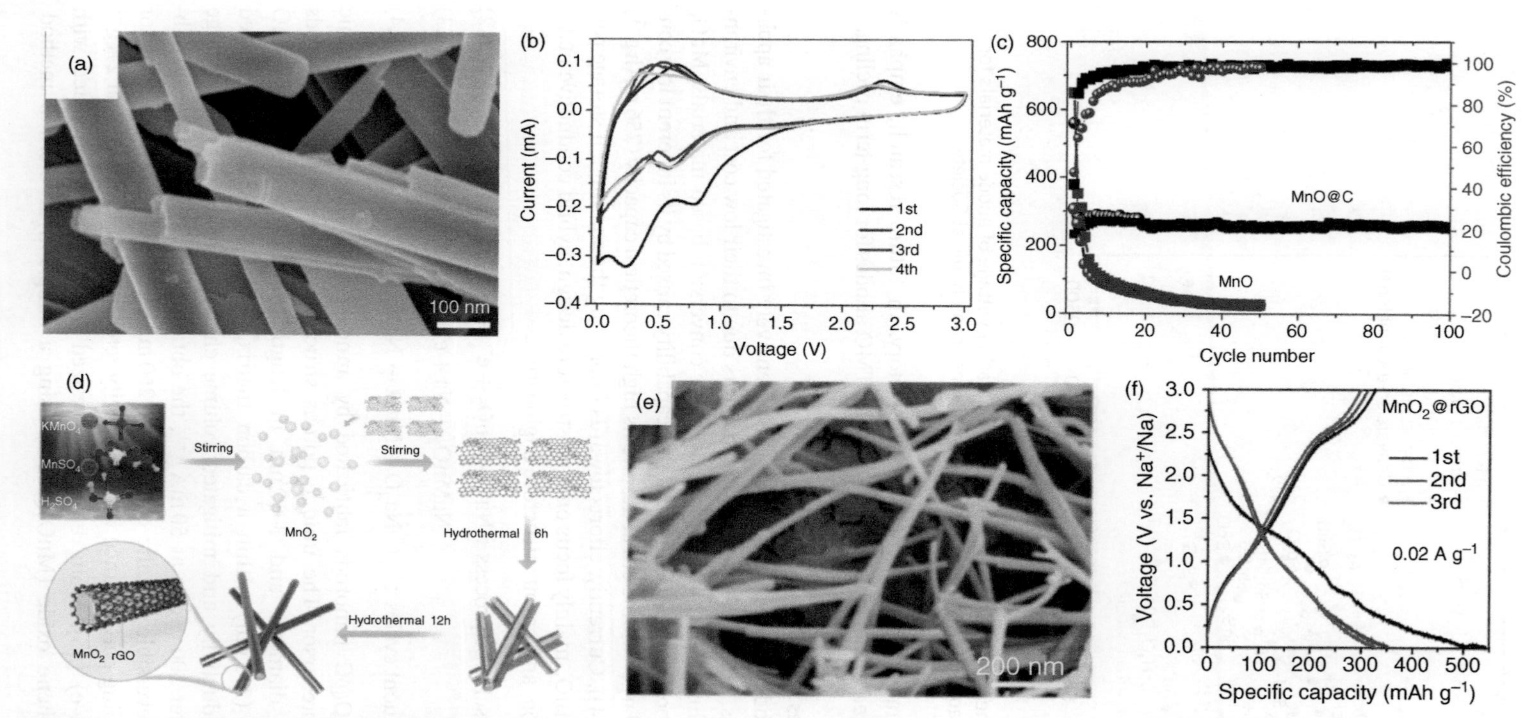

Figure 6.14 (a) SEM image and (b) CV curves of MnO@C composites. (c) Cycling stability of MnO@C and MnO electrodes. Source: Zhang et al. [95] / with permission of Elsevier. (d) Schematic illustration of the synthesis process, (e) SEM image, and (f) charge/discharge profiles of MnO_2@rGO nanowires. Source: Wang et al. [96] / with permission of John Wiley & Sons, Inc.

The incorporation of rGO not only induced lattice distortion to facilitate faster Na^+ transfer within highly crystalline MnO_2 nanowires but also created an interfacial structure between rGO and MnO_2. As an anode material for SIBs, the resulting MnO_2@rGO nanowires exhibited a high capacity of 228 mAh g^{-1} at 50 mA g^{-1}, with excellent capacity retention of 75% after 500 cycles (Figure 6.14d–f) [96].

Among various manganese oxides commonly studied for SIB anodes is MnO_2, which exhibits a discharge capacity of up to 1233 mAh g^{-1} when fully reduced to metallic Mn accompanied by Na^+ intercalation. The crystal structures of MnO_2 typically exist in α, β, γ, δ, and λ forms due to the different linkages of fundamental MnO_6 octahedra. Among these forms, α, β, and γ phases are usually characterized by a 1D structure, while δ is a 2D layered compound and λ exhibits a 3D spinel structure. Various MnO_2, α, β, and γ phases have been extensively studied for SIB anodes due to their large size channels that can provide a host for Na^+ insertion/extraction. The α-MnO_2 consists of (2 * 2) + (1 * 1) tunnel structures connected at the corners through shared MnO_6 octahedral double chains. While the β-MnO_2 features a 1D channel (1 * 1) structure composed of single chains of MnO_6 octahedral elements. As for γ-MnO_2, it possesses both ramsdellite-type (2 * 1) channels and pyrolusite-type (1 * 1) channels (Figure 6.15a) [97].

Furthermore, the properties of MnO_2 are significantly influenced by its crystal structure. By carefully controlling reaction conditions such as acidity and temperature during synthesis processes, specific polycrystals can be obtained. For instance, Su et al. reported the successful synthesis of α-MnO_2 and β-MnO_2 nanorods using hydrothermal methods. When utilized as anodes for SIBs, both α-MnO_2 and β-MnO_2 nanorods deliver high initial capacities of 278 and 298 mAh g^{-1} at 20 mA g^{-1}, respectively. The cycling performance and rate performance of β-MnO_2 is superior to α-MnO_2, possibly due to the compact and dense (1 * 1) tunnel structure in β-MnO_2 crystals. The two-tunnel structure of α-MnO_2 is not stable when Na^+ (de) intercalation, leading to a low capacity and poor cyclability than β-MnO_2 (Figure 6.15b–d) [98]. In addition, Su et al. fabricated β-MnO_2 nanorods with exposed tunnel structures by a hydrothermal method, leading to facile Na^+ insertion/extraction of the {111} crystal plane consisting of high-density (1 * 1) tunnels. The optimized β-MnO_2 nanorods deliver a high initial discharge capacity of 350 mAh g^{-1} and a good high-rate performance. The ex situ XRD and TEM results demonstrate that the gradual capacity decrease caused by the formation of $NaMn_2O_4$ inactive phase for β-MnO_2 nanorod electrodes (Figure 6.15e–g) [99].

CuO Nanowires

Cu_2O and CuO are two classical cupric oxides that have attracted increasing attention as anodes for SIBs due to their abundant resources, chemical stability, and non-toxic nature. In recent years, numerous studies have been conducted to investigate the electrochemical performance and sodium ion storage mechanism of Cu_2O and CuO. Wang et al. fabricated porous CuO nanowires by simply deleting a $Cu(OH)_2$ precursor. When used as a SIB anode, the porous CuO nanowires deliver a high discharge capacity of 640 mAh g^{-1}, which is very close to the theoretical specific capacity (674 mAh g^{-1}) (Figure 6.16a,b). Based on the CV, ex situ XRD,

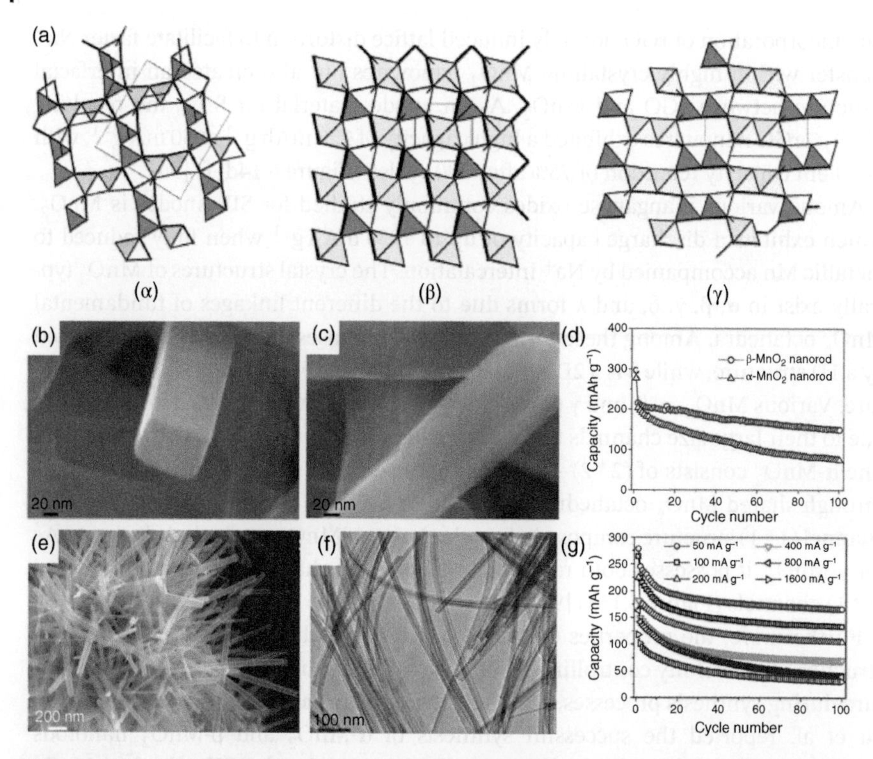

Figure 6.15 (a) Crystal structures of α-MnO_2, β-MnO_2, and γ-MnO_2. Source: Devaraj et al. [97] / with permission of American Chemical Society. SEM images of (b) α-MnO_2 nanorods and (c) β-MnO_2 nanorods. (d) Cycling performances of α-MnO_2 and β-MnO_2 nanorod anodes at a current density of 20 mA g^{-1}. Source: Su et al. [98] / with permission of Royal Society of Chemistry. (e) SEM image and (f) TEM image of β-MnO_2. (d) Cycling performances of β-MnO_2 nanorod at different current densities. Source: Su et al. [99] / with permission of Nature Publishing Group.

Raman spectroscopy, and HRTEM test, the discharge/charge processes can be explained by the following electrochemical conversion reactions (Eqs. (6.5)–(6.9), Figure 6.16c).

During the discharge process:

$$CuO + Na^+ + e^- \rightarrow Cu^{II}{}_{1-x}Cu^{I}{}_{x}O_{1-x/2} + Na_2O \tag{6.5}$$

$$2Cu^{II}{}_{1-x}Cu^{I}{}_{x}O_{1-x/2} + (2-2x)\,Na^+ + (2-2x)\,e^- \rightarrow Cu_2O + (1-x)\,Na_2O \tag{6.6}$$

$$Cu_2O + 2Na^+ + 2e^- \rightarrow 2Cu + Na_2O \tag{6.7}$$

During the charge process [100]:

$$2Cu + Na_2O \rightarrow Cu_2O + 2Na^+ + 2e^- \tag{6.8}$$

$$Cu_2O + Na_2O \rightarrow 2CuO + 2Na^+ + 2e^- \tag{6.9}$$

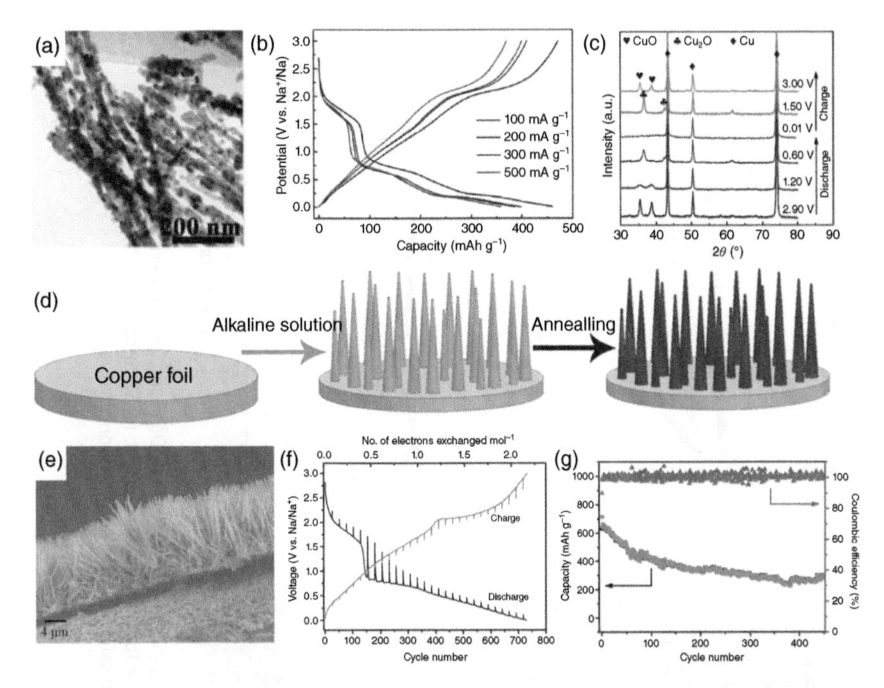

Figure 6.16 (a) The TEM image, (b) the charge/discharge profile, and (c) ex situ XRD patterns of CuO nanowires. Source: Wang et al. [100] / with permission of Springer Nature. (d) Schematic illustration of the fabrication for the in situ synthesis of CNAs on Cu foil. (e) SEM image of a CuO nanorod array. (f) GITT profile of CNA. (g) Cycling performance of binder-free CNA electrode at a high current density of 200 mA g^{-1}. Source: Yuan et al. [101] / with permission of John Wiley & Sons, Inc.

The flexible and porous CuO nanorod arrays (CNAs) were fabricated by in situ engraving commercial copper foils, as reported by Yuan et al. [101]. These CNAs can be directly utilized as anodes for SIBs without needing conductive agents or polymer binders. The binder-free porous CNA electrode delivers a stable discharge capacity of 640 mAh g^{-1} corresponding to 1.9 Na^+ intercalation per formula. Furthermore, a remarkable specific capacity of 290.6 mAh g^{-1} is maintained after 450 cycles at a current density of 200 mA g^{-1}. This exceptional electrochemical performance can be attributed to the unique array structure and absence of binders. Such strategies involving conductive carbon materials are widely recognized as effective means to enhance sodium-ion storage performance since they not only improve electrode conductivity but also mitigate volume changes during (de)-intercalation processes.

The excessive carbon content in composite materials may compromise their overall properties; therefore, future research should focus on reducing the carbon content without sacrificing electrical conductivity. Wang et al. prepared CuO quantum dots (~2 nm) uniformly encapsulated in carbon nanofibers (2-CuO@C) using a facile electrospinning method (Figure 6.17) [102]. This binder-free 2-CuO@C has a CuO content of 54 wt%. When used as an anode for SIBs, it demonstrates an initial capacity of 528 mAh g^{-1} with a high capacity retention of 83% after 100 cycles at 100 mA g^{-1} and high rate capability (even 250 mAh g^{-1} at 5000 mA g^{-1}). However,

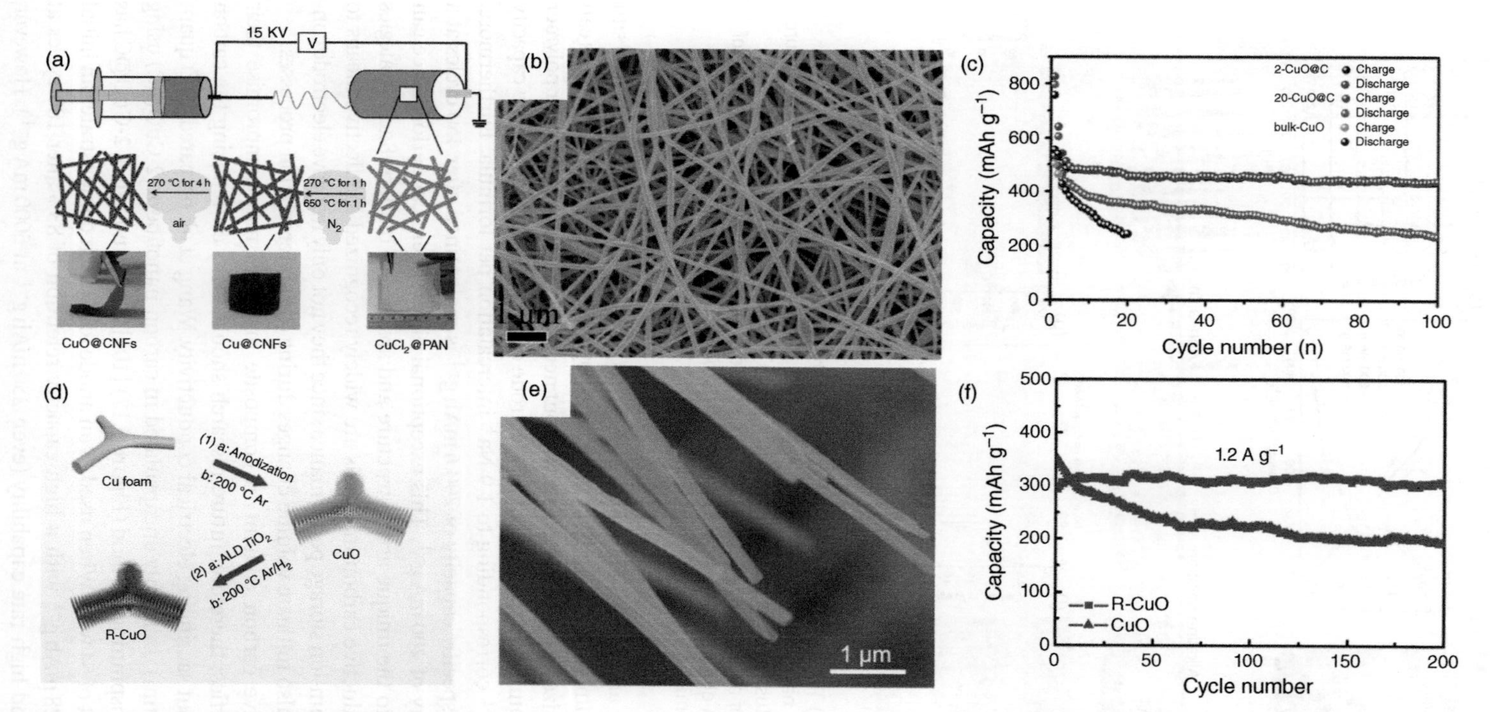

Figure 6.17 (a) Schematic illustration of the formation process for CuO@C nanofibers. (b) SEM image of 2-CuO@C. (c) Cycling performance of 2-CuO@C, 20-CuO@C, and bulk CuO at 100 mA g^{-1}. Source: Wang et al. [102] / with permission of John Wiley & Sons, Inc. (d) Schematic illustrating, (e) SEM image, and (f) cycling performance of R-CuO and CuO. Source: Refs. Ni et al. [103] / with permission of John Wiley & Sons, Inc.

the substantial volume change during Na intercalation and de-intercalation could disrupt the structural integrity (~170%) leading to poor electrochemical durability and low coulombic efficiency. Ni et al. presented a structural strategy involving atomic layer deposition to regulate the volume change in CuO nanoarray electrodes during Na storage by depositing cohesive TiO_2 thin films. The optimized CuO electrode enables large capacity performance reaching 592 mAh g^{-1} at 60 mA g^{-1} with stable cycling (82% after 1000 cycles), which may be due to the ductility of amorphous TiO_2 acting as nano glue to prevent particle fracture during Na^+ cycling [103].

6.3.4 Metal Sulfides

Metal sulfides (MSs) have attracted extensive attention due to their potential applications in fuel cells, SIBs, supercapacitors, and other energy devices. There are many kinds of MSs with relatively high theoretical specific capacity and energy density. They usually exist in nature in the form of molybdenite (MoS_2), sulfotungsten (WS_2), hexagonal nickel sulfide (Ni_3S_2), chalcocite (Cu2S), pyrite (FeS2), and other minerals [104], which also makes the sulfide cheaper. The sulfide used SIBs mainly includes sulfur-containing organic compounds and inorganic compounds. Most metal sulfides (MoS_2, SnS_2, WS_2, etc.) have a unique layered structure, which is unique in that the atoms in each layer exist stably through the action of covalent bonds, while the interlayer is formed only by the action of intermolecular forces such as van der Waals force. Due to the weak van der Waals force, under certain conditions, some polar molecules can enter the interlayer of layered compounds without destroying their layered structure by breaking the intermolecular force through adsorption, insertion, suspension, etc., so this type of transition metal sulfide can provide enough space for sodium-ion insertion. However, transition MSs also have some disadvantages, such as poor self-conductivity and low first Coulomb efficiency. During lithification, layered sulfides will be laminated and agglomerated, which will reduce the specific surface area of the material and the contact area with the electrolyte, limiting their application as electrode materials. For example, the cathode material NiS_2 has some problems during the battery charging and discharging processes: (i) Huge volume expansion; (ii) Poor electronic conductivity; (iii) Slow kinetics in the conversion reaction.

The commonly adopted methods to improve the electrochemical performance of MS as electrode materials include: (i) regulation of nano structure of materials, synthesis of 1D nanowires, hollow nanotubes, and 2D and 3D nano sheets [105]. Materials with nanostructures can shorten the transmission path of electrons and reduce the mechanical stress caused by excessive volume fluctuation after repeated natrification/sodium removal; Synthesis of hollow, core–shell, porous structural materials can solve the agglomeration of nanoparticles. (ii) Material composite, two or more materials can take advantage of multicomponent advantages to improve the electrochemical performance of materials. So far, the research on metal sulfide as lithium-ion electrode material has been relatively mature. Because lithium-ion and sodium-ion have similar embedding mechanism in the charge–discharge process, the development of new electrode materials for SIBs can be studied on the basis of

LIBs. The MS anode materials can be divided into layered structure and nonlayered structure according to the structural characteristics. The MS with layered structure includes molybdenum sulfide, tin sulfide, tungsten sulfide, and other materials, while the MS with nonlayered structure includes iron sulfide, nickel sulfide, and other materials.

6.3.4.1 Molybdenum Sulfide and Its Composites

The common anode material of molybdenum sulfide is molybdenum disulfide (MoS_2), which is a 2D layered structure composed of two S layers and one Mo layer (S–Mo–S) combined by van der Waals force. This configuration is conducive to the insertion/removal of lithium/sodium ions, and also plays a positive role in easing volume expansion and fixing polysulfides [106]. The theoretical capacity of pure MoS_2 can reach 670 mAh g^{-1}, which is much better than that of graphite anode (372 mAh g^{-1}). However, MoS_2 materials have the disadvantages of poor conductivity and structural stability, especially in SIBs. Tang et al. [107] took polystyrene nanospheres as templates, made chitosan and ammonium tetrathiomolybdate into lyophilized gels, treated them at 350 °C for 2 hours, and then annealed them at 800 °C for 2 hours. Using the electrostatic interaction between amino functional groups and molybdate ions, a hierarchical porous carbon framework embedded with MoS_2 nanosheets with a diameter of less than 10 nm was prepared. When it is used as the negative pole of SIB, its specific capacity reaches 330 mAh g^{-1} at 5 A g^{-1} current density, and its reversible specific capacity is 340 mAh g^{-1} after 550 cycles at 1 A g^{-1} high current. Xiong et al. [108] reported a flexible MoS_2/carbon nanofiber membrane while MoS_2 nanosheets were uniformly encapsulated in interconnected carbon nanofibers through a simple electrospinning method. The as-spun ammonium tetrathiomolybdate (ATTM) assisted polyacrylonitrile (PAN) ATTM-PAN and MoS_2-CNFs films exhibit excellent membrane flexibility and can be used as free-standing anode electrodes for SIBs (Figure 6.18a–c). The MoS_2-CNFs exhibit three reduction peaks (1.7, 0.92, and 0.2 V) in the first sodiation process. The reduction (1.7 V) and oxidation (1.75 V) peaks are attributed to the Na_1 ions insertion/deinsertion into MoS_2. The MoS_2-CNFs can maintain a reversible specific capacity of 283.9 mAh g^{-1} after 600 cycles at a current density of 100 mAh g^{-1} (Figure 6.18d–f).

6.3.4.2 Tungsten Sulfide and Its Composites

Tungsten disulfide (WS_2) has a layered structure similar to molybdenum disulfide (MoS_2), and its theoretical capacity is 433 mAh g^{-1}, which is considered to be a promising electrochemical active material. Tungsten disulfide-based anode materials mainly exist two problems: (i) its poor electronic/ionic conductivity, especially for the SIB system, the conductive response of pure WS_2 electrode materials is relatively low; (ii) WS_2 has poor crystallinity and unstable structure. These two factors together lead to poor cycle stability of WS_2 cathode material [109]. At present, the main way to solve these problems is to composite WS_2 with carbon-based materials, which can also adjust its structure to improve the conductivity and structural stability of WS_2. Liu et al. [110] obtained uniform WS_2 nanowires with

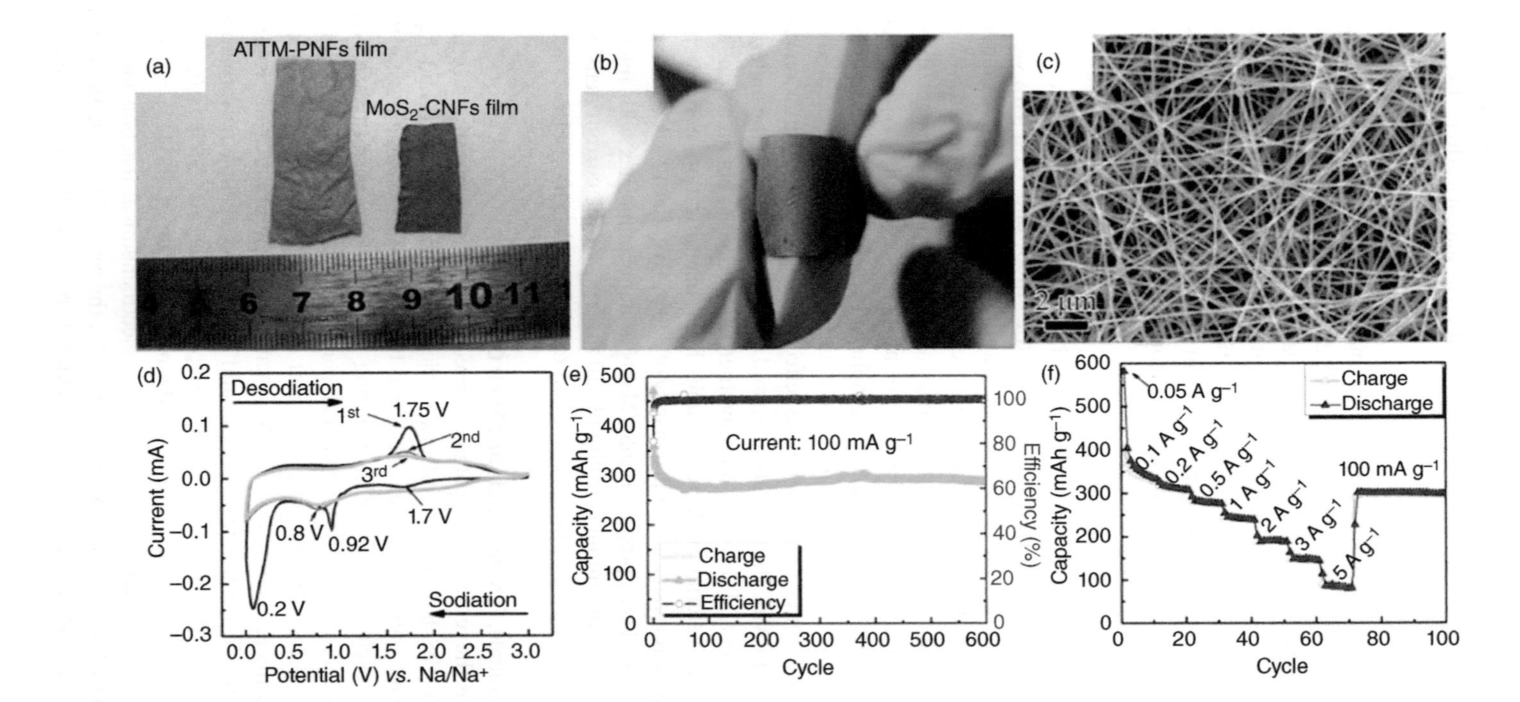

Figure 6.18 (a) Digital photo for the ATTM-PAN and MoS_2-CNFs films; (b) Digital photo shows the flexible property, (c) FESEM images, (d) initial three cycles of CV curves, (e) cycling performance, and (f) Rate capacity of the as-obtained MoS_2-CNFs film. Source: Hu et al. [108] / with permission of American Chemical Society.

a diameter of 25 nm by solvothermal method. The layer spacing was 0.83 nm. The layered structure of the material was protected by increasing the discharge cutoff potential (0.5–3 V). The material showed excellent cycle stability. After 1400 cycles at 1000 mA g^{-1} current density, the capacity remained at about 330 mAh g^{-1}. Xu et al. [111] prepared the DODA-WS_2 by a simple electro-spinning method, and used the obtained membrane woven as SIBs anode materials, as shown in Figure 6.19. The as-prepared material delivers large specific capacity of 377 mAh g^{-1} at current densities of 200 mA g^{-1}. After 400 cycles, the capacity still maintained 363 mAh g^{-1} with retention rate of 96%. The reversible capacities of DODA-WS_2 at 50, 100, 200, 500, 1000, 2000, 5000, and 10000 mA g^{-1} were 395, 365, 337, 305, 272, 230, 165, and 91 mAh g^{-1}, respectively. The unique results can be ascribed to the well-dispersed WS_2 cross-linked nanofiber framework and its conductive network.

6.3.4.3 Stannic Sulfide and Its Composites

The most widely studied tin disulfide material (SnS_2) has a higher theoretical capacity (1137 mAh g^{-1}) than 2D materials such as MoS_2 and WS_2. This can be attributed to its unique conversion reaction and Li/Na Sn alloying/dealloying mechanism [112]. In addition to these advantages, SnS_2 is currently faced with the main problems of poor material conductivity and large volume changes during charging and discharging. Prikhodchenko et al. [113] prepared SnS_2/graphene nanocomposites by H_2S vulcanizing SnO_2/graphene composites, and the reversible specific capacity can reach 650 mAh g^{-1}. In addition, the hydrothermal method can synthesize nanostructured SnS_2 composites, effectively reducing the diffusion path of Na^+, thereby improving the magnification and cycling properties of the materials. In addition, Zhang et al. [114] used the solvothermal method to obtain a thin layer of SnS_2/graphene material with remarkable high magnification performance, which still shows a reversible specific capacity of 330 mAh g^{-1} at 12.8 A g^{-1} current density. Guan et al. [115] prepared SnS_2/MWNTs nanocomposites with microporous structure, with the initial capacity of 910 mAh g^{-1} at the current density of 100 mA g^{-1}, and the capacity retention rate can reach 78% after 100 cycles.

6.3.4.4 Nickel Sulfide, Ferrous Sulfide and Their Composites

Ferrous sulfide is the representative of nonlayered sulfide. As the negative electrode material of SIB, FeS has the advantages of rich resources, low price and high theoretical specific capacity on the one hand, but its volume change during the electrochemical cycle reaches 200%, resulting in poor cycle performance. The researchers improved their electrochemical performance by controlling particle size and morphology. For example, Zhang et al. [116] prepared ultrathin carbon coated FeS microspheres in situ, which can effectively improve the volume expansion in the electrochemical process and have good cycling performance; The material also shows good low-temperature performance. At −25 °C, 0.05 A g^{-1} current density, the reversible capacity is 311 mAh g^{-1}. Wu et al. [117] reported that FeS/C nanocomposites can accelerate the electrochemical reaction kinetics. Under the current densities of 0.05, 35and 80 A g^{-1}, the reversible capacities of the nanocomposites are 547.1, 206.2 and 60.4 mAh g^{-1}, respectively, showing a good

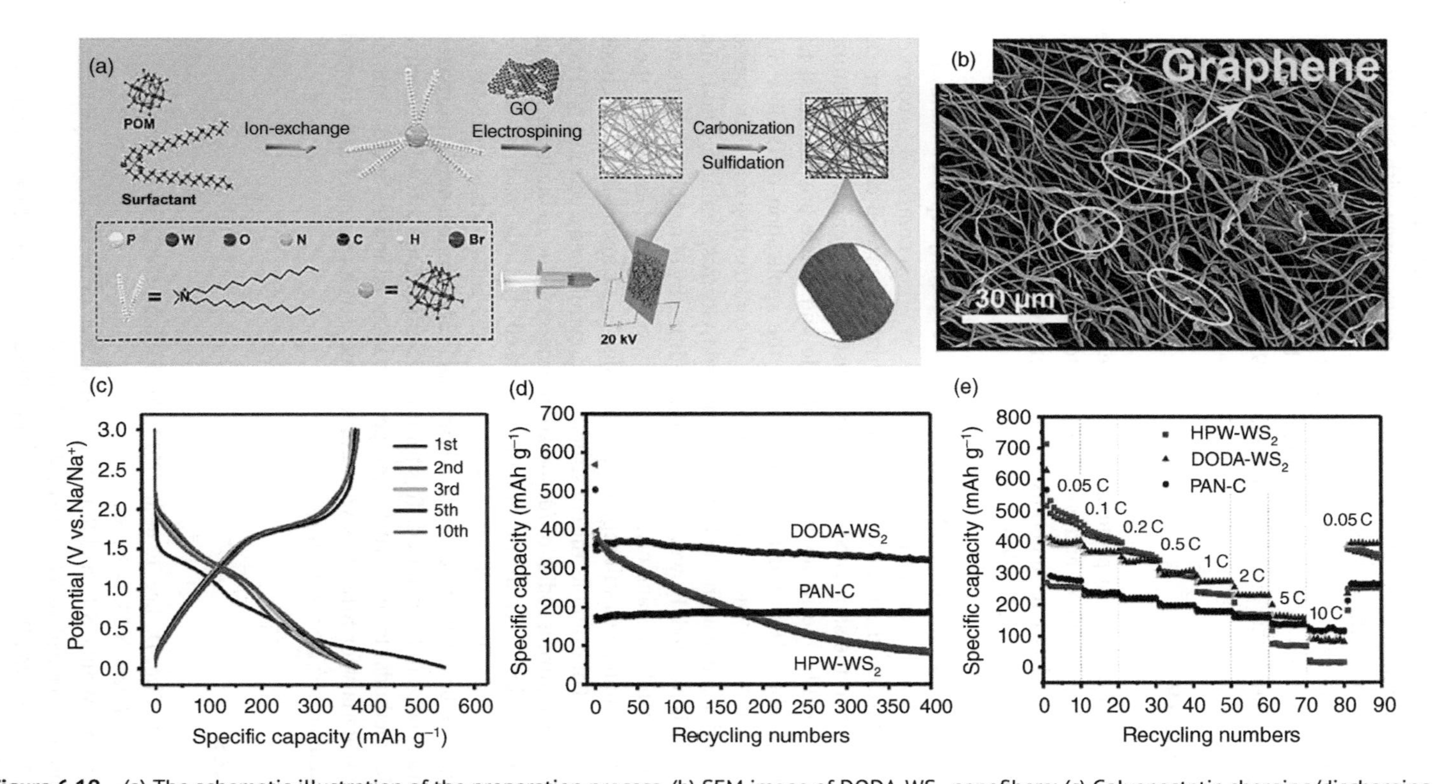

Figure 6.19 (a) The schematic illustration of the preparation process, (b) SEM image of DODA-WS_2 nanofibers; (c) Galvanostatic charging/discharging curves of DODA-WS_2 electrode at a current density of 200 m Ag^{-1}, (d) cycling performance comparison, and (e) rate performances of DODA-WS_2, HPW-WS_2 and PAN-C electrodes. Source: Li et al. [111] / with permission of Elsevier.

rate sodium storage performance. Similar to FeS, nickel-based sulfide (NiS_x) has also been widely reported as a negative electrode material for SIBs. For example, Jiang [118] and others prepared carbon coated NiS_x composites doped with N/S, which can effectively improve the electron and ion transport properties of the materials by using thiourea as raw materials. The reversible capacity under $2\,A\,g^{-1}$ current density can reach 338.4 $mAh\,g^{-1}$, which can still be maintained at 201 $mAh\,g^{-1}$ after 300 cycles, and the capacity retention rate is 99.5%.

6.4 Summary

SIBs are widely considered a promising candidate for large-scale energy storage devices because their low-cost and widely distributed sodium resources. However, the lack of proper electrode materials due to poor electrochemical performance hinders the practical application and development of SIBs. In this chapter, the nanowire-structured strategy is elucidated as an effective approach to the performance optimization of SIBs. We summarize the recently reported nanowires as cathode and anode materials for SIBs, respectively. When compared to conventional bulk electrodes, the nanowire structure exhibits improved electrochemical performance due to the shortened ion diffusion distance and enhanced structural stability. Additionally, the nanowire-based cathode and anode are beneficial for constructing flexible SIBs, which could broaden the application of flexible SIB devices. There are many typical synthesis methods for constructing nanowire-based electrode materials, such as the hydrothermal method, electrospinning method, and polymer-pyrolysis method. However, the synthesis difficulty of nanowires is related to the synthesis conditions and crystal structure of specific materials. Despite these advances, many challenges still remain in the design of nanowire-based electrode materials for SIBs. For example, the constructed nanowire-based electrodes still face structural damage during sodiation/desodiation processes due to the large volume changes. The reaction mechanisms of nanowire-based electrode materials need to be further systematically investigated. Additionally, the constructing methods for current nanowire-based electrode materials are single and limited; new synthesized methods are needed to be explored. To sum up, it is important to develop effective and simple methods to construct high-performance nanowire-based cathode and anode materials for SIBs; this will promote the development of SIBs toward flexibility and large-scale practicality.

References

1 Slater, M.D., Kim, D., Lee, E. et al. (2013). Sodium-ion batteries. *Advanced Functional Materials* 23: 947–958.
2 Hwang, J.Y., Myung, S.T., and Sun, Y.K. (2017). Sodium-ion batteries: present and future. *Chemical Society Reviews* 46: 3529–3614.
3 Chen, L., Fiore, M., Wang, J.E. et al. (2018). Readiness level of sodium-ion battery technology: a materials review. *Advanced Sustainable Systems* 2: 1700153.

4 Oshima, T., Kajita, M., and Okuno, A. (2004). Development of sodium-sulfur batteries. *International Journal of Applied Ceramic Technology* 1: 269–276.

5 Komaba, S. and Kubota, K. (2021). Layered $NaMO_2$ for the positive electrode. *Na-ion Batter* 1–46.

6 Stevens, D.A. and Dahn, J.R. (2000). High capacity anode materials for rechargeable sodium-ion batteries. *Journal of The Electrochemical Society* 147: 1271.

7 Jiang, M., Sun, N., Soomro, R.A. et al. (2021). The recent progress of pitch-based carbon anodes in sodium-ion batteries. *Journal of Energy Chemistry* 55: 34–47.

8 Yabuuchi, N., Kubota, K., Dahbi, M. et al. (2014). Research development on sodium-ion batteries. *Chemical Reviews* 114: 11636–11682.

9 Zhou, Y., Zhang, Z., Zhao, Y. et al. (2021). Cost-effective, long-term aqueous rechargeable hybrid sodium/zinc batteries based on Zn anode and $Na_3MnTi(PO_4)_3$ cathode. *Chemical Engineering Journal* 425: 130459.

10 Zhao, Y., Gao, X., Gao, H. et al. (2021). Elevating energy density for sodium-ion batteries through multielectron reactions. *Nano Letters* 21: 2281–2287.

11 Zhang, J., Lin, C., Xia, Q. et al. (2021). Improved performance of $Na_3TiMn(PO_4)_3$ using a non-stoichiometric synthesis strategy. *ACS Energy Letters* 6: 2081–2089.

12 Wu, C., Tong, J., Gao, J. et al. (2021). Studies on the sodium storage performances of $Na_3Al_xV_{2-x}(PO_4)_3$@C composites from calculations and experimental analysis. *ACS Applied Energy Materials* 4: 1120–1129.

13 Wang, S., Wang, G., He, C. et al. (2021). Enabling superior hybrid capacitive deionization performance in NASICON-structured $Na_3MnTi(PO_4)_3$/C by incorporating a two-species redox reaction. *Journal of Materials Chemistry A* 9: 6898–6904.

14 Singh, B., Wang, Z., Park, S. et al. (2021). A chemical map of NaSICON electrode materials for sodium-ion batteries. *Journal of Materials Chemistry A* 9: 281–292.

15 Shen, X., Zhou, Q., Han, M. et al. (2021). Rapid mechanochemical synthesis of polyanionic cathode with improved electrochemical performance for Na-ion batteries.Nature. *Communication* 12: 2848.

16 Rajagopalan, R., Zhang, Z., Tang, Y. et al. (2021). Understanding crystal structures, ion diffusion mechanisms and sodium storage behaviors of NASICON materials. *Energy Storage Materials* 34: 171–193.

17 Park, S., Chotard, J.-N., Carlier, D. et al. (2021). Crystal structures and local environments of NASICON-type Na3FeV(PO4)3 and Na4FeV(PO4)3 positive electrode materials for Na-ion batteries. *Chemistry of Materials* 33: 5355–5367.

18 Manna, S.C., Sandineni, P., and Choudhury, A. (2021). Low temperature hydrothermal synthesis of Na3Fe2(PO4)2F3 and its cathode electrochemistry in Na- and Li-ion batteries. *Journal of Solid State Chemistry* 295: 121922.

19 Ma, X., Wu, X., Liu, Y. et al. (2021). Toward a high-energy-density cathode with enhanced temperature adaptability for sodium-ion batteries: a case study of $Na_3MnZr(PO_4)_3$ microspheres with embedded dual-carbon networks. *ACS Applied Materials & Interfaces* 13: 21390–21400.

20 Liu, J., Lin, K., Zhao, Y. et al. (2021). Exceeding three-electron reactions in $Na_{3+2x}Mn_{1+x}Ti_{1-x}(PO_4)_3$ NASICON cathodes with high energy density for sodium-ion batteries. *Journal of Materials Chemistry A* 9: 10437–10446.

21 Li, X., Jiang, S., Li, S. et al. (2021). Overcoming the rate-determining kinetics of the $Na_3V_2O_2(PO_4)_2F$ cathode for ultrafast sodium storage by heterostructured dual-carbon decoration. *Journal of Materials Chemistry A* 9: 11827–11838.

22 Lee, J., Park, S., Park, Y. et al. (2021). Chromium doping into NASICON-structured $Na_3V_2(PO_4)_3$ cathode for high-power Na-ion batteries. *Chemical Engineering Journal* 422: 130052.

23 Lavela, P., Klee, R., and Tirado, J.L. (2021). On the benefits of Cr substitution on $Na_4MnV(PO_4)_3$ to improve the high voltage performance as cathode for sodium-ion batteries. *Journal of Power Sources* 495: 229811.

24 Klee, R., Lavela, P., and Tirado, J.L. (2021). Effect of the Mn/V ratio to optimize the kinetic properties of $Na_{3+x}Mn_xV_{1-x}Cr(PO_4)_3$ positive electrode for sodium-ion batteries. *Electrochimica Acta* 375: 137982.

25 Hou, J., Hadouchi, M., Sui, L. et al. (2021). Unlocking fast and reversible sodium intercalation in NASICON $Na_4MnV(PO4)_3$ by fluorine substitution. *Energy Storage Materials* 42: 307–316.

26 Han, K., An, F., Yan, F. et al. (2021). High-performance aqueous Zn–MnO_2 batteries enabled by the coupling engineering of K^+ pre-intercalation and oxygen defects. *Journal of Materials Chemistry A* 9: 15637–15647.

27 Hadouchi, M., Yaqoob, N., Kaghazchi, P. et al. (2021). Fast sodium intercalation in $Na_{3.41}£_{0.59}FeV(PO_4)_3$: a novel sodium-deficient NASICON cathode for sodium-ion batteries. *Energy Storage Materials* 35: 192–202.

28 Gu, Z.-Y., Guo, J.-Z., Sun, Z.-H. et al. (2021). Air/water/temperature-stable cathode for all-climate sodium-ion batteries. *Cell Reports Physical Science* 2: 100665.

29 Cui, G., Dong, Q., Wang, Z. et al. (2021). Achieving highly reversible and fast sodium storage of $Na_4VMn(PO_4)_3$/C-rGO composite with low-fraction rGO via spray-drying technique. *Nano Energy* 89: 106462.

30 Chen, M., Zhang, Y., Xing, G. et al. (2021). Electrochemical energy storage devices working in extreme conditions. *Energy & Environmental Science* 14: 3323–3351.

31 Chen, M., Hua, W., Xiao, J. et al. (2021). Activating a multielectron reaction of NASICON-structured cathodes toward high energy density for sodium-ion batteries. *Journal of the American Chemical Society* 143: 18091–18102.

32 Cao, Y., Yang, C., Liu, Y. et al. (2020). A new polyanion $Na_3Fe_2(PO_4)P_2O_7$ cathode with high electrochemical performance for sodium-ion batteries. *ACS Energy Letters* 5: 3788–3796.

33 Chen, M., Hua, W., Xiao, J. et al. (2019). NASICON-type air-stable and all-climate cathode for sodium-ion batteries with low cost and high-power density. *Nature Communications* 10: 1480.

34 Guo, J.Z., Wang, P.F., Wu, X.L. et al. (2017). High-energy/power and low-temperature cathode for sodium-ion batteries: in situ XRD study and superior full-cell performance. *Advanced Materials* 29: 1701968.

35 Deng, C. and Zhang, S. (2014). 1D nanostructured $Na_7V_4(P_2O_7)_4(PO_4)$ as high-potential and superior-performance cathode material for sodium-ion batteries. *ACS Applied Materials & Interfaces* 6: 9111–9117.

36 Kang, J., Park, H., Kim, H. et al. (2019). Development of a new mixed-polyanion cathode with superior electrochemical performances for Na-ion batteries. *ACS Sustainable Chemistry & Engineering* 8: 163–171.

37 Zhang, W., Wu, Y., Xu, Z. et al. (2022). Rationally designed sodium chromium vanadium phosphate cathodes with multi-electron reaction for fast-charging sodium-ion batteries. *Advanced Energy Materials* 12: 2201065.

38 Meng, Q., Lu, Y., Ding, F. et al. (2019). Tuning the closed pore structure of hard carbons with the highest Na storage capacity. *ACS Energy Letters* 4: 2608–2612.

39 Kim, H., Park, I., Lee, S. et al. (2013). Understanding the electrochemical mechanism of the new iron-based mixed-phosphate $Na_4Fe_3(PO_4)_2(P_2O_7)$ in a Na rechargeable battery. *Chemistry of Materials* 25: 3614–3622.

40 Li, H., Guan, C., Zhang, J. et al. (2022). Robust artificial interphases constructed by a versatile protein-based binder for high-voltage Na-ion battery cathodes. *Advanced Materials* 34: e2202624.

41 Li, Y., Lu, Y., Meng, Q. et al. (2019). Regulating pore structure of hierarchical porous waste cork-derived hard carbon anode for enhanced na storage performance. *Advanced Energy Materials* 9: 1902852.

42 Lu, Y., Zhao, C., Qi, X. et al. (2018). Pre-oxidation-tuned microstructures of carbon anodes derived from pitch for enhancing Na storage performance. *Advanced Energy Materials* 8: 1800108.

43 Wang, Y., Yu, X., Xu, S. et al. (2013). A zero-strain layered metal oxide as the negative electrode for long-life sodium-ion batteries. *Nature Communications* 4: 2365.

44 Wu, F., Zhao, C., Chen, S. et al. (2018). Multi-electron reaction materials for sodium-based batteries. *Materials Today* 21: 960–973.

45 Xu, C., Zhao, J., Wang, Y.A. et al. (2022). Reversible activation of V^{4+}/V^{5+} redox couples in NASICON phosphate cathodes. *Advanced Energy Materials* 12: 2200966.

46 Yuan, Y., Wei, Q., Yang, S. et al. (2022). Towards high-performance phosphate-based polyanion-type materials for sodium-ion batteries. *Energy Storage Materials* 50: 760–782.

47 Zhao, A., Yuan, T., Li, P. et al. (2022). A novel Fe-defect induced pure-phase $Na_4Fe_{2.91}(PO_4)_2P_2O_7$ cathode material with high capacity and ultra-long lifetime for low-cost sodium-ion batteries. *Nano Energy* 91: 106680.

48 Zhu, T., Hu, P., Cai, C. et al. (2020). Dual carbon decorated $Na_3MnTi(PO_4)_3$: a high-energy-density cathode material for sodium-ion batteries. *Nano Energy* 70: 104548.

49 Senthilkumar, B., Murugesan, C., Sada, K. et al. (2020). Electrochemical insertion of potassium ions in $Na_4Fe_3(PO_4)_2P_2O_7$ mixed phosphate. *Journal of Power Sources* 480: 228794.

50 Tang, L., Liu, X., Li, Z. et al. (2019). CNT-decorated $Na_4Mn_2Co(PO_4)_2P_2O_7$ microspheres as a novel high-voltage cathode material for sodium-ion batteries. *ACS Applied Materials & Interfaces* 11: 27813–27822.

51 Au, H., Alptekin, H., Jensen, A.C.S. et al. (2020). A revised mechanistic model for sodium insertion in hard carbons. *Energy & Environmental Science* 13: 3469–3479.

52 Kim, H., Yoon, G., Park, I. et al. (2015). Anomalous Jahn–Teller behavior in a manganese-based mixed-phosphate cathode for sodium ion batteries. *Energy & Environmental Science* 8: 3325–3335.

53 Liang, L., Sun, X., Denis, D.K. et al. (2019). Ultralong layered $NaCrO_2$ nanowires: a competitive wide-temperature-operating cathode for extraordinary high-rate sodium-ion batteries. *ACS Applied Materials & Interfaces* 11: 4037–4046.

54 Liu, Y., Shen, Q., Zhao, X. et al. (2019). Hierarchical engineering of porous P_2-$Na_{2/3}Ni_{1/3}Mn_{2/3}O_2$ nanofibers assembled by nanoparticles enables superior sodium-ion storage cathodes. *Advanced Functional Materials* 30: 1907837.

55 Shen, Q., Zhao, X., Liu, Y. et al. (2020). Dual-strategy of cation-doping and nanoengineering enables fast and stable sodium-ion storage in a novel Fe/Mn-based layered oxide cathode. *Advancement of Science* 7: 2002199.

56 He, H., Jin, G., Wang, H. et al. (2014). Annealed NaV_3O_8 nanowires with good cycling stability as a novel cathode for Na-ion batteries. *Journal of Materials Chemistry A* 2: 3563.

57 Li, Y. and Wu, Y. (2010). Formation of $Na_{0.44}MnO_2$ nanowires via stress-induced splitting of birnessite nanosheets. *Nano Research* 2: 54–60.

58 Hosono, E., Saito, T., Hoshino, J. et al. (2012). High power Na-ion rechargeable battery with single-crystalline $Na_{0.44}MnO_2$ nanowire electrode. *Journal of Power Sources* 217: 43–46.

59 Cao, Y., Xiao, L., Wang, W. et al. (2011). Reversible sodium ion insertion in single crystalline manganese oxide nanowires with long cycle life. *Advanced Materials* 23: 3155–3160.

60 Jin, T., Han, Q., Wang, Y. et al. (2018). 1D Nanomaterials: design, synthesis, and applications in sodium–ion batteries. *Small* 14: 1703086.

61 Jiang, Y., Yao, Y., Shi, J. et al. (2016). One-dimensional $Na_3V_2(PO_4)_3$/C nanowires as cathode materials for long-life and high rate Na-ion batteries. *ChemNanoMat* 2: 726–731.

62 Wang, X., Niu, C., Meng, J. et al. (2015). Novel $K_3V_2(PO_4)_3$/C bundled nanowires as superior sodium-ion battery electrode with ultrahigh cycling stability. *Advanced Energy Materials* 5: 1500716.

63 Peng, M., Li, B., Yan, H. et al. (2015). Ruthenium-oxide-coated sodium vanadium fluorophosphate nanowires as high-power cathode materials for sodium-ion batteries. *Angewandte Chemie, International Edition* 54: 6452–6456.

64 Chen, Y., Fan, Q., Li, J. et al. (2022). Preparation and electrochemical performance of nanowire-shaped $Na_3Mn_{2-x}Fe_x(P_2O_7)(PO_4)$ for sodium-ion and lithium-ion batteries. *Dalton Transactions* 51: 4173–4181.

65 Wu, C., Kopold, P., Ding, Y.L. et al. (2015). Synthesizing porous NaTi2(PO4)3 nanoparticles embedded in 3D graphene networks for high-rate and long cycle-life sodium electrodes. *ACS Nano* 9: 6610–6618.

66 Cao, Y., Xiao, L., Sushko, M.L. et al. (2012). Sodium ion insertion in hollow carbon nanowires for battery applications. *Nano Letters* 12: 3783–3787.

67 Anji Reddy, M., Helen, M., Groß, A. et al. (2018). Insight into sodium insertion and the storage mechanism in hard carbon. *ACS Energy Letters* 3: 2851–2857.

68 Kim, H., Yoon, G., Park, I. et al. (2016). Highly stable iron- and manganese-based cathodes for long-lasting sodium rechargeable batteries. *Chemistry of Materials* 28: 7241–7249.

69 Li, Y., Hu, Y.-S., Titirici, M.-M. et al. (2016). Hard carbon microtubes made from renewable cotton as high-performance anode material for sodium-ion batteries. *Advanced Energy Materials* 6: 1600659.

70 Liu, J., Huang, Y., Zhao, Z. et al. (2021). Yeast template-derived multielectron reaction NASICON structure $Na_3MnTi(PO_4)_3$ for high-performance sodium-ion batteries. *ACS Applied Materials & Interfaces* 13: 58585–58595.

71 Luo, W., Schardt, J., Bommier, C. et al. (2013). Carbon nanofibers derived from cellulose nanofibers as a long-life anode material for rechargeable sodium-ion batteries. *Journal of Materials Chemistry A* 1: 10662.

72 Sheng, J., Peng, C., Xu, Y. et al. (2017). $KTi_2(PO_4)_3$ with large ion diffusion channel for high-efficiency sodium storage. *Advanced Energy Materials* 7.

73 Sun, N., Guan, Z., Liu, Y. et al. (2019). Extended "adsorption–insertion" model: a new insight into the sodium storage mechanism of hard carbons. *Advanced Energy Materials* 9: 1901351.

74 Cho, J.S., Lee, J.K., and Kang, Y.C. (2016). Graphitic carbon-coated $FeSe_2$ hollow nanosphere-decorated reduced graphene oxide hybrid nanofibers as an efficient anode material for sodium ion batteries. *Scientific Reports* 6: 23699.

75 Zhu, Q., Nan, B., Shi, Y. et al. (2017). $Na_3V_2(PO_4)_3$/C nanofiber bifunction as anode and cathode materials for sodium-ion batteries. *Journal of Solid State Electrochemistry* 21: 2985–2995.

76 Zhu, T., Hu, P., Wang, X. et al. (2019). Realizing three-electron redox reactions in NASICON-structured $Na_3MnTi(PO_4)_3$ for sodium-ion batteries. *Advanced Energy Materials* 9: 1803436.

77 Hu, P., Zhu, T., Cai, C. et al. (2022). Sodium ion storage in $Na_4MnV(PO_4)$3@C free-standing electrode. *Advanced Functional Materials* 32: 2208051.

78 Li, W.-J., Chou, S.-L., Wang, J.-Z. et al. (2013). Simply mixed commercial red phosphorus and carbon nanotube composite with exceptionally reversible sodium-ion storage. *Nano Letters* 13: 5480–5484.

79 Wang, J.W., Liu, X.H., Mao, S.X. et al. (2012). Microstructural evolution of tin nanoparticles during in situ sodium insertion and extraction. *Nano Letters* 12: 5897–5902.

80 Darwiche, A., Marino, C., Sougrati, M.T. et al. (2012). Better cycling performances of bulk Sb in Na-ion batteries compared to Li-ion systems: an unexpected electrochemical mechanism. *Journal of the American Chemical Society* 134: 20805–20811.

81 Xu, Y., Zhu, Y., Liu, Y. et al. (2013). Electrochemical performance of porous carbon/tin composite anodes for sodium-ion and lithium-ion batteries. *Advanced Energy Materials* 3: 128–133.

82 Ko, Y.N. and Kang, Y.C. (2014). Electrochemical properties of ultrafine Sb nanocrystals embedded in carbon microspheres for use as Na-ion battery anode materials. *Chemical Communications* 50: 12322–12324.

83 Zhang, C., Wang, X., Liang, Q. et al. (2016). Amorphous phosphorus/nitrogen-doped graphene paper for ultrastable sodium-ion batteries. *Nano Letters* 16: 2054–2060.

84 Liu, Y., Xu, Y., Zhu, Y. et al. (2013). Tin-coated viral nanoforests as sodium-ion battery anodes. *ACS Nano* 7: 3627–3634.

85 Baggetto, L., Ganesh, P., Sun, C.-N. et al. (2013). Intrinsic thermodynamic and kinetic properties of Sb electrodes for Li-ion and Na-ion batteries: experiment and theory. *Journal of Materials Chemistry A* 1: 7985–7994.

86 Luo, W., Ren, J., Feng, W. et al. (2021). Engineering nanostructured antimony-based anode materials for sodium ion batteries. *Coatings* 11 (10).

87 Liang, L., Xu, Y., Wang, C. et al. (2015). Large-scale highly ordered Sb nanorod array anodes with high capacity and rate capability for sodium-ion batteries. *Energy & Environmental Science* 8: 2954–2962.

88 Zhu, Y., Han, X., Xu, Y. et al. (2013). Electrospun Sb/C fibers for a stable and fast sodium-ion battery anode. *ACS Nano* 7: 6378–6386.

89 Liu, Z., Yu, X.-Y., Lou, X.W. et al. (2016). Sb@C coaxial nanotubes as a superior long-life and high-rate anode for sodium ion batteries. *Energy & Environmental Science* 9: 2314–2318.

90 Liu, S., Feng, J., Bian, X. et al. (2016). The morphology-controlled synthesis of a nanoporous-antimony anode for high-performance sodium-ion batteries. *Energy & Environmental Science* 9: 1229–1236.

91 Zhou, G., Xu, L., Hu, G. et al. (2019). Nanowires for electrochemical energy storage. *Chemical Reviews* 119: 11042–11109.

92 Wu, L., Hu, X., Qian, J. et al. (2014). Sb–C nanofibers with long cycle life as an anode material for high-performance sodium-ion batteries. *Energy & Environmental Science* 7: 323–328.

93 Li, Q., Zhang, W., Peng, J. et al. (2021). Metal–organic framework derived ultrafine Sb@Porous carbon octahedron via in situ substitution for high-performance sodium-ion batteries. *ACS Nano* 15: 15104–15113.

94 Qiao, S., Zhou, Q., Ma, M. et al. (2023). Advanced anode materials for rechargeable sodium-ion batteries. *ACS Nano* 17: 11220–11252.

95 Zhang, X., Zhu, G., Yan, D. et al. (2017). MnO@C nanorods derived from metal-organic frameworks as anode for superiorly stable and long-life sodium-ion batteries. *Journal of Alloys and Compounds* 710: 575–580.

96 Wang, Z., Yan, X., Wang, F. et al. (2021) Reduced graphene oxide thin layer induced lattice distortion in high crystalline MnO_2 nanowires for high-performance sodium-and potassium-ion batteries and capacitors. *Carbon* 174: 556–566.

97 Devaraj, S. and Munichandraiah, N. (2008). Effect of crystallographic structure of MnO_2 on its electrochemical capacitance properties. *Journal of Physical Chemistry C* 112: 4406–4417.

98 Su, D., Ahn, H.-J., and Wang, G. (2013). Hydrothermal synthesis of α-MnO_2 and β-MnO_2 nanorods as high capacity cathode materials for sodium ion batteries. *Journal of Materials Chemistry A* 1: 4845–4850.

99 Su, D., Ahn, H.-J., and Wang, G. (2013). β-MnO_2 nanorods with exposed tunnel structures as high-performance cathode materials for sodium-ion batteries. *NPG Asia Materials* 5: 70.

100 Wang, L., Zhang, K., Hu, Z. et al. (2014). Porous CuO nanowires as the anode of rechargeable Na-ion batteries. *Nano Research* 7: 199–208.

101 Yuan, S., Huang, X.-l., Ma, D.-l. et al. (2014). Engraving copper foil to give large-scale binder-free porous CuO arrays for a high-performance sodium-ion battery anode. *Advanced Materials* 26: 2273–2279.

102 Wang, X., Liu, Y., Wang, Y. et al. (2016). CuO quantum dots embedded in carbon nanofibers as binder-free anode for sodium ion batteries with enhanced properties. *Small* 12: 4865–4872.

103 Ni, J., Jiang, Y., Wu, F. et al. (2018). Regulation of breathing CuO nanoarray electrodes for enhanced electrochemical sodium storage. *Advanced Functional Materials* 28: 1707179.

104 Zhao, W., Ci, S., Hu, X. et al. (2019). Highly dispersed ultrasmall NiS_2 nanoparticles in porous carbon nanofiber anodes for sodium ion batteries. *Nanoscale* 11: 4688–4695.

105 Zhu, T., Chen, J.S., and Lou, X.W. (2011). Glucose-assisted one-pot synthesis of FeOOH nanorods and their transformation to Fe_3O_4@ carbon nanorods for application in lithium ion batteries. *Journal of Physical Chemistry C* 115: 9814–9820.

106 Kang, W., Wang, Y., and Xu, J. (2017). Recent progress in layered metal dichalcogenide nanostructures as electrodes for high-performance sodium-ion batteries. *Journal of Materials Chemistry A* 5: 7667–7690.

107 Tang, Y., Zhao, Z., Wang, Y. et al. (2017). Ultrasmall MoS, nanosheets mosaiced into nitrogen-doped hierarchical porous carbon matrix for enhanced sodium storage performance. *Electrochimica Acta* 225: 369–377.

108 Xiong, X., Luo, W., Hu, X. et al. (2015). Flexible membranes of MoS_2/C nanofibers by electrospinning as binder-free anodes for high-performance sodium-ion batteries. *Scientific Reports* 5: 9254.

109 Zeng, X., Ding, Z., Ma, C. et al. (2016). Hierarchical nanocomposite of hollow N-doped carbon spheres decorated with ultrathin WS_2 nanosheets for high-performance lithium-ion battery anode. *ACS Applied Materials & Interfaces* 8: 18841–18848.

110 Liu, Y., Zhang, N., Kang, H. et al. (2015). WS_2 nanowires as a high-performance anode for sodium-ion batteries. *Chemistry: A European Journal* 21: 11878–11884.

111 Xu, X., Li, X., Zhang, J. et al. (2019). Surfactant assisted electrospinning of WS_2 nanofibers and its promising performance as anode material of sodium-ion batteries. *Electrochimica Acta* 302: 259–269.

112 Yu, X.Y., Yu, L., and Lou, X.W. (2016). Metal sulfide hollow nanostructures for electrochemical energy storage. *Advanced Energy Materials* 6: 1501333.

113 Prikhodchenko, P.V., Denis, Y., Batabyal, S.K. et al. (2014). Nanocrystalline tin disulfide coating of reduced graphene oxide produced by the peroxostannate deposition route for sodium ion battery anodes. *Journal of Materials Chemistry A* 2: 8431–8437.

114 Zhang, Y., Zhu, P., Huang, L. et al. (2015). Few-layered SnS_2 on few-layered reduced graphene oxide as Na-ion battery anode with ultralong cycle life and superior rate capability. *Advanced Functional Materials* 25: 481–489.

115 Zhao, Y., Guo, B., Yao, Q. et al. (2018). A rational microstructure design of SnS_2–carbon composites for superior sodium storage performance. *Nanoscale* 10: 7999–8008.

116 Fan, H.-H., Li, H.-H., Guo, J.-Z. et al. (2018). Target construction of ultrathin graphitic carbon encapsulated FeS hierarchical microspheres featuring superior low-temperature lithium/sodium storage properties. *Journal of Materials Chemistry A* 6: 7997–8005.

117 Hou, B.-H., Wang, Y.-Y., Guo, J.-Z. et al. (2018). Pseudocapacitance-boosted ultrafast Na storage in a pie-like FeS@ C nanohybrid as an advanced anode material for sodium-ion full batteries. *Nanoscale* 10: 9218–9225.

118 Tao, H., Zhou, M., Wang, K. et al. (2018). N/S co-doped carbon coated nickel sulfide as a cycle-stable anode for high performance sodium-ion batteries. *Journal of Alloys and Compounds* 754: 199–206.

7

Application of Nanowire Materials in Metal-Chalcogenide Battery

Rechargeable batteries are of great significance for solving the current prominent energy and environmental problems. At present, traditional lithium-ion batteries occupy a large share of the portable electronic equipment market. However, due to their high price, low energy density, and limited adaptability to the environment, their applications in electric vehicles and smart grids are not as wide as imagined. In order to achieve higher energy density, higher battery safety, and adapt to more stringent use environments, various new energy storage systems are being developed all over the world [1]. At present, research on new batteries is being carried out in the laboratory. Although there is still a long way to go before large-scale applications, the development of various new battery systems provides more imagination and possible space for exploration of new energy conversion and storage. This chapter will introduce the principles and applications of various new batteries from the perspective of nanowire electrode materials, and explain the characteristics and advantages of nanowire electrode materials in new batteries.

The idea of using sulfur as a cathode material for batteries was initially proposed by Herbert and Ulam in 1962 [2]. In a typical lithium–sulfur (Li–S) battery, the electrode reaction of the entire battery can be described as: $S + 2Li \leftrightarrow Li_2S$. The theoretical specific capacities corresponding to elemental sulfur and lithium metal are 1675 and 3861 mAh g^{-1}, and the average voltage corresponding to the full battery is 2.15 V. As a result, the energy density achieved by these advancements in Li–Se batteries is approximately 2500 Wh kg^{-1} and 2800 Wh l^{-1}, which is significantly higher than the energy density of commercialized lithium-ion battery systems [3].

Of course, there are also many technical problems in the commercialization of Li–S batteries. For example, sulfur and discharge product lithium sulfide's electronic conductivity is about 5×10^{-30} and 1×10^{-13} S cm^{-1}, respectively, and lithium sulfide is not easy to oxidize, which will lead to a low discharge rate. In addition, problems such as volume expansion and structural deterioration of discharge product (lithium sulfide) are also prominent. In recent years, the elemental selenium (Se) with the same main group as sulfur and the elemental sodium (Na) with the same main group as lithium have gradually entered the field of vision. The development

Nanowire Energy Storage Devices: Synthesis, Characterization, and Applications, First Edition.
Edited by Liqiang Mai.

trend of lithium–selenium (Li–S) batteries and sodium–sulfur (Na–S) batteries is also gradually rising, and their respective performance and cost advantages also make it expected to replace Li–S batteries in the future.

7.1 Lithium–Sulfur Battery

At present, the conventional Li–S battery under research is mainly composed of a composite material prepared from sulfur, conductive additives, and a binder as the cathode, and metallic lithium as the anode (Figure 7.1a). Elemental sulfur is generally composed of sulfur molecules composed of eight sulfur atoms. The discharge process of the corresponding sulfur element undergoes the following multistep reaction [5]:

$$S_8 + 2e^- \rightarrow S_8^{2-} \tag{7.1}$$

$$3S_8^{2-} + 2e^- \rightarrow 4S_6^{2-} \tag{7.2}$$

$$2S_6^{2-} + 2e^- \rightarrow 3S_4^{2-} \tag{7.3}$$

$$S_4^{2-} + 4Li^+ + 2e^- \rightarrow 2Li_2S_2 \tag{7.4}$$

$$Li_2S_2 + 2Li^+ + 2e^- \rightarrow 2Li_2S \tag{7.5}$$

The first three reaction steps correspond to the higher voltage plateau (2.15–2.4 V) in the charge–discharge curve in Figure 7.1b. The polysulfides obtained in these three reactions are all soluble liquid-phase polysulfides. The latter two reactions correspond to a lower and long voltage plateau (2.1 V), resulting in insoluble Li_2S_2 and Li_2S solids deposited on the surface of the cathode. The whole process of sulfur reduction is a "solid-liquid-solid" three-phase conversion process [4]. When the Li_2S is all over the electrode, the voltage will drop rapidly, leading to the termination of the final reaction.

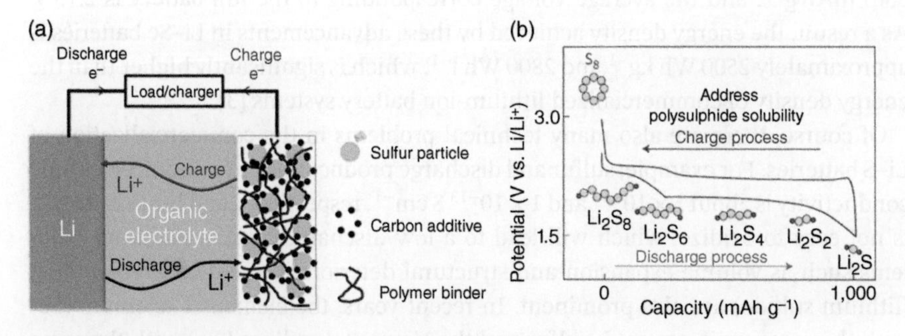

Figure 7.1 (a) Schematic diagram of lithium–sulfur battery structure; (b) Typical lithium-sulfur battery charge and discharge curve. Source: Manthiram et al. [4]/with permission of American Chemical Society.

While batteries offer high energy density and attractive raw material prices, there are still numerous challenges that must be addressed in order for them to meet the requirements of commercialization [1]. One major challenge is the significant volume change that occurs in sulfur during charging and discharging. The density of elemental sulfur and Li_2S is 2.03 and 1.66 g cm^{-3}. As a result of the significant volume change, the active material may become pulverized leading to a rapid attenuation of the final capacity. Second, sulfur and Li_2S are both electronic and ionic insulators. The electronic impedance of Li_2S is much higher than that of 1014 Ω cm. The diffusion rate of lithium ions in Li_2S is as low as 10–15 cm^2 S^{-1}. Once the thin layer of Li_2S is deposited on the whole electrode surface, further lithium intercalation will be seriously hindered, and the voltage will decrease sharply. Therefore, it is very difficult for sulfur to be completely converted to lithium sulfide, and the discharge capacity reported in most studies is less than 80% of the theoretical capacity. A third challenge is that during discharge, the intermediate discharge product, polysulfide, can migrate to the negative electrode where it may react with the negative metal lithium. In this way, the passivation of the negative electrode not only results in the loss of active material but also leads to an increase in impedance. Moreover, sulfur will change the morphology of the cathode material in the process of repeated dissolution and deposition, which will lead to stress in the electrode and shorten the cycle life of the battery. The fourth is the safety problem caused by the use of lithium metal in the negative electrode, and excessive use of lithium metal will lead to a decrease in energy density.

In order to solve the above challenges, the reasonable construction of electrode materials is particularly important. The construction of sulfur cathode materials should have the following requirements [6–8]: (i) enough space to accommodate the volume expansion of sulfur; (ii) fast ion and electron transfer rate and short transmission path; (iii) enough conductive surfaces for depositing Li_2S; (iv) effective physical and chemical restrictions of polysulfides; and (v) higher sulfur loading and utilization.

Sulfur presents in the form of irregular particles, which is difficult to make one-dimensional (1D) nanostructures, but it can be effectively loaded on nanowires to form 1D nanostructure composites. Up to now, the sulfur cathode materials based on nanowires can be divided into the following types according to the types of composites: (i) sulfur–carbon nanowire composites; (ii) sulfur/conductive polymer nanowire composites; and (iii)sulfur/metal compound nanowire composites. This chapter will introduce in detail the development of three kinds of nanowires constructed sulfur cathode materials.

7.1.1 Sulfur–Carbon Nanowire Composite Cathode Materials

When applied to the cathode of lithium–sulfur battery, nanowires have more continuous electron transport paths and structural advantages that can be braided compared to structures such as nanoparticles and nanosheets. The traditional sulfur cathode film is usually composed of active sulfur element, conductive carbon, and binder. Early research mainly used carbon fibers (CNFs), multi-walled carbon nanotubes (MWCNTs) and other highly conductive nanowires as conductive additives to be added to the sulfur cathode film to improve the rate performance of the sulfur cathode and prolong the cycle life of electrode materials [9]. The results show that using carbon nanowires to construct conductive network framework is more effective than using traditional 0D materials (Super P, carbon black, and acetylene black). However, it is difficult to obtain a uniformly dispersed MWCNT electrode through this traditional paste mixing coating method, and it is difficult to ensure good contact between the sulfur nanoparticles and CNFs or carbon nanotubes. In subsequent studies, high-temperature melting or chemical synthesis was often used to allow carbon nanotubes to cover the surface of the sulfur element well, so that effective contact between sulfur and carbon nanotubes could be obtained (Figure 7.2a) [10, 11]. High-temperature melting mainly refers to mixing elemental sulfur and conductive carbon matrix uniformly, sealing it in a reactor or crucible, heating up to 155–500 °C, and allowing elemental sulfur to melt or gasify through diffusion and adsorption into the pores and cavities of the conductive carbon matrix. In the process, a uniform sulfur–carbon composite is obtained. The main method of chemical synthesis is to dissolve polysulfides or thiosulfate in the dispersed aqueous solution of carbon matrix, and then add a certain concentration of acid to form sulfur elements in situ attached to the surface or pores of carbon matrix. The electron conduction capacity and magnification performance of the sulfur–carbon composite electrodes synthesized by these two methods have been effectively improved. However, their cycle life is always limited to 100 cycles of charge and discharge. This is because simple sulfur–carbon compounding makes it difficult to limit the shuttle effect and volume expansion. Cui et al. [12] synthesized a hollow carbon fiber array using anodized aluminum oxide (AAO) template and then melted and infiltrated elemental sulfur into the cavity of the carbon fiber array by capillary force at 155 °C to form a core-shell sulfur carbon nanowire structure (Figure 7.2b). This structure can physically limit the loss and overflow of liquid polysulfide to a certain extent, alleviate the shuttle effect and volume expansion, and extend the cycle life [17]. After assembling the button cell, the specific capacity remains at 730 mAh g^{-1} after 160 cycles at a current density of 0.2 C (1 C = 1675 mA g^{-1}).

Subsequent studies have focused on the inhibition of the dissolution and shuttling of polysulfides by surface modification or structural design of 1D carbon nanostructures. In terms of surface modification, the surface of conventional CNFs and CNTs is SP_3 or SP_2 hybrid C—C bond, which is difficult to chemically interact with polysulfides, so the dissolution of polysulfides can only be alleviated by the restriction of physical structure [18]. However, CNFs or CNTs doped with N, O, and S have suspended C—N, C—O, and C—S bonds, and the positive and negative charge centers of these valence bonds do not overlap, resulting in a dipole moment and showing a certain polarity. The surface with polar bonds can form Li–N, O, S bonds with the same polar polysulfide and play the effect of chemisorption. Some

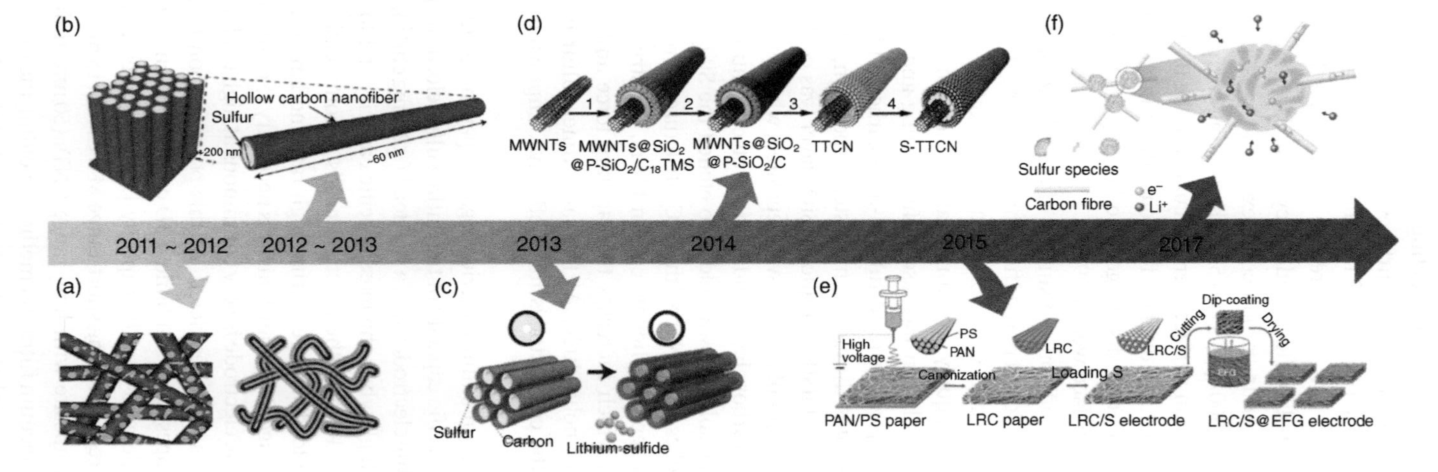

Figure 7.2 Development route of nanowire sulfur–carbon composite cathode materials. (a) Composite structure of simple sulfur carbon nanowires and nanotubes; Source: Refs. [10, 11]/with permission of Royal Society of Chemistry and Nature Publishing Group. (b) Sulfur carbon core–shell nanowires [12]. (c) The structure of sulfur carbon core–shell nanowires modified by organic matter; Source: Zheng et al. [13]/with permission of American Chemical Society. (d) Tube-in-tube carbon nanostructures as sulfur-fixing substrates; Source: Zhao et al. [14]/with permission of Wiley-VCH. (e) Multichannel ultralong carbon nanofibers/sulfur composite cathode films; Source: Li et al. [15]/with permission of Nature Publishing Group. (f) Unrestricted carbon fiber networks as sulfur-fixing substrates; Source: Pan et al. [16]/with permission of Nature Publishing Group. MWNTs: multi-walled carbon nanotubes; TTCN: Multilayer Carbon Nanotubes with Tube-in-Tube-Structure.

organics also have abundant suspended polar bonds, which means they can be used as modifiers to modify the surface of conducting carbon matrix. For example, Zheng et al. [13] modified the inner and outer surfaces of hollow CNFs with PVP, and then injected sulfur into them to obtain PVP modified sulfur carbon core–shell nanowire structure (Figure 7.2c). This structure can be used for both physical restriction and chemisorption of polysulfides, and the cycle life is greatly improved compared to that of unmodified sulfur–carbon complexes. The capacity retention rate is maintained above 80% after 300 cycles at 0.5 C. The binding energies between S, Li_2S_2, Li_2S, and graphitized carbon surface were calculated to be 0.79, 0.21, and 0.29 eV, respectively, indicating that it is difficult for elemental sulfur to chemisorb with graphitized carbon surface during discharge. The binding energies of Li_2S_2, Li_2S, and PVP modified carbon surface were 1.29 and 1.01 eV, respectively, which assisted to prove that the polar bonds on the surface of organic compounds promoted the adsorption and redeposition of polysulfide, thus mitigating its dissolution and transit effects. The delicate design of nanostructures is another way to limit the dissolution of polysulfides. Uniquely structured carbon nanowires with more pores or smaller confining channels can increase sulfur loading and better limit polysulfide loss. Zhao et al. [14] constructed tube-in-tube carbon (TTC) nanostructures as sulfur-fixing substrates through a multistep deposition, calcination, and etching method (Figure 7.2d). Firstly, a uniform SiO_2 layer was coated on the surface of MWCNTs, and then a layer of silicone was deposited, which was calcined to produce SiO_2 and pyrolytic carbon layer under the protective atmosphere of Argon. Then, sodium hydroxide solution was added to etched the SiO_2 layer, and the multilayer carbon nanotubes (TTCNs) with TTC structure were obtained. Compared with the previously reported hollow carbon fibers, the structure has a smaller 1D cavity (10–20 nm), which increases the capillary force and promotes the physical confining effect of polysulfide. At the same time, sulfur elemental and polysulfide contact with the conductor is more effective, improving the rate performance.

After 2014, researchers of lithium-sulfur batteries began to pay more attention to the loading capacity and area specific capacity of the entire sulfur electrode [19]. In earlier studies, most of the sulfur electrodes had a low load ($<2.0\,mg\,cm^{-2}$), but the area specific capacity was not high when the mass specific capacity and long cycle life were obtained. It is difficult to meet the practical needs of commercialization. Increasing the sulfur electrode load while ensuring high rate and long life performance is a very important challenge. To achieve this, it is necessary to ensure that the electron transport throughout the electrode is very continuous at high loads and that the dissolution of polysulfides is more restricted. At this point, carbon nanowires show unique structural advantages; they can build a 3D connected conductive network and improve the electrode area-specific capacity [20]. Zhang et al. Flexible thin film electrodes were prepared by mixing short-range MWCNTs (15 nm in diameter, 10–50 μm in length)/S composite with long-range CNTs (50 nm in diameter, 1000–2000 μm in length) using vacuum filtration method, avoiding the use of binder and conductive additives in traditional electrodes. It shows that the 3D conductive network constructed by the composite structure of short-range and long-range carbon nanotubes greatly promotes the rapid electron transport under high load. If the ultralong hollow carbon fiber with physical limiting effect is applied to the construction of 3D high-load sulfur electrode, better cyclic stability can be obtained. Lou et al. [15] synthesized an ultra-long nanofiber with multiple parallel hollow channels by electrospinning (Figure 7.2e). First, polyacrylonitrile (PAN) and polystyrene (PS) spheres were mixed, and the PAN/PS were prepared into nanowires interwoven organic films by electrospinning. Then, the PAN was annealed at high temperature and pyrolyzed to form carbon fiber, while the PS spheres in the middle were decomposed and escaped to form parallel 1D pores. Then the elemental sulfur was injected into the pores by high-temperature melting method. Finally, the film was immersed in the amine reduced graphene oxide (EFG) dispersion solution to form the graphene coating layer. The sulfur–carbon composite film obtained by this method has a sulfur content of 72.3%. Moreover, the loading capacity was gradually increased to 7.2 and 10.8 mg cm^{-2}, and the specific area capacity reached 7.2 and 10.7 mAh cm^{-2}, respectively. The excellent high load performance shows that the structure combines the domain-limiting effect of multichannel hollow nanofibers and the braiding property, realizes the construction of 3D continuous conductive framework, and well limits the dissolution and shuttle of polysulfide.

All the studies reviewed above pay more attention to the use of chemisorption or physical structure. Liu et al. recently proposed a different strategy [16]. They found that when the carbon fiber network was compounded with the high temperature molten sulfur. Because this structure will generate continuous Li_2S insulating layer in the discharge process, subsequent Li^+ is difficult to further diffuse into the internal polysulfide, the capacity is greatly lost, and in the charging process, Li_2S insulating layer is difficult to have effective conductive contact with the carbon fiber, so the formation of "dead sulfur," cycle performance decline. However, if the cathode electrolyte droplets formed by Li_2S_6 dissolved in ethylene glycol dimethyl ether (DME)/1,3-dioxy-penyl ring (DOL) are placed in the carbon fiber network and the discharge process is carried out, Li_2S will gather at the intersection points of several CNFs in the form of nano-flowers (Figure 7.2f). During the charging process, solid Li_2S can be effectively transferred into polysulfides and sulfur elementals. The specific capacity of the structure is up to the theoretical specific capacity when tested at the current density of 0.1 C, and the sulfur utilization rate is about 99%. This study shows that the diffusion and deposition of sulfur discharge products in the reaction process play an important role in the performance of lithium– sulfur batteries.

1D carbon nanostructures have unique advantages in the application of highly loaded sulfur cathode. Their high conductivity can provide efficient electron transport for insulating sulfur and lithium sulfide, and can greatly reduce the introduction of conductive additives and significantly increase sulfur content. However, it could not adsorb polysulfide well, which limits its further development. If the problem of cycle life can be effectively solved by regulating the deposition form of lithium sulfide, the commercial application of high-load carbon sulfur nanowire composite electrode will be a big step forward. Therefore, further structural design and experimental exploration are essential.

7.1.2 Conductive Polymer Nanowire/Sulfur Composite Cathode Materials

As mentioned in Section 7.1.1, organic materials have been widely studied as sulfides due to their abundant polar bonds on the surface. These include dilute hydrocarbons, thiols, nitriles, lipids, amines, thiophene, and other short- or long-chain polymers. Short-chain organic matter is usually added to the sulfur-carbon composite electrode sheet as additive, modifier, or binder to enhance the effect of chemical sulfur fixation [21]. In addition to being used as additives, some long-chain polymer materials are usually synthesized into nanoparticles, nanowires, nanotubes, and other structures as substrates for sulfur filling. However, because most of the organic matter is an insulator, it cannot achieve smooth electron transfer, so it is difficult to use as a sulfide substrate alone. Some conductive polymers can be combined with sulfur elemental substance for physical sulfur recombination and can also undergo chemical reactions to form sulfur-conductive polymer copolymers. Liu et al. [22] synthesized S-PANI copolymer by reaction of PANI nanotubes with elemental sulfur (Figure 7.3a). At 280 °C, PANI chain can undergo dehydrogenation and hybridization reaction with elemental sulfur to form

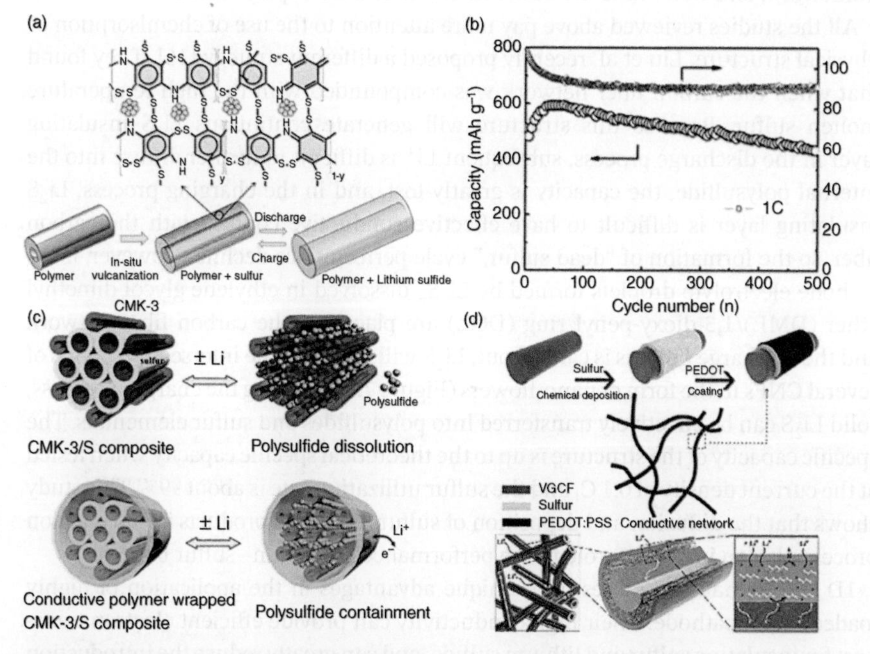

Figure 7.3 (a) Schematic illustration of the construction and discharge/charge process of the SPANI- NT/S composite. (b) Prolonged cycling performance of the electrode up to 500 cycles at 1 C rate; Source: Xiao et al. [22]/with permission of Wiley-VCH. (c) Schematic of PEDOT:PASS-coated CMK-3/sulfur composite for improving the cathode performance; Source: Reproduced with permission Yang et al. [23]/with permission of American Chemical Society. (d) Schematic illustration for the fabrication of CF/S/PEDOT composite; Source: He et al. [24]/with permission of American Chemical Society. PEDOT:Poly, 3,4-ethylene dioxythiophene; PEDOT-PASS:Poly, 3,4-ethylene dioxythiophene-Sodium Polystyrene Sulfonate.

heterocyclic polymers connected by disulfide bonds, restricting elemental sulfur in polymer cavities and three-dimensional chain networks in both physical and chemical ways. During the discharge process, on the one hand, the polysulfide can form Li–N and Li–S bonds with N and S atoms, and on the other hand, the 3D polymer chain network slows down the loss of polysulfide, thus limiting the dissolution of polysulfide well. The capacity retention rate is 76% after 500 cycles at the current density of 1 C (Figure 7.3b), showing good cycling stability. However, compared with inorganic carbon matrix, the conductive polymer matrix still cannot compare with its higher conductivity. Another strategy is to combine the conductive polymer with a 1D inorganic carbon material to form a sulfide fixer with both high conductivity and chemical limitation. For example, Cui et al. at Stanford University [23] coated a layer of poly (3,4-ethylenedioxythiophene) -poly (sodium styrene sulfonate) (PEDOT:PSS) on the surface of 1D mesoporous carbon/sulfur (CMK-3/S) composite electrode. Compared with uncoated CMK-3/S, the composite exhibited better cycle stability while maintaining good rate performance (Figure 7.3c) Yang et al. [24] synthesized three layers of CNFS/S/PEDOT composite fiber cathode material through multiple deposition (Figure 7.3d). The innermost CNFs provide a 1D continuous conductive path for the intermediate sulfur elements, while the outermost PEDOT physically limits and chemically adsorbs the dissolution shuttle of polysulfides. This multilayer fiber structure can also be used as a building block for self-supporting sulfur film anode, providing support for highly loaded sulfur anode.

Although polymers are difficult to achieve fine nanostructure design, they have good processability, flexibility, and stability, and subsequent studies mostly compound them with conductive carbon. At the same time, a variety of copolymers were formed by some high temperature sulfurization reactions with sulfur elementary to achieve the construction of high-performance composite electrode. However, the research on high loads still needs to be strengthened, and more research is needed on how to build a better conductive network after increasing the load. If a composite positive electrode with high load, long cycle, and high magnification performance can be prepared, and the processing and synthesis of polymer and carbon materials have the characteristics of low cost, simplicity, and low temperature, this kind of composite electrode will have great potential on the road to commercialization.

7.1.3 Metal Compound Nanowires/Sulfur Composite Cathode Materials

With the application of carbon materials and conducting polymers in sulfur-fixing substrates, the mass-specific capacity, rate performance and cycle stability of lithium–sulfur batteries have been improved. The sulfur content of the positive electrode is mostly between 40 and 65 wt% in this studies. At the current density of 0.1 C, the average level of the first discharge specific capacity reaches 900 mAh g^{-1}, and the capacity retention rate reaches more than 75% after 100 cycles. The number of cycles can reach about 200 times and the capacity retention rate is about 80% at 1 C. The volume energy density is about 800 Wh l^{-1}, the energy density is similar to that of $LiMO_2$-graphene (M is a transition metal) system [3]. To further improve the energy density, more efficient sulfur-fixing substrates need to be developed, with mass specific capacity approaching 1500 mAh g^{-1} and sulfur content increasing to 75%, while maintaining good cycling stability.

In 2013, metal oxide TiO_2 was developed into sulfur-fixing substrates, and a series of TiO_2 substrates with different structure and phases were reported, leading to a series of research upsurge in metal compound/sulfur composite cathode materials. This is because metallic compounds tend to form stronger absorption with polysulfide. α-TiO_2 and β-TiO_2 in TiO_2 can show better cycle stability. Later, it was found that titanium oxide with lower price could show stronger absorption abilities. This is mainly due to the fact that low-titanium ions are more likely to bond with a polysulfide ion in a polysulfide. The developed low-price titanium oxide matrix mainly includes Ti_nO_{2n-1} (Ti_4O_7), and TiO, Ti_6O_{13}. At the same time, these oxide substrates are usually compounded with carbon to ensure the electrical conductivity of the entire electrode. Li et al. [25] designed a "grape cluster" shaped TiO/C porous nanofibers as sulfur-fixing substrates by electrospinning (Figure 7.4a–e). They first coated SiO_2 nanospheres with a diameter of about 100 nm with a TiO_2 layer of about 10 nm (SiO_2@TiO_2) and then SiO_2@TiO_2 was

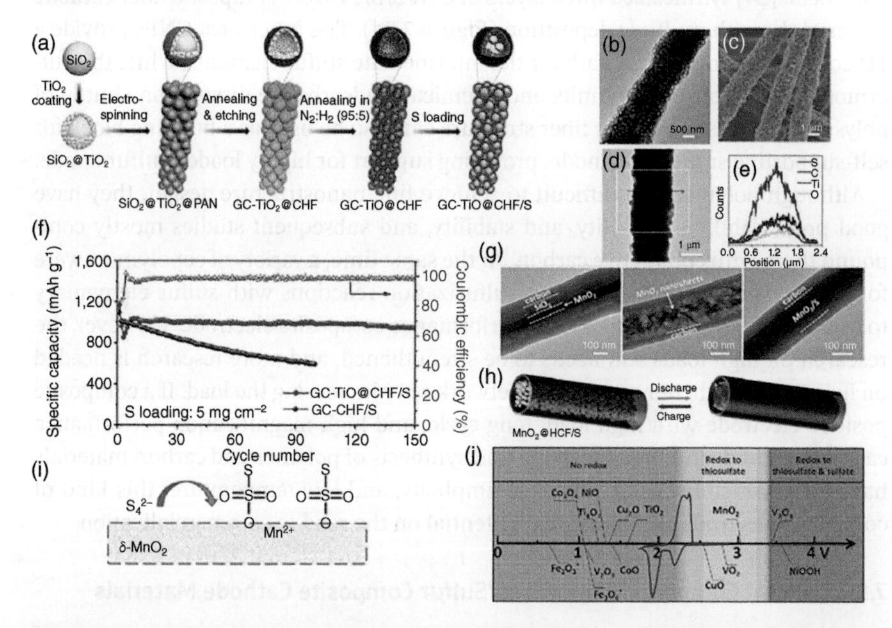

Figure 7.4 Grape clusters of porous TiO/C nanofibers are used as sulfur-fixing substrates. (a) Schematic of the grape-cluster-inspired design. (b) SEM images of sulfur host materials. (c) SEM images of GC-TiO@CHF/S. (d) TEM images of GC-TiO@CHF/S. (e) EDX elemental mappings. (f) Prolonged cycle performance of GC-TiO@CHF/S at 0.2 C; Source: Li et al. [25] / Reproduced with permission of Cell Press. (g) MnO_2@KCF SEM image of the synthesis process; (h) Schematic of morphology evolution during discharge; Source: Li et al. [26]/with permission of John Wiley & Sons, Inc. (i) The schematic illustrates the surface-level oxidation of polysulfide initially formed by δ-MnO_2, leading to the formation of thiosulfate, and the simultaneous reduction of Mn^{4+} to Mn^{2+}; Source: Reproduced with permission Liang et al. [27]/with permission of Nature Publishing Group. (j) The oxides with different voltages form different media with the positive electrode of sulfur; Source: Reproduced with permission Liang et al. [28]/with permission of Wiley-VCH. CG: Grape clusters; PAN: Polypropylene cyanide; CHF: Hollow carbon fiber.

dispersed in PAN solution for electrospinning to obtain composite fibers assembled by the nanospheres. After carbonization and etching, the nanofibers assembled by carbon-coated TiO_2 hollow spheres were obtained. Finally, the "grape" string GC-TiO@CHF matrix was obtained by calcination in reducing atmosphere. The matrix of the nanowires is mainly composed of carbon coated TiO hollow spheres. After filling with elemental sulfur, each hollow nanowire unit can play a physical limiting effect. At the same time, the inner layer of TiO has a strong absorb ability to polysulfide, and the outer layer of carbon can provide good electron conduction. The load of TiO/C positive electrode is $5\,mg\,cm^{-2}$ and the sulfur content is 73 wt%. Compared with the porous carbon fiber without TiO, this structure shows excellent cycling stability (Figure 7.4f). At a high specific capacity of $810\,mAh\,g^{-1}$ and the cycle retention rate was 79% after 400 cycles at 2 C. At the same time, the area specific capacity of the electrode is more than $3\,mAh\,cm^{-2}$ below the rate of 0.5 C, which is very close to the level of the positive electrode of commercial lithium battery.

In addition to titanium oxide, other oxide sulfide substrates have also been developed in large numbers, such as Co_3O_4, NiO, Fe_2O_3, Cu_2O, CuO, MnO_2, and VO_2. Among them, MnO_2 is one of the popular sulfur-fixing substrates, which shows good sulfur fix ability. For example, Li et al. synthesized a hollow MnO_2 nanosheet filled carbon fiber sulfur-fixing substrate (MnO_2@KCF) through a few design, which was melted into sulfur elemental substance at high temperature to form composite sulfur cathode material (Figure 7.4g,h) [26]. The organic combination of MnO_2 and hollow carbon nanofibers provides support for rate performance of sulfur elemental substance. Moreover, the appearance of MnO_2 initiates a new type of polysulfide interfacial reaction, So that different potential oxides can be selected as sulfur fixers with certain regularity. This mechanism was mainly proposed by Liang et al. [27]. They synthesized a water–sodium–manganese mineral-type layered δ-MnO_2 sulfur with a certain amount of oxygen defects at the edges. At the same time, there are water molecules or alkali potassium ions between the layers. The cathode material with sulfur composite showed very good cyclic stability. Non-in-situ XPS studies have shown that the Mn^{4+} and Mn^{3+} on the surface of δ-MnO_2 undergo a redox reaction with polysulfides during the discharge process to form a thiosulfate medium (Figure 7.4i). Sulfate thiosulfate reacts with polysulfide with high sulfur content (Li_2S_x, $4 \leq x \leq 8$) are further redox to form polysulfide with low sulfur content (Li_2S_x, $1 \leq x \leq 2$). This reaction actually promotes the gradual transformation of polysulfide, which improves the conversion of polysulfide, enhances the rate performance and specific capacity, and reduces the probability of polysulfide being dissolved and improves the cycling stability. In order to further know the mechanism of δ-MnO_2 generation of thiosulfate, Nazar et al. studied the change of the valence state of various oxide matrix during sulfur discharge and the change of the valence state of sulfur using non-in-situ XPS. They found that the formation of thiosulfate was related to the reduction potential of the oxide itself (Figure 7.4j) [28]. In the voltage range of 2.4–3.05 V (vs Li^+/Li), the oxide matrix undergoes redox reaction. at the same time, the polysulfide is oxidized to thiosulfate as the medium. The typical materials are MnO_2, VO_2, and CuO.

If the reduction potential of the oxide matrix is lower than 2.4 V, the formation of thiosulfate is difficult, The polysulfide is adsorbed by the polar bonds on the surface of the oxide matrix. The previously mentioned TiO_2, Ti_4O_7, Co_3O_4, NiO, Fe_2O_3, and so on are this kind of oxide sulfur-fixers. However, when the reduction voltage is greater than 3.05 V, the polysulfide will be further oxidized to sulfate, sulfate ions are difficult to further react with the polysulfide, but will cover the surface of the oxide matrix, hindering the adsorption of polysulfide, which is not conducive to the improvement of cyclic stability. High voltage materials such as V_2O_5 and NiOOH are typical.

The mechanism mentioned above can also be regarded as a catalytic effect, which promotes the conversion of lithium polysulfide through the mediating action of thiosulfate. Some sulfur fixers also exhibit similar catalytic effects, such as CoS_2, Co_3S_4, MoS_2, WS_2, and TiS_2 [29]. However, the average potential of these sulfide substrates is below 2.0 V, which makes it difficult to react with polysulfide to form thiosulfate. It is found that they can directly catalyze the transformation of multisulfide lithium without the formation of the medium, and promote the reaction kinetics, so can reduce the dissolution probability of polysulfide; At the same time, it can also prevent the dissolution of polysulfides by strong polar adsorption of metal sulfur bonds on the surface or edge. For example, Pu et al. at Tsinghua University synthesized hollow Co_3S_4 nanotube matrix and applied it to the cathode of lithium-sulfur batteries (Figure 7.5a,b) [31]. They used DFT to calculate the chemical binding energies between the (111) crystal planes of Co_3S_4 and Li_2S_4, Li_2S_6 and Li_2S_8, which were 2.26, 1.61, and 1.68 eV, respectively (Figure 7.5c). This indicates that there is strong adsorption between Co_3S_4 and lithium polysulfide. To further investigate the catalytic effect of Co_3S_4, they assembled a symmetric cell with a Co_3S_4 composite electrode containing Li_2S_6 and then tested the cyclic voltage curves (Figure 7.5d,e). It was found that the Co_3S_4 symmetrical battery containing Li_2S_6 had a strong current response, indicating the presence of charge transfer, but the acetylene black symmetrical battery containing Li_2S_6 and the Co_3S_4 symmetrical battery without Li_2S_6 had almost no current response, indicating that Co_3S_4 promoted the oxidation reduction of Li_2S_6 and accelerated its reaction kinetics.

In addition, researchers also found that some sulfide can be used as a catalyst, which can relieve the accumulation of bulk Li_2S and the difficulty of oxidation [32]. In addition to sulfide, some traditional metal electrocatalysis have also been studied as sulfide fixers, such as Pt, Co, and Ni. These metallic materials also exhibit similar catalytic effects [30]. This strategy of introducing catalysts into electrode materials to speed up the reaction is interesting and effective, and it may be further applied to other transformational battery reaction systems. However, more research is needed on how to further fabricate 1D nanostructures through synthesis techniques to achieve high load performance.

With the gradual application of metal oxide and sulfide matrix to lithium–sulfur cathode materials, research on improving the cycle life, load capacity, and specific capacity of lithium–sulfur battery cathodes has reached a golden age. All kinds of metal compound matrix have been developed successively. In 2016, Professor Goodenough, the father of lithium electricity, reported the work of TiN hollow nanotube

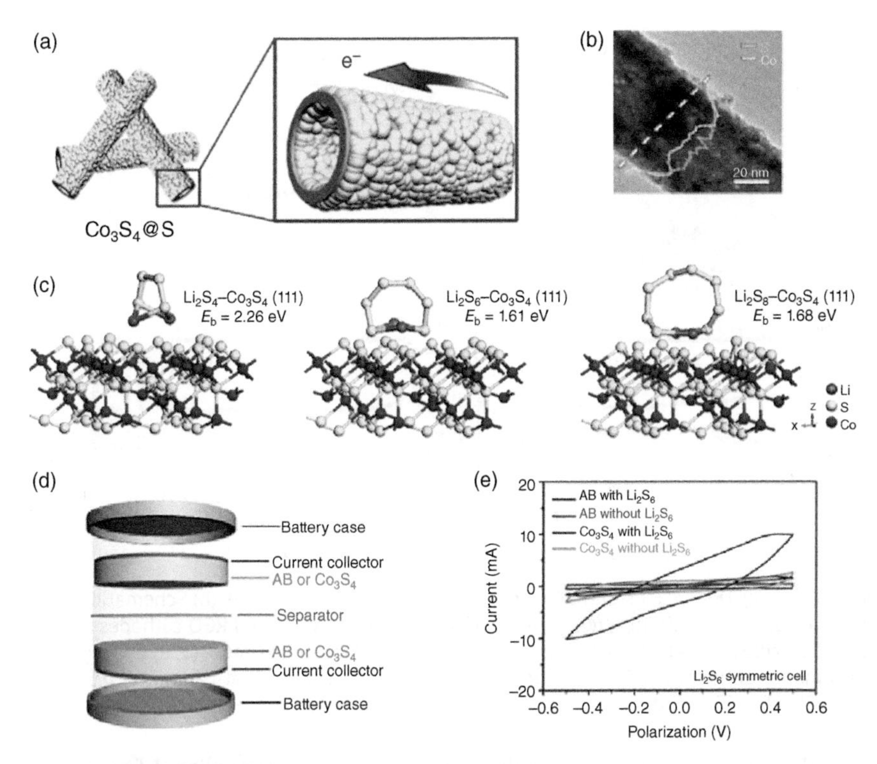

Figure 7.5 (a) Schematic illustration of Co_3S_4@S nanotubes. (b) TEM and energy spectrum analysis of Co_3S_4@S nanotubes. (c) Typical binding geometries and energies of three polysulfide molecules (Li_2S_4, Li_2S_6, Li_2S_8) on Co_3S_4 (111) surface. (d) Schematic illustration of a symmetric cell. (e) CV curves of the Li_2S_6 and AB symmetrical cells. E_b: binding energy. Source: Al Salem et al. [30]/with permission of with permission of Elsevier.

arrays as sulfur-fixing substrate, which initiated the upsurge in the application of materials such as nitride and even carbide to the positive electrode of lithium sulfur batteries [33]. After that, Vn, Co_4N, WN, and other substrates were also studied successively. Compared with sulfide and oxides, the main advantages of nitride and carbide lie in their high electronic conductivity, which can reduce the use of conductive carbon, promote the diffusion and transfer of polysulfide on the substrate, and also have a certain chemisorption effect. Sun et al. [34] prepared a porous VN nanowire and graphene composite gel matrix (Figure 7.6a). Then a certain amount of Li_2S_6 cathode electrolyte was dropped into the gel to form a composite electrode for testing (Figure 7.6b). The loading capacity was 3 mg cm^{-2}. At the current density of 0.2 C, its initial specific mass capacity is as high as 1471 mAh g^{-1} (Figure 7.6c). At a high rate of 2 C, the first specific discharge capacity is 1128 mAh g^{-1}, and the capacity retention rate is 81% after 200 cycles. The excellent electrochemical performance of the electrode is not only due to the high conductivity and adsorption caused by VN but may also be related to the deposition behavior of cathode electrolyte. Nitride and carbide in the cathode materials of lithium–sulfur batteries may also act as catalysts, but the relevant reports have not yet appeared. At the same

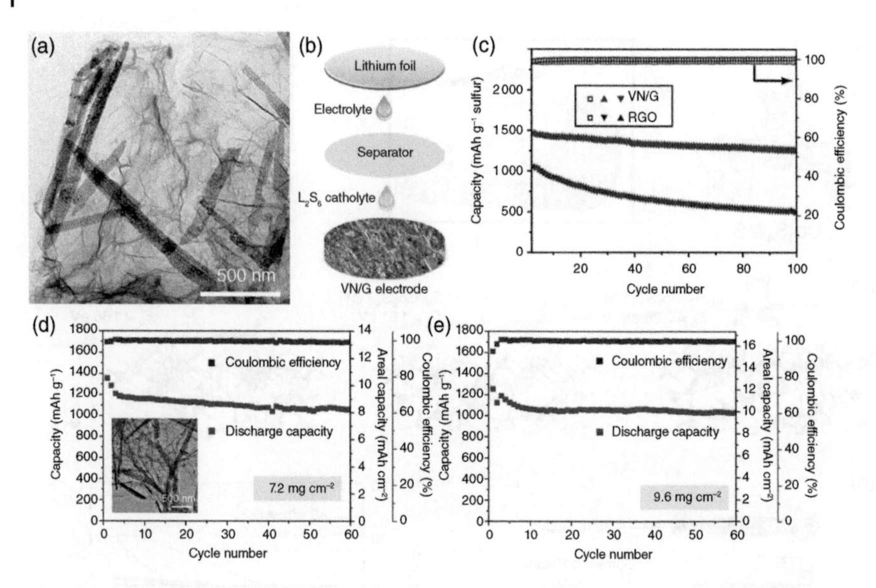

Figure 7.6 (a) TEM images of the as-prepared porous VN/G composite; (b) Schematic of the assembly of button batteries; (c) Cycling performance of the VN/G and RGO cathodes at 0.2 C; Source: Sun et al. [34]/with permission of Nature Publishing Group. (d, e) Cycling performance of the 3DNG/TiN cathode with 7.2 and 9.6 mg cm^{-2} sulfur loadings at 0.5 C. Source: Li et al. [35]/with permission of John Wiley & Sons, Inc.

time, this kind of nitride and graphene composite three-dimensional framework has a lot of pores, which is conducive to the realization of high load performance. Li et al. designed TiN/N-doped graphene-based composite sulfates, which could achieve high area-specific capacity of 9.94 and 12.0 mAh cm^{-2} at high loads of 7.2 and 9.6 mg cm^{-2}, respectively (Figure 7.6d,e) [35].

From the development of various sulfur cathode materials, we can see that before 2014, the main line of research was to first solve the problem of conductivity and then concentrate a lot of research on the topic of limiting the dissolution of polysulfide shuttle. Although this has made great progress, the real commercialization of lithium–sulfur batteries is still not mature. Some of these key issues remain unresolved. Composites, for example, are complicated and expensive to synthesize and are not conducive to mass production. In addition, the electrode studied in the laboratory has a low sulfur content, which reduces the overall energy density of the electrode. Therefore, the overall energy density will not significantly exceed that of the lithium-ion battery when it is assembled into a commercial battery. Therefore, the main direction of future research is to develop simple and effective composite materials with high sulfur injection volume, high capacity, and long life performance.

In this regard, nanowire material is a very potential sulfur cathode composite element. For example, the preparation of carbon fiber interwoven films for high-load sulfur positive electrode reduces the use of adhesive and is suitable for mass production. The 3D continuous conductive frame provides a very effective conductive contact for elemental sulfur and lithium sulfide under high loads, improving power density. At the same time, this structure is conducive to the diffusion of polysulfide and deposition of lithium sulfide, improving the sulfur utilization rate and making the specific capacity close to the theoretical capacity, so as to improve the energy density. If we want to achieve a long cycle life, we must carry out fine structure design and surface modification of nanowire monomers, or introduce some metal compound units with chemical adsorption or catalytic effect, to ensure the rapid transformation and slow dissolution of polysulfide. It is expected that nanowires can solve more problems and play a greater role in the future research and application of lithium–sulfur batteries.

7.2 Sodium–Sulfur Battery and Magnesium–Sulfur Battery

7.2.1 Sodium–Sulfur Battery

Compared to lithium metal, sodium metal resources are widely available and cost-effective, which has led to increased attention toward sodium batteries in recent years. Compared with lead-acid batteries at that time, it has higher energy density, and can discharge at high power, and the coulomb efficiency is close to 100%. All of this is due to the fact that sodium–sulfur batteries were a high temperature battery in the early days of their development. As shown in Figure 7.7, the high-temperature sodium–sulfur battery employs molten metal sodium as the negative electrode and molten sulfur as the positive electrode. These two components are separated by an all-solid electrolyte, sodium β-Al_2O_3 ($NaAl_{11}O_{17}$) [36]. At 300–350 °C, sodium β-Al_2O_3 can realize the rapid conduction of sodium ions, and at the same time, sodium ions, and sulfur elemental quickly sodium reaction, to form liquid sodium polysulfide. The reaction equation is: $x\text{S} + 2\text{Na} \rightarrow \text{Na}_2\text{S}x$ ($x = 3$–5). Because sodium and sulfur are in the molten state in this state, the reaction kinetics is faster, and there are not too many side reactions, and there will not be sodium polysulfide solution problem. Therefore, high temperature sodium–sulfur batteries have high Coulomb efficiency and cycle life. However, it has not been applied on a large scale, mainly due to the following reasons: (i) High working temperature limits its application range, and the use of heat source also makes its cost higher; (ii) The strong activity of metal sodium is easy to lead to fire, and the safety is low. However, early studies on high-temperature sodium sulfur batteries mainly focused on improving the ionic conductivity of solid electrolyte $NaAl_{11}O_{17}$ thin film and improving the battery configuration to improve safety. To enhance the energy density, expand the application range, and ensure the safety of sodium–sulfur batteries, reducing their operating temperature has become a primary concern.

At room temperature, the solid electrolyte $NaAl_{11}O_{17}$ cannot conduct sodium ions, so the first problem to be solved is to develop an electrolyte (compatible with sulfur) that can reversibly dissolve the deposited Na^+. At the same time, the reaction kinetics between sulfur elemental and metal sodium ions are slow, which leads to incomplete reaction of sodium polysulfide and low capacity. The Yuguo Guo team synthesized

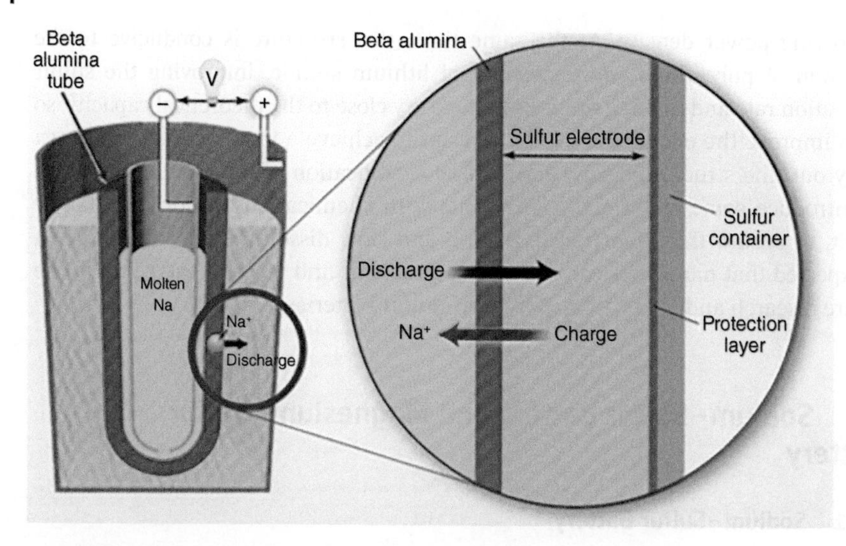

Figure 7.7 Schematic of the high-temperature Na/S battery. Source: Dunn et al. [36]/with permission of American Association for the Advancement of Science.

two to four sulfur atom connections in small molecules for sodium sulfur batteries, greatly promoting the positive reaction between sodium ion and sulfur dynamics, enabling the S reaction to form the final product Na_2S, and getting an energy density of 955 Wh kg^{-1} [37]. First, carbon nanotubes coated with microporous carbon (MPC) were immersed in the molten sulfur element. At this time, S_8 molecules would break into small molecular chains S_{2-4} and diffuse into the micropores (~0.5 nm). When the temperature was reduced to room temperature, the S_{2-4} diffused into the micropores was restricted and did not revert to S_8 molecules. To prove this, the authors calculated the molecular sizes of S_{2-4}, Na_2S_2, and Na_2S, all of which are only allowed to exist as monomolecules in micropores. This S/(CNT@MPC) is used as the positive electrode of the sodium–sulfur battery (Figure 7.8a). Its initial discharge capacity is as high as 1610 mAh g^{-1}, and the average voltage is about 1.4 V, indicating that S_{2-4} is almost completely converted to Na_2S. After the first charge, the first Coulomb efficiency was 71.3%, and after 20 cycles, the capacity remained at more than 1000 mAh g^{-1}. Further analysis of the reaction mechanism (Figure 7.8b,c) revealed that the first platform at 1.55 V was mainly generated by the reaction of Na_2S_2, while the second platform at 1.2 V was derived from the formation of Na_2S. The application of small-molecule sulfur to the cathode of sodium sulfur battery can achieve higher specific capacity and fast reaction kinetics, but the voltage lag is a problem that needs to be further solved. The theoretical reaction potential of room-temperature sodium–sulfur batteries is between 2.2 and 1.6 V, while the reaction potential of small molecules of sulfur and sodium ions is relatively low, and the platform is inclined. In addition, the deposition and oxidation of sodium sulfide also need to be further optimized to improve the coulomb efficiency.

Yu et al. used the liquid intermediate Na_2S_6 cathode electrolyte to drop into MWCNT film as the cathode material [38]. This method can provide uniformly

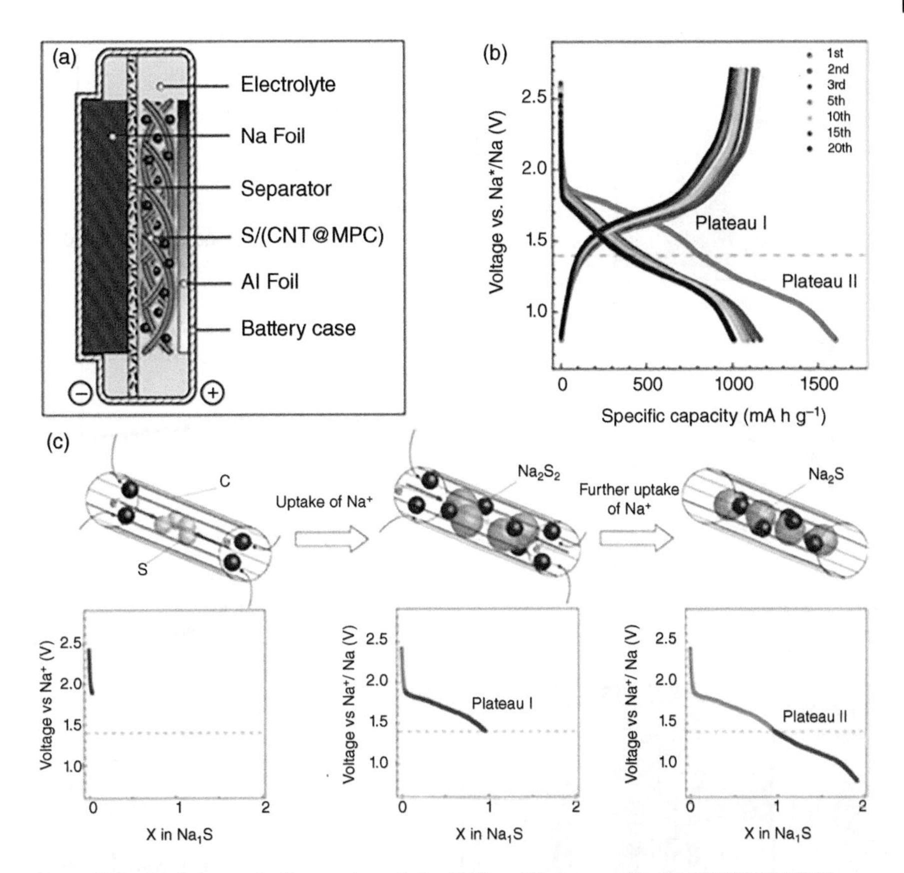

Figure 7.8 (a) Schematic illustration of the RT Na–S battery with the S/(CNT@MPC) cathode; (b) GDC voltage profiles of the S/(CNT@MPC) cathode at 0.1 C; (c) Study on the structure change and mechanism of S/(CNT@MPC) cathode during the first discharge. Source: Xin et al. [37]/with permission of Wiley-VCH.

dispersed active sulfur in the conductive carbon network (Figure 7.9a,b), and is conducive to the diffusion and deposition of Na_2S in the conductive matrix, thus improving the first Coulomb efficiency. Its initial capacity is around 930 mAh g^{-1} in the voltage range of 1.2–2.8 V, representing a typical 2.2 and 1.6 V voltage platforms (Figure 7.9c). However, due to the solubility of sodium polysulfide, its cycling performance is poor. The capacity decays to about 400 mAh g^{-1} after 20 cycles (Figure 7.9d). The configuration of the battery can be further optimized. In terms of the anode, activated carbon was introduced into the MWCNT film to make contacts with conductive network Na_2S_6 better, while at the same time using the Nafion membrane of sodium (tetrafluoroethylene perfluoroalkyl vinyl ether copolymer) as a diaphragm to limit sodium polysulfide solution problem (Figure 7.9e) [39]. The main reason is that organic Nafion membrane has certain chemisorption effect on sodium polysulfide, which slows down its dissolution shuttle.

They further optimized the battery's configuration. On the positive electrode, activated carbon was introduced into the MWCNT film to make the contact between

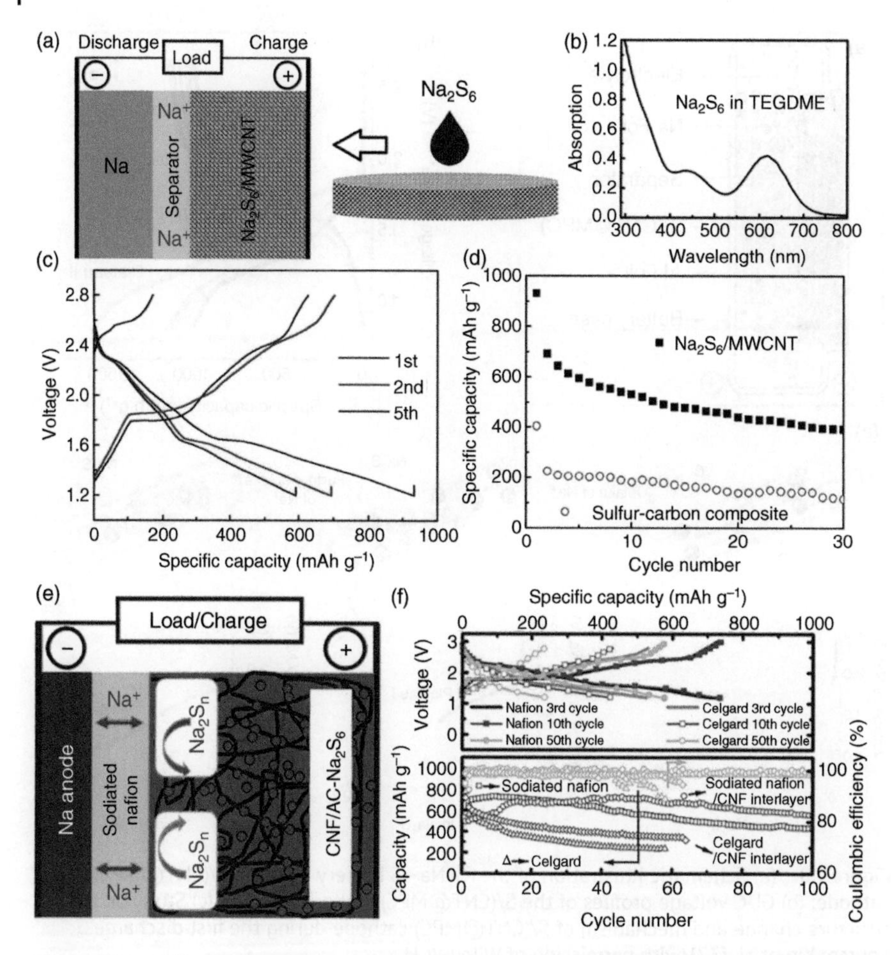

Figure 7.9 (a) Na_2S_6 dissolved in tetraethylene glycol dimethyl ether as the catholyte application and the structure of sodium–sulfur battery; (b) the Raman spectrum of the catholyte; Na_2S_6/MWCNT charge–discharge curve; (c) and cycle life; (d) in the voltage range of 1.2–2.8 V; Source: Yu and Manthiram [38]/with permission of American Chemical Society. (e) Schematic diagram of room temperature sodium–sulfur battery with CNF/AC-Na2S6 as the positive electrode and Nafion membrane as the separator; (f) Charging curve and cycle performance at 0.5 C; Source: Yu and Manthiram [39]/with permission of Wiley-VCH. MWCNT: multi-walled carbon nanotube; CNF: carbon nanofiber; AC: activated carbon.

conductive network with Na_2S_6 better. At the same time, the sodiumized Nafion membrane was used as a separator to limit the dissolution of sodium polysulfide (Figure 7.9f) [39]. This is mainly because the organic Nafion membrane has a certain chemical adsorption effect on sodium polysulfide, which slows down its dissolution and shuttle.

Currently, room temperature sodium–sulfur batteries are still in the nascent stage of research and development. Although usable electrolyte and electrode materials have been developed, it will take a long time to achieve high energy density and low cost. How to better solve the problem of slow dynamics and dissolution of sodium polysulfide is a topic that researchers should pay attention to. For instance, similar to lithium–sulfur batteries, certain matrix materials with catalytic properties can be incorporated into the cathode material of sodium–sulfur batteries to enhance their kinetics and simultaneously adsorb sodium polysulfides; or small organic molecules are added to the composite electrode to chemically connect the sodium polysulfide molecules to prevent its dissolution. It is crucial to address the challenges of high load and commercial-scale production for practical applicability. Meantime, sodium metal is more active than lithium metal, and the guarantee of safety is also a major problem for sodium–sulfur battery. It can be seen that the commercial production is still difficult.

7.2.2 Magnesium–Sulfur Battery

Compared with the high activity of lithium and sodium metal negative electrodes, magnesium metal negative electrodes show good stability. Incorporating magnesium as the negative electrode in the metal chalcogenide battery system can significantly enhance the safety of the battery. In addition, compared to lithium and sodium metals, magnesium metal has a higher capacity energy density ($3833\,mAh\,cm^{-3}$), no obvious dendritic behavior, and rich surface content. These advantages make magnesium–sulfur batteries a promising high-energy-density battery system with potential applications in various fields. However, the development of magnesium–sulfur batteries is relatively late, and there are still many unsolved or even unknown problems. Among them, the slow reaction dynamics of magnesium ions and sulfur elements will lead to extreme battery polarization; furthermore, non-nucleophilic magnesium electrolyte compatible with element sulfur is also urgently needed to develop. In 2011, Professor John Muldoon of Toyota North America Research Institute developed an electrolyte system with bis(hexamethyldisilazide) magnesium, magnesium chloride, and aluminum chloride salt ($HMDSMgCl\text{-}AlCl_3$) as the main components for the first time, which achieve the rechargeable normal-temperature magnesium–sulfur battery. And it has a first discharge capacity of $1200\,mAh\,g^{-1}$ and a stable voltage plateau of 0.89 V [40]. Nevertheless, the primary challenge lies in the low Coulombic efficiency, substantial polarization, and high cost, which hinder the practical implementation of magnesium–sulfur batteries in commercial settings. Therefore, to develop no nucleophilic magnesium electrolytic solution with high coulombic efficiency, high ionic conductivity and low cost is one of the important problems in the development of magnesium–sulfur battery.

In addition, magnesium–sulfur batteries also suffer from the problem of easily dissolving magnesium polysulfide in organic electrolyte. The nanofilm made of nanowire materials can play a great role in fixing sulfur. Yu et al. [41] covered a layer of carbon fiber interwoven film on the surface of ordinary glass fiber film to prevent the dissolution and loss of magnesium polysulfide (Figure 7.10a–c). It was discovered that the battery coated with carbon fiber exhibited higher charge and discharge capacities, achieving a discharge capacity of approximately $1200\,mAh\,g^{-1}$ at a current density of C/50 ($1\,C = 1675\,mA\,g^{-1}$). The discharge capacity decreases with the increase of rate (Figure 7.10d,e). In addition, after

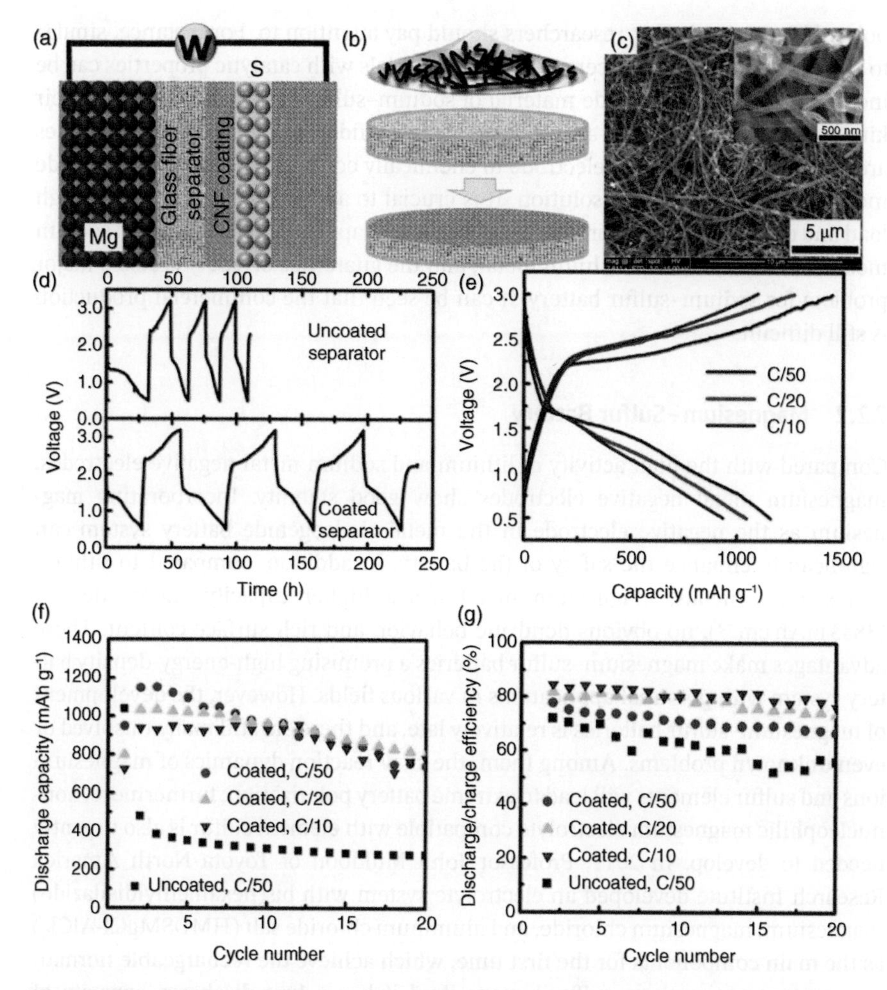

Figure 7.10 (a) Schematic diagram of magnesium–sulfur battery assembled with carbon fiber membrane covered with glass fiber membrane; (b) Schematic diagram of carbon fiber covered with glass fiber membrane; (c) SEM image of carbon fiber; (d) constant current charge and discharge test curve; (e) Charge and discharge curves at different rates; (f) Cycle performance graph; (g) Coulomb efficiency graph; Source: Yu and Manthiram [41]/with permission of the American Chemical Society.

20 cycles, the magnesium–sulfur battery coated with carbon fiber can still output about 800 mAh g^{-1}, which is much better than the magnesium–sulfur battery without carbon fiber coating (Figure 7.10f). At the same time, it shows good coulombic efficiency (Figure 7.10g). The outcomes suggest that the carbon fiber coating effectively mitigates the dissolution of magnesium polysulfide and decreases the polarization of the battery to a certain degree. But the existing performance is still far from commercialization.

We can find that the Coulombic efficiency of the currently studied magnesium-sulfur battery cannot reach 90%, which can be cycled 20 times under the condition of excessive magnesium in the negative electrode, but in commercial batteries, such a large irreversible reaction will cause rapid consumption of magnesium. The capacity decays further rapidly. Therefore, the rapid development of magnesium–sulfur battery has to rely on a lot of research and optimized design.

7.3 Lithium–Selenium Battery

Presently, conventional lithium-ion batteries are constrained by the theoretical capacity of their positive and negative materials, hence the pursuit of new energy storage devices/systems has become a prominent research focus [5, 42–47]. In the previous chapters, we have introduced lithium–sulfur battery as promising new generation of high-performance energy storage devices with significant potential. Researchers have widely used carbon/sulfur composite, titanium dioxide-coated sulfur, electrolyte modification, and separator modification to improve Li–S batteries electrochemical performance. The development and research of a new type of secondary battery system with high conductivity electrode materials and higher energy density is of irreplaceable value for solving people's increasing energy demand.

Selenium (Se), an element of the same main group as sulfur, has gradually entered people's vision in recent years and is expected to become a strong competitor of Li–S battery. According to Table 7.1, although selenium has a lower theoretical mass specific capacity relative to sulfur (selenium: 675 $mAh\,g^{-1}$), it has a relatively high tap density (4.81 $g\,cm^{-3}$), which means it have a volumetric capacity as high as 3265 $mAh\,cm^{-3}$, a higher volumetric capacity will be of great significance in the future to solve space-constrained power battery. Based on the above-mentioned excellent physical properties of selenium, to a certain extent, it has a better development prospect and potential than sulfur as a high-energy battery in the future, and has made major breakthroughs in the research in recent years, gradually showing better performance and application prospects.

Table 7.1 Comparison of physical and chemical properties of sulfur and selenium.

Index	Li-S	Li-Se
Cell voltage (V)	2.2	2.0
Specific capacity of cathode ($mA\,h\,g^{-1}$)	1675	675
Capacity density of cathode ($mAh\,cm^{-3}$)	3467	3265
Theoretical specific energy ($W\,h\,Kg^{-1}$)	2567	1155
Theoretical capacity energy density ($Wh\,cm^{-3}$)	5314	5544
Price($\$\,kg^{-1}$)	0.124	77
Density ($g\,cm^{-3}$)	2.07	4.81
Conductivity ($S\,cm^{-1}$)	$>10^{-30}$	$>10^{-5}$

Source: Yang et al. [61]/American Chemical Society.

7.3.1 Reaction Mechanism of Lithium–Selenium Battery

Abouimrane et al. group conducted the first study on the electrochemical reaction mechanism of Li–Se batteries in carbonate electrolytes. The molecular structure and charge–discharge curves of the final product of Li–Se battery charge and discharge are shown in Figure 7.11a–c [48]. The lithiation process of the first circle of Li–Se batteries follows different patterns. This is mainly due to the electrochemical properties of chain-shaped Se_n molecules, which can interact with the electrolyte. The carbonyl group in it forms Se-C and Se-O bonds and finally reacts completely with lithium ions to form Li_2Se, thereby providing a higher capacity. Figure 7.11d shows typical CV curves of Li–Se batteries in carbonate electrolyte. From the first discharge curve, it can be seen that there is an obvious discharge platform near 2 V, which indicates that a single-phase transformation reaction has occurred in the cathode material, namely $Se + 2Li^+ + 2e^- \leftrightarrow Li_2Se$. During the charging process, Li_2Se is involved in the oxidation reaction to form more Selenide (Li_2Se_n [$n \geq 4$],) is finally oxidized

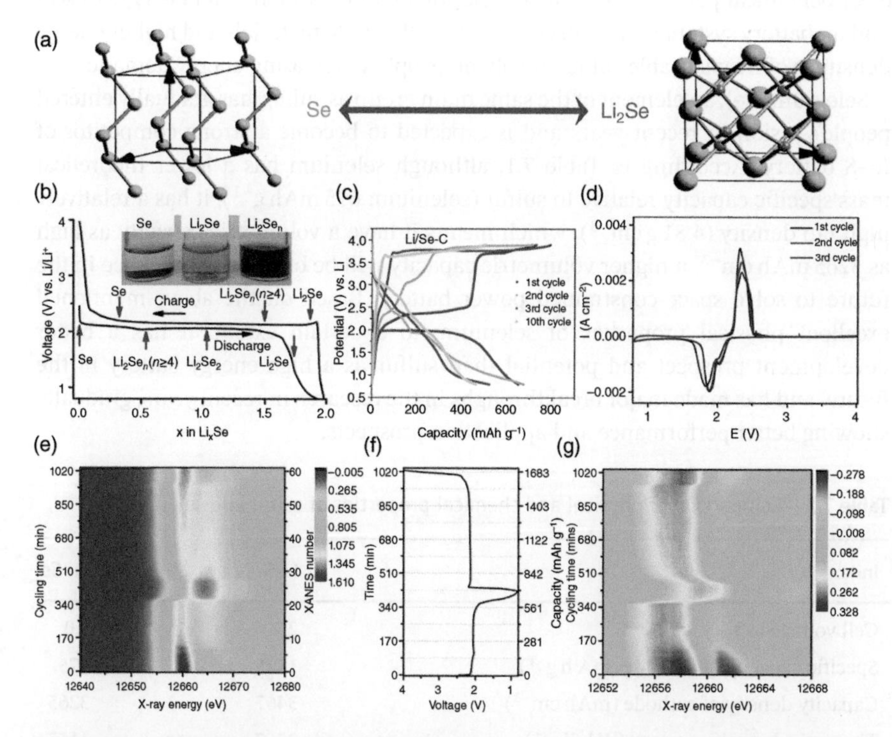

Figure 7.11 (a) Schematic diagram of the molecular structure of Li–Se battery charge and discharge products; (b) Representation of cathode phase evolution during charge and discharge of a Li/Se cell in an ether-based electrolyte; (c) Li–Se battery in the first, second, third, and tenth charge and discharge curve; Source: Abouimrane et al. [48]/with permission of American Chemical Society. (d) Li–Se battery first three cycles voltammetry curve. (e) XANES pattern of Li–Se battery cycle. (f) voltage curve during cycle. (g) The derivative of the standardized XANES spectrum of the battery cycling process; Source: Cui et al. [49]/with permission of Royal Society of Chemistry.

into selenium. The atomic skeleton structure of the charge and discharge products of the Li–Se battery is shown in Figure 7.11a. The elemental selenium is transformed from the initial triangular selenium crystals to the inverse fluorite structure Li_2Se during the discharge process. This reaction mechanism was further confirmed by in situ X-ray diffraction and in situ X-ray absorption spectroscopy [50].

At the same time, Cui et al. studied Li–Se batteries in ether electrolytes [49]. Analogous to the electrochemical reaction of Li–S batteries, the electrochemical reaction corresponding to 2.02 V is that Se combines with lithium ions and is reduced to polylithium selenide Li_2Se_n ($n \geq 4$). 1.83 V corresponds to Li_2Se_n as the discharge deepens, and finally was reduced to Li_2Se_2 and Li_2Se. In order to further verify the above analysis results, the author performed the X-ray absorption near edge structure (XANES) 2D contour map characterization of the circulating Li–Se battery (Figure 7.11e), And get the corresponding voltage curve (Figure 7.11f) and the first derivative diagram (Figure 7.11g) at the same time. The results found that the Se diffraction peak intensity gradually weakened during the discharge process, while the Li_2Se peak gradually became stronger.

Li et al. observed the structural changes of selenium in the electrochemical process in real time (Figure 7.12) [51]. According to the in situ TEM image and the corresponding electron diffraction pattern, we can clearly observe that as the lithium intercalation continues, the characteristic diffraction peaks of single crystal selenium gradually disappear, and the corresponding Li_2Se diffraction peaks begin to appear and gradually increase, indicating that Li. In the electrochemical reaction of the Se battery, elemental selenium reacts with lithium ions to gradually transform into Li_2Se, and this result is also consistent with the above-mentioned electrochemical reaction mechanism. In summary, it can be seen that elemental selenium exhibits two electrochemical reaction mechanisms in carbonate- and ether-based electrolytes. This may be because the intermediate phase product lithium selenide is insoluble in carbonate-based electrolytes, and soluble in ether electrolyte.

7.3.2 Selenium-Based Cathode Materials

It is evident that these materials still face the problem of Li_2Se_n dissolution in ether electrolytes, leading to the formation of a shuttle effect in the battery. Materials have similar life and efficiency issues; therefore, the use of low-cost carbonate-based electrolytes based on the physics and electrochemical reaction characteristics of selenium and the structural design and preparation of selenium cathode materials have become hot spots in Li–Se battery research. In response to the above problems, Yang et al. used ordered mesoporous carbon (CMK-3) as the matrix and prepared Se/CMK- by filling nano-selenium particles in the mesopores of CMK-3 through high-speed ball milling and melt diffusion. The composite material shows a higher discharge capacity and excellent cycle stability (Figure 7.13) [52]. At the same time, it is proven that the Se/CMK-3 composite material has no polyselenide produced during the electrochemical reaction. The initial amorphous Se_8 molecular ring

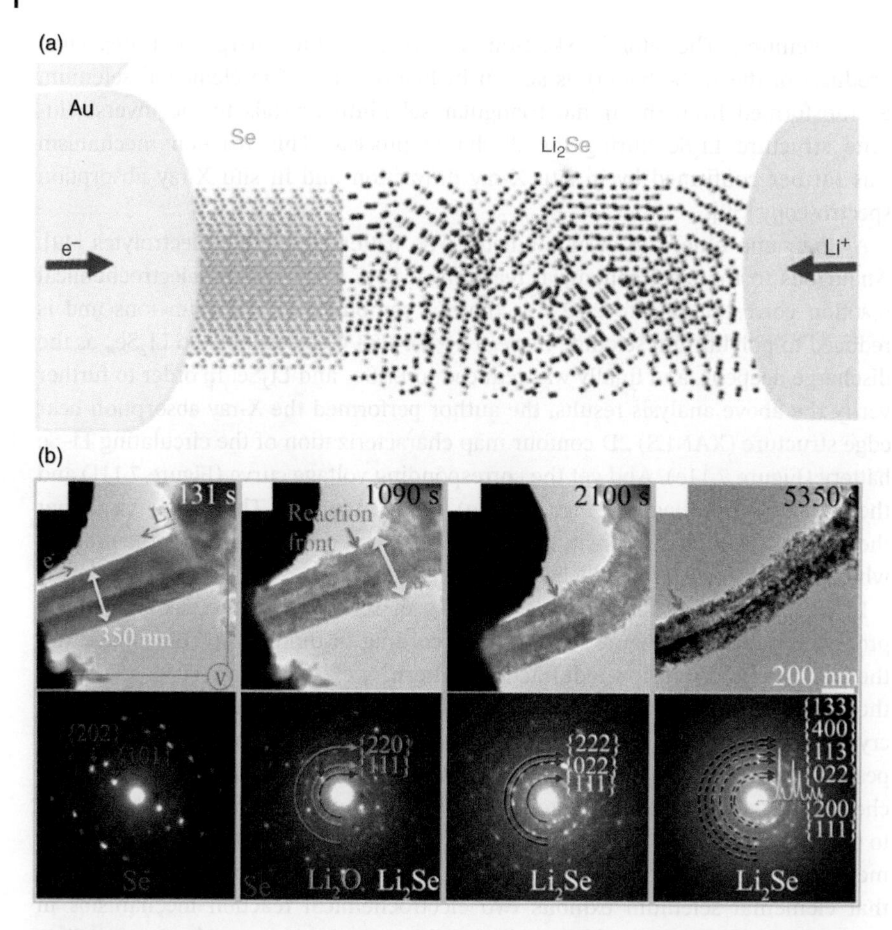

Figure 7.12 (a) Schematic illustration of sodiation and lithiation of selenium nanotubes. Three different phases appeared in sodiation due to the alloying reaction. In lithiation, selenium transforms into polycrystalline Li_2Se phase. (b) Evolution of morphology and microstructure of a selenium nanotube in the lithiation process. Source: Li et al. [51]/with permission of American Chemical Society.

is transformed into Se_n molecular chain, and finally it is directly transformed into Li_2Se, followed by lithium insertion-delithiation. The occurrence of this electrochemical reaction mechanism is mainly due to the fact that the nano-sized Se in the Se/CMK-3 composite material is confined in the mesopores of CMK-3, the large-sized Se is suppressed, and the chain-like Se_n is relatively small compared with the carbon matrix. This unique mesoporous composite structure exhibits excellent rate and cycle performance. It is cycled 500 times at a current density of 1 C, and the capacity is still 45% of the theoretical capacity. In view of the unique advantages of nanowire 1D structures in electrochemistry, researchers also hope to solve a series of problems with Li–Se batteries by constructing nanowire structures or introducing other nanowire structures. Zhang et al. reported that using selenium nanowires as the matrix is shown in Figure 7.14a–d [53]. The electrochemical performance

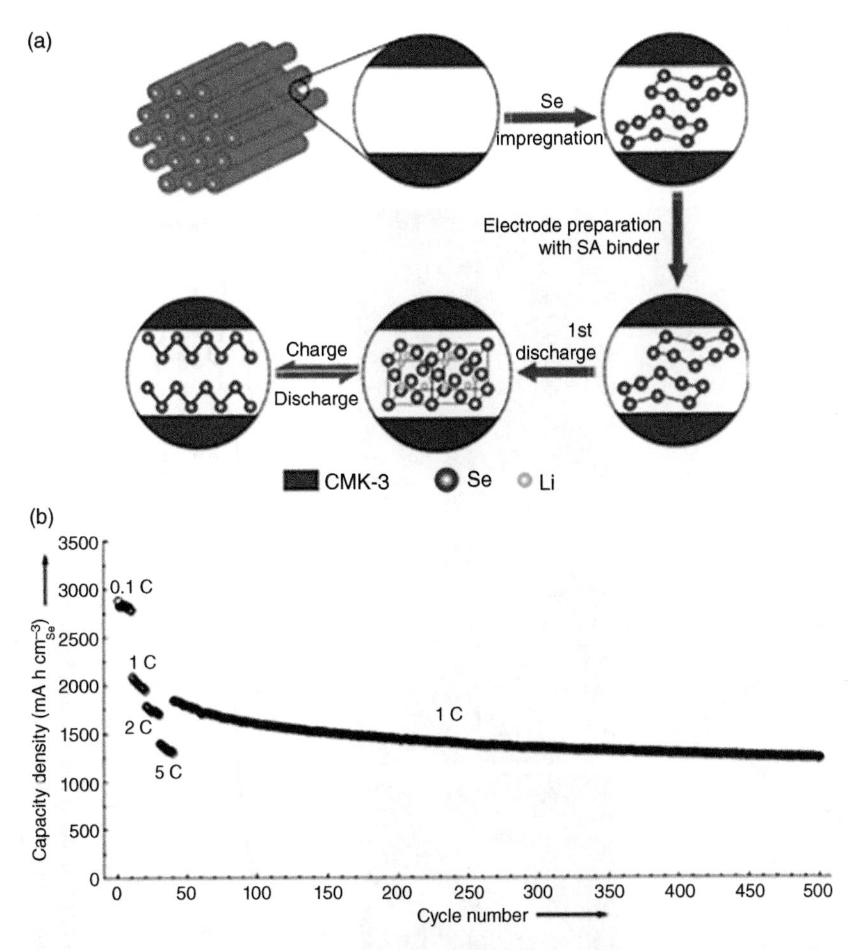

Figure 7.13 (a) proposed lithiation/delithiation processes of Se/CMK-3. (b) Reversible capability density of Se/CMK-3 at 0.1, 1, 2, and 5 C, and long cycling performance at 1 C. Source: Yang et al. [52]/with permission of Wiley-VCH.

test G@Se/PANI shows good cycle stability and excellent rate performance. The outstanding electrochemical performance of the material can be mainly attributed to the presence of highly conductive graphene, a unique polyaniline shell, and one-dimensional selenium nanowires.

Zeng et al. used electrospinning and subsequent heat treatment to prepare bendable, self-supporting selenium/porous carbon nanofiber composites (Se/PCNFs refer to selenium embedded within 3D interconnected mesoporous carbon nanofibers) [54]. The results of electrochemical lithium storage show that the Se/PCNFs composite electrode undergoes 900 charge–discharge cycles at $0.5\,Ag^{-1}$, and the specific capacity value is almost not lost, remaining at $516\,mAh\,g^{-1}$. Han et al. reported on a 3D self-supporting composite film made of graphene, selenium, and carbon nanotubes (G-Se@CNT) [55]. Compared with the graphene–selenium composite material, the ternary G-Se@CNT composite

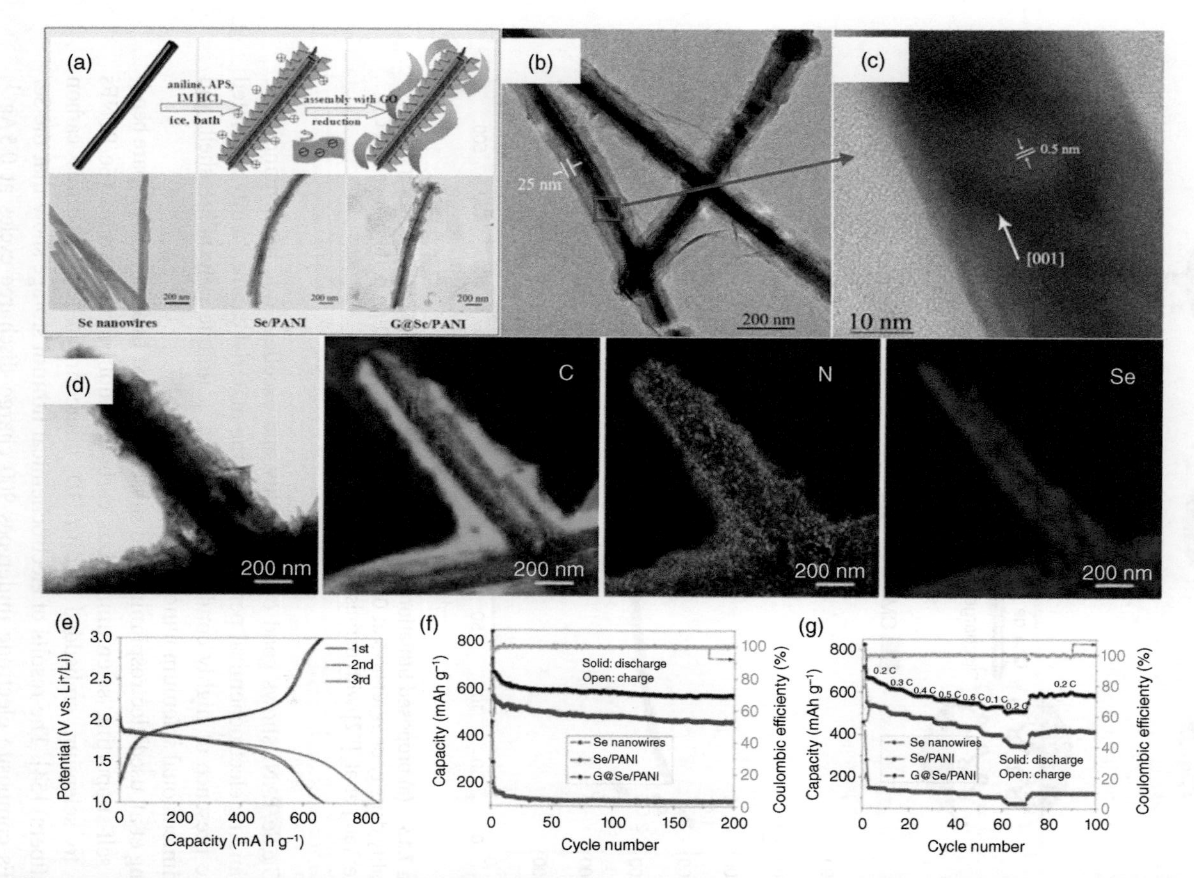

Figure 7.14 (a) Schematic illustration for the formation of G@Se/PANI. (b) High magnification TEM image. (c) HRTEM image of G@Se/PANI. (d) Galvanostatic charge–discharge curves of G@Se/PANI electrode. (f) cycle performance at a current density of 0.2 C between 1.0 and 3.0 V and (g) rate capability at various current densities between 1.0 and 3.0 V. Source: Zhang et al. [53]/with permission of Elsevier.

film electrode maintains a capacity of 315 mAh g^{-1} after 100 cycles of 0.1 C (1 C = 678 mA g^{-1}). At the same time, Han et al. also prepared selenium-mesoporous carbon nanoparticles-graphene (Se/MCN-RGO) [56]. The composite material has the ability to accommodate up to 62% selenium, resulting in a remarkable reversible specific capacity value of 655 mAh g^{-1} at 0.1 C (1 C = 678 mA g^{-1}). Even after undergoing 1300 cycles at 1 C, it experiences a mere 0.008% capacity decay rate per cycle. This high capacity utilization rate is mainly attributed to this unique 3D layered structure, in which the layered graphene structure increases the conductivity of the composite material and facilitates the diffusion of lithium ions. The mesoporous structure of mesoporous carbon nanoparticles limit the shuttle effect of polyselenide.

Balakumar et al. used the hard template method to prepare porous carbon-coated MWCNTs to form a TTC structure, which was used as a carrier to support selenium and applied to Li–Se batteries (Figure 7.15) [57]. This 1D tube nanostructure has a large number of voids and free volumes, with a pore volume as high as 2.167 cc g^{-1}

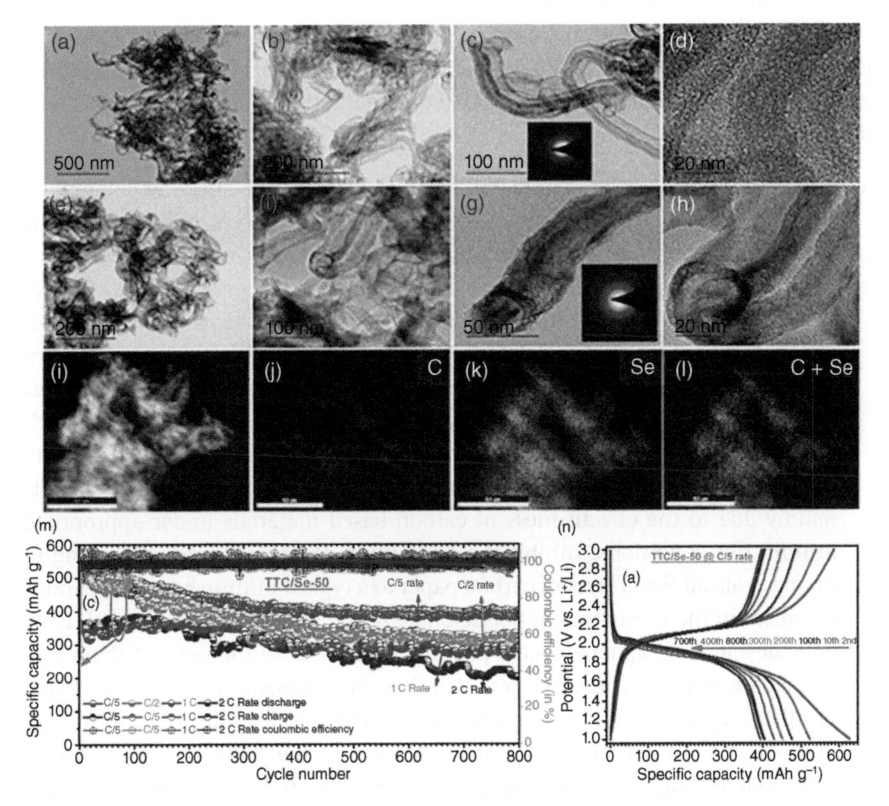

Figure 7.15 HRTEM images of (a–d) TTC at different magnification, (e–h) TTC/Se-50 at different magnification (inset of (c) and (g): SAED pattern of TTC and TTC/Se-50 respectively) and (i) STEM image, (j–k) individual and (l) cumulative elemental mapping of TTC/Se-50 composite. (m) consolidated cycling performance with Coulombic efficiency of TTC/Se-50 cathode under the influence of C/5, C/2, 1 C and 2 C rate. (n) profile of TTC/Se-50 cathode under the influence of C/5 rate. Source: Balakumar and Kalaiselvi [57]/with permission of Elsevier.

and an ultra-high specific surface area of 1131 m^2 g^{-1}. This unique structure is not only conducive to buffer electrode materials. The volume expands and shrinks in the process of deintercalating lithium, and it can load a large amount of active material, selenium, so that the mass ratio of selenium in the composite material can reach 70%. Secondly, the core MWCNT framework provides a gentle electron transfer path. In addition, the TTC 1D nanostructure has a short lithium ion diffusion path. Even when charging and discharging at ultra-high current density, it also ensures the long-term cycle stability of the electrode material and has a close 100% excellent Coulomb effect. Through electrochemical tests, the electrode material with a mass ratio of 50% selenium still has a high capacity close to 400 mAh g^{-1} at a current density of 0.2 C after 800 cycles.

7.3.3 Existing Problems and Possible Solutions

Since the current research on Li–Se batteries is still in its early stages, some speculations and inferences are still based on Li–S batteries. Therefore, there may be several shortcomings and problems in the development of Li–Se batteries. However, addressing these issues is crucial for the advancement of Li–Se batteries. Finding solutions to these problems will undoubtedly facilitate the development of Li–Se batteries.

(1) The production of polyselenide will bring about a shuttle effect, which greatly affects the cycle life of Li–Se batteries. How to suppress polyselenide has become a current research hotspot and difficulty. The traditional, effective method is to load Se on carbon materials to inhibit the dissolution of polyselenide. Lee et al. changed the concentration of lithium bis(trifluoromethanesulfonyl) imide (LiTFSI) (1, 3, 5, 7 mol l^{-1}) to study the electrochemical lithium storage performance of micro mesoporous carbide-derived carbon/sulfur composite (selenium/carbide-derived carbon, Se-CDC) [58]. The research results show that the Se-CDC composite electrode has the highest capacity retention rate after 150 cycles at a LiTFSI concentration of 5 mol l^{-1}. This is mainly due to the effectiveness of carbon-based materials under appropriate lithium salt conditions. Inhibit the dissolution of polyselenide. According to Zhang et al., adding a layer of carbon paper to a typical lithium battery separator can restrain the movement of polyselenide to the negative electrode [59]. Fang et al. developed graphene-coated polymer separators and integrated them into Li–Se batteries. Their research revealed that the composite separator is capable of effectively curbing the diffusion of polyselenide, leading to a substantial improvement in the electrochemical stability of the battery [60]. It can be seen that through carbon coating, electrolyte modification, and membrane modification, the shuttle effect of soluble polyselenide can be effectively suppressed. Even so, there is currently no perfect strategy that can effectively and completely solve the shuttle effect caused by polyselenide.
(2) The high cost of selenium is a significant constraint on the wide-scale application of Li–Se batteries. It is essential to reduce their cost while retaining

Table 7.2 Selenium and sulfur cost and energy density comparison table.

Material	Price($ kg^{-1})	Theoretical mass energy (Wh kg^{-1})
$S_{0.3}Se_{0.7}$	53.9372	1578.6
$S_{0.5}Se_{0.5}$	38.562	1861
$S_{0.7}Se_{0.3}$	23.1868	2143.4

their excellent electrochemical performance. According to the current development of Li–Se batteries, researchers have summarized the following two effective improvement methods: (i) Changing the selenium cathode into a sulfur–selenium mixture cathode can effectively reduce the cost of the entire Li–Se battery and increase the battery (Table 7.2); (ii) Compared the ether electrolyte used in Li–S batteries, low-cost carbonate electrolytes can be used in Li–Se batteries.

Li–Se batteries have gained significant attention as a research hotspot due to their remarkable advantages, particularly their high volumetric specific capacity. Currently, research on Li–Se batteries is predominantly focused on enhancing and innovating the concept of sulfur-based cathode materials in Li–S batteries, as well as developing and modifying cathode materials based on selenium. Among them, selenium nanowires and their composite materials have a unique one-dimensional microstructure compared with other selenium cathodes, and they also exhibit excellent electronic and ion conductivity. They are designed to achieve a new type of energy storage with high capacity. The device provides a very promising development direction.

7.4 Summary and Outlook

Here, we present a summary of recent research progress on cathode materials for lithium–sulfur batteries, room temperature sodium–sulfur battery, magnesium–sulfur battery, and lithium–selenium battery. This chapter introduces 1D carbon nanostructures, conductive polymers as well as all kinds of metal compounds in solid application in the lithium–sulfur battery. The introduction of these materials increases the material conductivity, at the same time also limits the volume expansion of electrodes. The application of nanowire materials in room temperature sodium–sulfur battery and magnesium–sulfur battery are still in the nascent stage of research and development. It is also necessary to develop composite materials with catalytic properties to solve the problem of slow dynamics and dissolution of polysulfides. As for the lithium–sulfur battery, due to the different characteristics between selenium and sulfur, we also focused on introducing the different reaction mechanism, problems faced in large-scale applications, and solutions. It is worth mentioning that, from a longer-term perspective, due to the shortage of lithium resources and increasing prices, sodium–selenium (Na–Se) batteries and magnesium–selenium (Mg–Se) batteries are also expected to become future research directions.

References

1 Evers, S. and Nazar, L.F. (2013). New approaches for high energy density lithium-sulfur battery cathodes. *Accounts of Chemical Research* 46 (5): 1135–1143.

2 Han, S.C., Song, M.S., Lee, H. et al. (2003). Effect of multiwalled carbon nanotubes on electrochemical properties of lithium/sulfur rechargeable batteries. *Journal of the American Chemical Society* 150 (7): A889.

3 Yang, Y., Zheng, G., and Cui, Y. (2013). Nanostructured sulfur cathodes. *Chemical Society Reviews* 42 (7): 3018–3032.

4 Manthiram, A., Fu, Y., Chung, S.H. et al. (2014). Rechargeable lithium-sulfur batteries. *Chemical Reviews* 114 (23): 11751–11787.

5 Bruce, P.G., Freunberger, S.A., Hardwick, L.J. et al. (2012). Li-O_2 and Li-S batteries with high energy storage. *Nature Materials* 11 (1): 19–29.

6 Li, W., Zheng, G., Yang, Y. et al. (2013). High-performance hollow sulfur nanostructured battery cathode through a scalable, room temperature, one-step, bottom-up approach. *Proceedings of the National Academy of Sciences* 110 (18): 7148–7153.

7 Zhou, G. (2017). Graphene-pure sulfur sandwich structure for ultrafast, long-life lithium-sulfur batteries. In: *Design, Fabrication and Electrochemical Performance of Nanostructured Carbon Based Materials for High-Energy Lithium-Sulfur Batteries*, 75–94. Singapore: Springer.

8 Zhou, G., Paek, E., Hwang, G.S. et al. (2015). Long-life Li/polysulphide batteries with high sulphur loading enabled by lightweight three-dimensional nitrogen/sulphur-codoped graphene sponge. *Nature Communications* 6 (1): 1–11.

9 Rosenman, A., Elazari, R., Salitra, G. et al. (2014). Li-S cathodes with extended cycle life by sulfur encapsulation in disordered micro-porous carbon powders. *Journal of Electrochemical Science and Technology* 161 (5): A657.

10 Ji, L., Rao, M., Aloni, S. et al. (2011). Porous carbon nanofiber-sulfur composite electrodes for lithium/sulfur cells. *Energy & Environmental Science* 4 (12): 5053–5059.

11 Ji, X. and Lee, K.T. (2009). A highly ordered nanostructured carbon-Sulphur cathode for lithium-Sulphur batteries. *Nature Materials* 8 (6): 500–506.

12 Zheng, G., Yang, Y., Cha, J.J. et al. (2011). Hollow carbon nanofiber-encapsulated sulfur cathodes for high specific capacity rechargeable lithium batteries. *Nano Letters* 11 (10): 4462–4467.

13 Zheng, G., Zhang, Q., Cha, J.J. et al. (2013). Amphiphilic surface modification of hollow carbon nanofibers for improved cycle life of lithium sulfur batteries. *Nano Letters* 13 (3): 1265–1270.

14 Zhao, Y., Wu, W., Li, J. et al. (2014). Encapsulating MWNTs into hollow porous carbon nanotubes: a tube-in-tube carbon nanostructure for high-performance lithium-sulfur batteries. *Advanced Materials* 26 (30): 5113–5118.

15 Li, Z., Zhang, J.T., Chen, Y.M. et al. (2015). Pie-like electrode design for high-energy density lithium–sulfur batteries. *Nature Communications* 6 (1): 1–8.

16 Pan, H., Chen, J., Cao, R. et al. (2017). Non-encapsulation approach for high-performance Li-S batteries through controlled nucleation and growth. *Nature Energy* 2 (10): 813–820.

17 Moon, S., Jung, Y.H., Jung, W.K. et al. (2013). Encapsulated monoclinic sulfur for stable cycling of Li-S rechargeable batteries. *Advanced Materials* 25 (45): 6547–6553.

18 Guo, J. and Xu, Y. (2011). Sulfur-impregnated disordered carbon nanotubes cathode for lithium-sulfur batteries. *Nano Letters* 11 (10): 4288–4294.

19 Fang, R., Zhao, S., Hou, P. et al. (2016). 3D interconnected electrode materials with ultrahigh areal sulfur loading for Li-S batteries. *Advanced Materials* 28 (17): 3374–3382.

20 Yuan, Z., Peng, H.J., Huang et al. (2014). Hierarchical free-standing carbon-nanotube paper electrodes with ultrahigh sulfur-loading for lithium-sulfur batteries. *Advanced Functional Materials* 24 (39): 6105–6112.

21 Liu, J., Wang, M., Xu, N. et al. (2018). Progress and perspective of organosulfur polymers as cathode materials for advanced lithium-sulfur batteries. *Energy Storage Materials* 15: 53–64.

22 Xiao, L., Cao, Y., Xiao, J. et al. (2012). A soft approach to encapsulate sulfur: polyaniline nanotubes for lithium-sulfur batteries with long cycle life. *Advanced Materials* 24 (9): 1176–1181.

23 Yang, Y., Yu, G., Cha, J.J. et al. (2011). Improving the performance of lithium–sulfur batteries by conductive polymer coating. *ACS Nano* 5 (11): 9187–9193.

24 He, F., Ye, J., Cao, Y. et al. (2017). Coaxial three-layered carbon/Sulfur/polymer nanofibers with high Sulfur content and high utilization for lithium-Sulfur batteries. *ACS Applied Materials & Interfaces* 9 (13): 11626–11633.

25 Li, Z., Guan, B.Y., Zhang, J. et al. (2017). A compact nanoconfined sulfur cathode for high-performance lithium-sulfur batteries. *Joule* 1 (3): 576–587.

26 Li, Z., Zhang, J., and Lou, X.W. (2015). Hollow carbon nanofibers filled with MnO_2 nanosheets as efficient sulfur hosts for lithium-sulfur batteries. *Angewandte* 127 (44): 13078–13082.

27 Liang, X., Hart, C., Pang, Q. et al. (2015). A highly efficient polysulfide mediator for lithium-sulfur batteries. *Nature Communications* 6 (1): 1–8.

28 Liang, X., Kwok, C.Y., Lodi-Marzano, F. et al. (2016). Tuning transition metal oxide-sulfur interactions for long life lithium sulfur batteries: the "goldilocks" principle. *Advanced Energy Materials* 6 (6): 1501636.

29 Liu, D., Zhang, C., Zhou, G. et al. (2018). Catalytic effects in lithium-sulfur batteries: promoted sulfur transformation and reduced shuttle effect. *Advancement of Science* 5 (1): 1700270.

30 Al Salem, H., Babu, G., Rao, C. et al. (2015). Electrocatalytic polysulfide traps for controlling redox shuttle process of Li-S batteries. *Journal of the American Chemical Society* 137 (36): 11542–11545.

31 Pu, J., Shen, Z., Zheng, J. et al. (2017). Multifunctional Co_3S_4@ sulfur nanotubes for enhanced lithium-sulfur battery performance. *Nano Energy* 37: 7–14.

32 Zhou, G., Tian, H., Jin, Y. et al. (2017). Catalytic oxidation of Li_2S on the surface of metal sulfides for Li-S batteries. *Proceedings of the National Academy of Sciences* 114 (5): 840–845.

33 Cui, Z., Zu, C., Zhou, W. et al. (2016). Mesoporous titanium nitride-enabled highly stable lithium-sulfur batteries. *Advanced Materials* 28 (32): 6926–6931.

34 Sun, Z., Zhang, J., Yin, L. et al. (2017). *Conductive Porous Vanadium Nitride/Graphene Composite as Chemical Anchor of Polysulfides for Lithium-Sulfur Batteries*, 14627. Commun: Nat.

35 Li, Z., He, Q., Xu, X. et al. (2018). A 3D nitrogen-doped graphene/TiN nanowires composite as a strong polysulfide anchor for lithium-Sulfur batteries with enhanced rate performance and high areal capacity. *Advanced Materials* 30 (45): 1804089.

36 Dunn, B., Kamath, H., and Tarascon, J.M. (2011). Electrical energy storage for the grid: a battery of choices. *Science* 334 (6058): 928–935.

37 Xin, S., Yin, Y.X., Guo, Y.G. et al. (2014). A high-energy room-temperature sodium-sulfur battery. *Advanced Materials* 26 (8): 1261–1265.

38 Yu, X. and Manthiram, A. (2014). Room-temperature sodium-sulfur batteries with liquid-phase sodium polysulfide catholytes and binder-free multiwall carbon nanotube fabric electrodes. *Journal of Physical Chemistry C* 118 (40): 22952–22959.

39 Yu, X. and Manthiram, A. (2015). Ambient-temperature sodium-Sulfur batteries with a Sodiated Nafion membrane and a carbon nanofiber-activated carbon composite electrode. *Advanced Energy Materials* 5 (12): 1500350.

40 Kim, H.S., Arthur, T.S., Allred, G.D. et al. (2011). Structure and compatibility of a magnesium electrolyte with a Sulphur cathode. *Nature Communications* 2 (1): 1–6.

41 Yu, X. and Manthiram, A. (2016). Performance enhancement and mechanistic studies of magnesium-sulfur cells with an advanced cathode structure. *ACS Energy Letters* 1 (2): 431–437.

42 Armand, M. and Tarascon, J.M. (2008). Building better batteries. *Nature* 451 (7179): 652–657.

43 Sun, Y.K., Myung, S.T., Park, B.C. et al. (2009). High-energy cathode material for long-life and safe lithium batteries. *Nature Materials* 8 (4): 320–324.

44 Poizot, P.L.S.G., Laruelle, S., Grugeon, S. et al. (2000). Nano-sized transition-metal oxides as negative-electrode materials for lithium-ion batteries. *Nature* 407 (6803): 496–499.

45 Goodenough, J.B. and Kim, Y. (2010). Challenges for rechargeable Li batteries. *Chemistry of Materials* 22 (3): 587–603.

46 Ji, X., Nazar, L.F., Hardwick, L.J. et al. (2010). Advances in Li-S batteries. *Journal of Materials Chemistry A* 20 (44): 9821–9826.

47 Barghamadi, M., Kapoor, A., Wen, C. et al. (2013). A review on Li-S batteries as a high efficiency rechargeable lithium battery. *Journal of Electrochemical Science and Technology* 160 (8): A1256.

48 Abouimrane, A., Dambournet, D., Chapman, K.W. et al. (2012). A new class of lithium and sodium rechargeable batteries based on selenium and selenium–sulfur as a positive electrode. *Journal of the American Chemical Society* 134 (10): 4505–4508.

49 Cui, Y., Abouimrane, A., Sun, C.J. et al. (2014). Li–Se battery: absence of lithium polyselenides in carbonate based electrolyte. *ChemComm* 50 (42): 5576–5579.

50 Cui, Y., Abouimrane, A., Lu, J. et al. (2013). (De) Lithiation mechanism of Li/SeS_x (x = 0–7) batteries determined by in situ synchrotron X-ray diffraction and X-ray absorption spectroscopy. *Journal of the American Chemical Society* 135 (21): 8047–8056.

51 Li, Q., Liu, H., Yao, Z. et al. (2016). Electrochemistry of selenium with sodium and lithium: kinetics and reaction mechanism. *ACS Nano* 10 (9): 8788–8795.

52 Yang, C.P., Xin, S., Yin, Y.X. et al. (2013). An advanced selenium–carbon cathode for rechargeable lithium–selenium batteries. *Angewandte Chemie, International Edition* 52 (32): 8363–8367.

53 Zhang, J., Xu, Y., Fan, L. et al. (2015). Graphene–encapsulated selenium/polyaniline core–shell nanowires with enhanced electrochemical performance for Li–Se batteries. *Nano Energy* 13: 592–600.

54 Zeng, L., Zeng, W., Jiang, Y. et al. (2015). A flexible porous carbon nanofibers-selenium cathode with superior electrochemical performance for both Li-Se and Na-Se batteries. *Advanced Energy Materials* 5 (4): 1401377.

55 Han, K., Liu, Z., Ye, H. et al. (2014). Flexible self-standing graphene–Se@ CNT composite film as a binder-free cathode for rechargeable Li–Se batteries. *Journal of Power Sources* 263: 85–89.

56 Han, K., Liu, Z., Shen, J. et al. (2015). A free-standing and ultralong-life lithium-selenium battery cathode enabled by 3D mesoporous carbon/graphene hierarchical architecture. *Advanced Functional Materials* 25 (3): 455–463.

57 Balakumar, K. and Kalaiselvi, N. (2017). Selenium containing tube-in-tube carbon: a one dimensional carbon frame work for selenium cathode in Li-Se battery. *Carbon* 112: 79–90.

58 Lee, J.T., Kim, H., Oschatz, M. et al. (2015). Micro-and mesoporous carbide-derived carbon–selenium cathodes for high-performance lithium selenium batteries. *Advanced Energy Materials* 5 (1): 1400981.

59 Zhang, Z., Zhang, Z., Zhang, K. et al. (2014). Improvement of electrochemical performance of rechargeable lithium–selenium batteries by inserting a free-standing carbon interlayer. *RSC Advances* 4 (30): 15489–15492.

60 Fang, R., Zhou, G., Pei, S. et al. (2015). Localized polyselenides in a graphene-coated polymer separator for high rate and ultralong life lithium–selenium batteries. *ChemComm* 51 (17): 3667–3670.

61 Yang, C.P., Yin, Y.X., and Guo, Y.G. (2015). Elemental selenium for electrochemical energy storage. *Journal of Physical Chemistry Letters* 6 (2): 256–266.

8

Application of Nanowires in Supercapacitors

Among the electrochemical energy storage (EES) devices, secondary supercapacitors and batteries are two crucial units that are deft at releasing the stored electric energy with long discharge time and high-power output, respectively. In the previous chapters of 5th, 6th, and 7th, the one-dimensional (1D) nanomaterials developed for various alkali metal ion-based batteries and high-energy batteries are discussed systematically. This unique structure provides a new idea for development of secondary batteries with high power density. Similarly, the supercapacitor, first developed in 1957 [1], is another extensively used EES. The nanowire structure possesses huge potential to promote the inferior energy density and maintain its native high cycle life and fast charge–discharge capability.

Supercapacitors include pseudocapacitors, electric double-layer capacitors (EDLC), and so on, all of which are interface or near-interface reaction between electrode material and electrolyte. Although there are some differences in energy storage mechanism between EDLC and pseudocapacitances, the interface state of electrode material is closely related to the specific capacitance. When increasing the specific surface area (SSA) of the material, there is a higher requirement for the "ability" of the ions to diffuse from bulk phase to the energy storage electrolyte interface. The transfer of electrons from the external circuit to the reaction interface is also different between the two energy storage systems. The transformation of the bulk phase structure leads to changes in the electrical conductivity of the secondary battery materials, and more charge transfer also requires better electron transport capacity of the electrode materials, which is also applicable to the pseudocapacitors based on oxide semiconductors. Therefore, high power density is derived from rapid surface reaction, ion transport, and electron conduction of supercapacitor.

The surface state, ion diffusion, and electronic conductivity of supercapacitor electrode materials are critical to their performance, and massive research is also carried out around these three points. In 2005, Wang Xianyou et al. used the anodic aluminum oxide (AAO) template method to electrodeposit MnO_2 nanowire arrays and applied them to supercapacitors, showing the possibility of

Nanowire Energy Storage Devices: Synthesis, Characterization, and Applications, First Edition.
Edited by Liqiang Mai.

application of nanowire materials in EES devices [2]. On the whole, nanowire materials assembled in electrochemical capacitors and secondary batteries are not strictly distinguished. Compared with bulk materials, they have unique physical and chemical properties: larger SSA can provide more adsorption sites, and linear structure benefits a variety of structural design possibilities. The one-dimensional cross-linked network is conducive to ion diffusion, which is still suitable in the supercapacitor. The difference is that, due to the characteristics of surface energy storage, there is few ion diffusion in the bulk phase.

In the EDLC and pseudocapacitors, there are some differences in designing one-dimensional materials for practical applications. When EDLC charges, along with the transfer of electrolyte ions, voltage between cathodes and anodes is generated. When the supercapacitor discharges, electrons travel through an appliance from anode to cathode, generating a current, and the positive and negative ions are released from electrodes and enter the electrolyte. 1D nanocarbon materials such as carbon nanotubes (CNTs) and carbon nanofibers (CNFs), have unique advantages in structural design. Obviously, the pore structure is the research core of electrode materials for EDLC. 1D nanowire electrode materials are easy to construct porous nanowire structure due to their large aspect ratio, which accelerates the diffusion of ions in pore structure and increases the diversity of pore structure design in carbon-based electrode material. In addition, the cross-linked network in nanowire structure can realize the continuity of the macroporous structure and further promote the slow transport process of electrolyte ions between the interface and electrolyte during the charge and discharge.

Pseudocapacitive supercapacitors are energy storage devices with electrochemical performance between EDLC and lithium-ion batteries. Whether they are underpotential deposition, redox faradaic reaction, or intercalation reaction mechanism, rapid electrochemical reaction steps require electrode materials to have corresponding electron conductivity and ion diffusion ability (especially in porous electrode materials). Similar to the 1D EDLC electrode material, the abundant pore structure and cross-linked network facilitate the diffusion of ions. Pseudocapacitor electrode materials based on transition metal oxides have semiconductor properties and are often not conducive to electrons. Therefore, composite material has become a better choice for application in pseudocapacitive supercapacitors. The one-dimensional nanowire materials with pseudocapacitance can be in closer contact with the conductive materials to further enhance their electrical conductivity. At the same time, their unique 1D structure can alleviate the expansion of electrode material in the charging/discharging processes, and effectively enhance the cycling stability of supercapacitors.

This chapter will focus on 1D nanomaterials applied to three typical capacitors, including EDLCs, pseudocapacitive supercapacitors, and hybrid supercapacitors. The relationship between 1D structure and performance of capacitors is explored, which could provide theoretical guidance for the development of nanowire-based supercapacitors.

8.1 Nanowire Electrode Material for Electrochemical Double-Layer Capacitor

EDLCs play an irreplaceable role in energy storage owing to their unique merits, such as high power density, short charging time, and long service life. The EDLCs operate on ion adsorption/desorption at the electrode/electrolyte interface, which is highly reversible owing to their physical charge storage mechanisms. Basically, when an external voltage is applied between the EDLC electrodes, the electronic charge that accumulates at electrode surface is balanced through the adsorption of cations and anions from the electrolyte, and the electrolyte concentration remains roughly constant.

Nanowires, CNTs, and CNFs as 1D porous carbon materials possess anisotropy and large aspect ratio, benefiting rapid axial electron transmission and radial ion diffusion in energy storage [3]. As a member of the carbon family, 1D porous carbon materials have enriched the pore structure design of carbon-based electrode materials, receiving extensive attention. In addition, the 1D porous carbon materials are characterized by excellent thermochemical stability, and can avoid the agglomeration problem of nanomaterials. The 1D carbon materials rely on electrode/electrolyte interfaces to charge storage rather than interior parts of electrode materials, the effective SSA of electrolyte infiltration that forming the electrochemical double layer determines the capacitance. The SSA and pore structure are two principal parameters that make a difference to the performance of carbon materials. Engineering the pore volume, microstructure, crystallinity, and surface functionality of one-dimensional porous carbon materials hold the key to boosting their capacitive performance [4].

Generally, the larger the specific surface on the carbon material, the more charge is stored at the interface. However, improving the SSA does not generate a linear increase in capacitance because not all pores can be infiltrated by electrolyte ions [5]. In addition, pores with different sizes also contribute quite different capacitance. According to the formula $C = \varepsilon A/d$, where C is capacitance created by this charge separation at the interface between electrode and electrolyte, ε is dielectric constant of electrolyte, d (m) is average approaching distance of ions to the electrode surface, and A (m^2) is the accessible SSA of electrode. Pores with similar size to ion diameter can contribute the most capacitance, which decreases in the slightly largerer pores and macropores 0.7 and 0.8 nm are the best-suited pore diameters in regular aqueous and organic electrolytes, respectively [6]. Therefore, microporous carbon materials are important in developing EDLSs with high specific capacitance. The Gogotsi group found a relationship between the pore size of the carbon material and the ion size of the electrolyte in the organic electrolyte, and believed that the material's specific capacitance can be maximized when the pores of the material and the electrolyte ions are similar in size, which also applies to aqueous electrolytes. The demonstration of charge storage in pores with smaller size than that of ion in electrolyte will improve our understandings of ionic transport in porous media. These

findings should also permit the design of application-specific supercapacitors: for longer discharge times where energy density is at a premium, such as in hybrid electric vehicles, extremely narrow pores should prove optimal, but for pulse power applications, increasing the pore size might be beneficial. Further tuning the carbon porosity and designing the carbon materials with a large volume of narrow but short pores may allow both energy and power characteristics to be improved.

8.1.1 The Application of Carbon Nanotubes in EDLCs

CNTs are large cylindrical carbon materials consisting of a hexagonal arrangement of sp^2 hybridized carbon atoms, which are formed by rolling up a single sheet of graphene (single-walled carbon nanotubes, SWCNTs) or by rolling up multiple sheets of graphene (multiwalled carbon nanotubes, MWCNTs) (Figure 8.1) [7]. To date, the commonly used synthesis techniques for CNTs are arc discharge, laser ablation, and chemical vapor deposition (CVD) [8]. The SSA of CNTs ranges from 100 to 1000 $m^{-2}\ g^{-1}$. CNTs can exhibit narrow pore diameter range, but the internal pores are unlikely to contribute to EDL capacitance due to the ion diffusion limitation inside the carbon walls and because of the absence of the inner electric field in regular operating conditions. However, the outer surface of CNTs is convex, exhibiting a strong affinity with electrolyte ions. Generally, convex surfaces with higher curvature can significantly increase the specific capacitance [9]. CNTs possess brilliant electrical conductivity and can maintain the dominant position even under extreme climatic conditions, which makes them suitable candidates for high-power devices.

When applied as electrode materials in EDLCs, the research point of CNTs is generally not on the structural design of a single-carbon nanotube, but on the integrated design of multiple carbon nanotubes. Hata et al. synthesized high-density stacked SWCNTs by employing the zipper effect of the liquid [10]. When assembled in EDLCs, the power density reached 43 $kW\ kg^{-1}$. Pan et al. successfully synthesized a MWCNT film through electrophoretic deposition technology with a power density of 0.02 $kW\ g^{-1}$ [11]. In view of their excellent conductivity, CNT-based EDLCs demonstrate an ultrahigh frequency response. CNTs are generally employed as

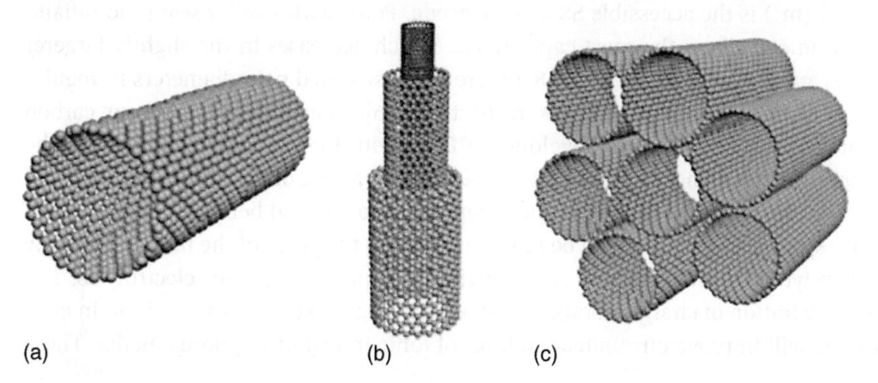

Figure 8.1 Schematic diagram of different types of CNTs: (a) SWCNTs; (b) MWCNTs; (c) CNT bundles. Source: Adapted from Baughman et al. [7].

model materials to study the formation of electrochemical double layers inside and outside the tubes in simulation experiments. Although CNT-based electrode materials exhibit multiple merits, the expensive synthesis process limits their application. Therefore, developing facile CNT synthesis technology is an extremely important research direction.

8.1.2 The Application of Carbon Nanofibers in EDLCs

CNFs are 1D carbon nanomaterials composed of stable continuous carbon molecules with high strength and high modulus. Commonly used CNFs preparation techniques include template method, sol–gel method, electrostatic spinning method, and CVD method [12]. Although the conductivity of CNFs is inferior to that of CNTs, CNFs possess abundant defect reactive sites, flexible designability, and cost-effective production, which are more prospective for EDLCs. The 1D structure of CNFs guarantees efficient mass transfer, and the abundant defective active sites on the surface provide a sufficient electrode/electrolyte contact interface. Carbon precursors tend to form 3D spherical structures with lower surface energy rather than 1D structures during spontaneous polymerization. Therefore, 1D phenolic resin-based porous carbon materials are mainly synthesized by template method including soft template and hard template and electrospinning method.

Lu's research group has made achievements in 1D phenolic resin-based carbon fibers [13]. They fabricated the uniform polymer nanofibers (PNFs) via a controlled hydrothermal approach by using resorcinol and hexamethylenetetramine (HMTA) as carbon precursors and Pluronic F127 as morphology-directing agents. The morphology of the end product is closely related to the adding amount of F127. By controlling the amount of F127 from 0.0159 to 0.0636 mmol, the morphology of the phenolic resin was transformed from nanospheres to nanofibers (Figure 8.2) [14]. Subsequently, they obtained mesoporous phenolic resin-based CNFs with a radius of 20 nm and micrometers long by introducing SiO_2 in CNFs. After steam activation, activated CNFs with a SSA of 1218 m^{-2} g^{-1} was obtained. CNFs-I-A/NF displayed a capacitance of 276 F g^{-1} under 0.5 A g^{-1}, and retained a capacitance of 206 F g^{-1} under 5 A g^{-1}, indicating a good capacitance retention capability of such an electrode [15].

Apart from template method, electrospinning technology is another common mean for preparing 1D resin-based porous carbon materials [13]. Bai et al. immersed resole-type phenolic resin and polyvinyl butyral (PVB) in ethanol or ethanol/*N*,*N*-dimethylformamide (DMF) to prepare microporous carbon ultra-fine fibers with different diameters by electrospinning [16]. Average diameter of as-spun fibers changed from 1.1 to 0.33 μm with increasing dimethylformamide content in the spinning solution, caused by more fiber divisions. Li et al. prepared mesoporous CNFs by a sol–gel/electrospinning process using phenolic resin precursor as carbon source and triblock copolymer Pluronic F127 as template. And the mesopore volumes depended on the amount of tetraethyl orthosilicate (TEOS) in the spinnable sols [17]. Mai et al. proposed a low-cost, one-step synthesis of magnesium hydroxide ($Mg(OH)_2$) that was deposited into polyacrylonitrite (PAN) nanofibers by electrospinning [18]. After removing the MgO template, the

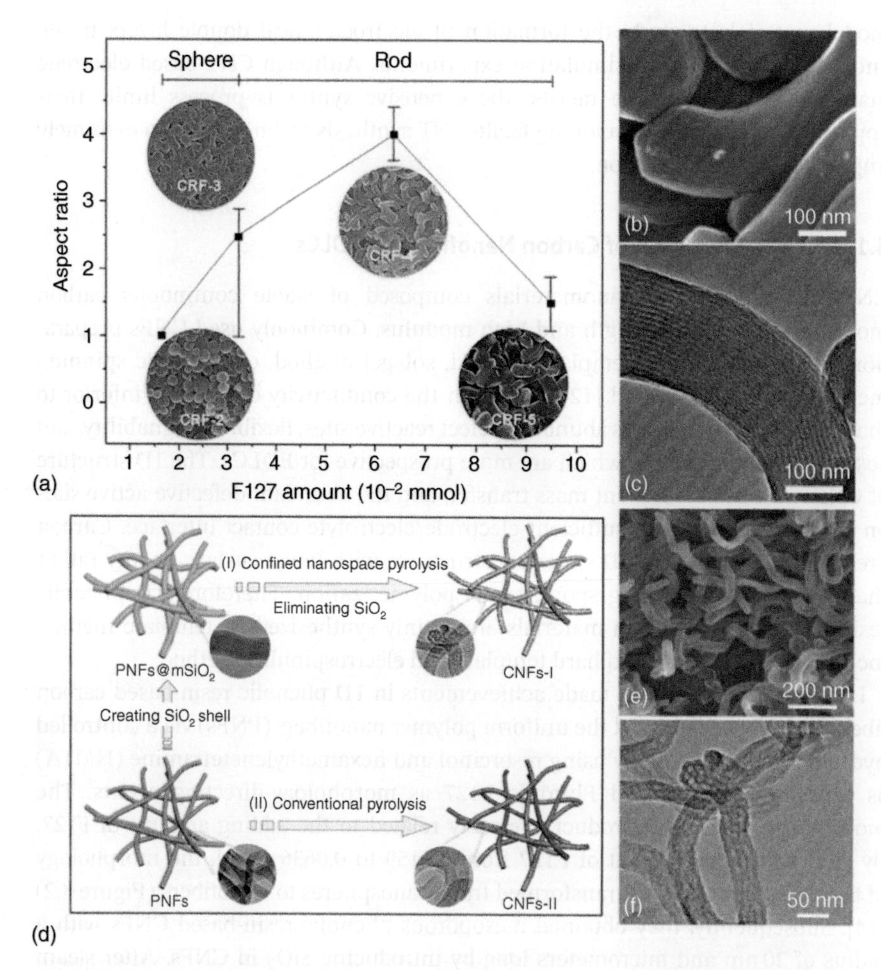

Figure 8.2 Variation in morphology and aspect ratio of the as-synthesized carbon materials along with F127 amount (a); SEM (b) and TEM (c) images of CRF-4. Source: Sun et al. [14] / with permission of Royal Society of Chemistry. Schematic illustration for the formation process of mesoporous carbon nanofibers (CNFs-I) and microporous carbon nanofibers (CNFs-II) (d); SEM (e) and TEM (f) images of CNFs-I-A in water. Source: Zhang et al. [15] / with permission of Royal Society of Chemistry.

mesopores were formed in the CNFs. The as-constructed supercapacitor based on N-doped mesoporous CNFs can deliver excellent performance with an ultrahigh capacitance of 327.3 $F g^{-1}$ under 1.0 $A g^{-1}$, and a remarkable stability, e.g. only about 7% loss after 10 k cycles under constant charging–discharging current of 20 $A g^{-1}$ in 6 M KOH electrolyte.

Currently, the preparation of CNFs is more convenient than CNTs. However, the commercialization evolution is still hampered, not only because most of the synthesis procedure requires templates but also the later template removal is cumbersome and energy-consuming. In addition, the synthesis process employs considerable environmentally unfriendly organic solvents. Therefore, the large-scale preparation of cheap 1D porous carbon materials in aqueous phase for supercapacitors is the principal challenge for researchers.

8.2 Nanowire Electrode Materials for Pseudocapacitive Supercapacitors

EES technology is very important for 3C electronic products, transportation, as well as huge energy storage systems. Porous 1D nanomaterial combines the advantages of 1D nanostructure and porous structure, and greatly promotes the development of EES field.

The active materials of pseudocapacitive supercapacitors are mainly transition metal compounds and conducting polymers [19–24]. Through the rapid surface redox reaction of electrode material, it can maintain a high power density while obtaining a high energy density. The key factors relating to the properties of pseudocapacitive supercapacitors mainly include SSA, electrical conductivity, structural stability, and so on.

8.2.1 Metal Oxide Nanowire Electrode Materials

The most widely studied 1D transition metal oxide is RuO_2, which possesses high specific capacitance, good conductivity, relative stability in the electrolyte, and is a kind of excellent pseudocapacitive material in supercapacitors. Buzzanca et al., an Italian electrochemical laboratory, first proposed and used RuO_2 as electrode material for electrochemical capacitors [25]. Zhou et al. prepared RuO_2 nanowires/SWCNTs composite membrane through a simple method [26]. After the introduction of RuO_2 nanowires, the charge transfer impedance of the composite film decreased by 48%, the specific capacity increased to 138 F g^{-1}, and the energy density was 0.0188 Wh g^{-1}.

However, the high price of precious metal Ru in RuO_2 restricts its further development. In other studies of cheap transition metal oxides, MnO_2 is cheap, abundant in nature, and has a variety of oxidation states. It shows good Shim capacitance characteristics in neutral electrolyte, and its voltage window is wide, generally up to more than 1 V. It is a kind of new capacitor material with a low price and good effect. However, its electrical conductivity is low, and the electrochemical reaction rate is also slow. By compositing with carbon materials and conductive polymers, the electrical conductivity of the composite material can be enhanced, and good electrochemical performance can be obtained at high rate. Nguyen et al. synthesized reduced GO, Co_3O_4, and ultralong MnO_2 nanowires (5–10 μm) ternary composites through hydrothermal method [27], the specific capacity reached 1560 F g^{-1} under 2 A g^{-1}, and the specific capacity retention reached 96% at 1500th cycle. In addition, Lin et al. synthesized a coaxial MnO_2/PANI super-long nanowire structure [28], whose cross-connected network structure could increase the electronic conductivity as well as specific capacity. When the mass fraction of MnO_2 is 62.5%, the material presented an area capacitance of 346 mF cm^{-2} at a scanning speed of 5 mV s^{-1}.

Increasing SSA and reducing diffusion distance are other ways to get superior pseudocapacitive materials. Among these researches, Mai et al. realized the heterotrophic nucleation growth of cobalt molybdate nanowires based on manganese molybdate nanowires through "self-assembly orientation overlapping" process by matching the crystal plane of cobalt molybdate and manganese molybdate [29]. The electrochemical properties of composite isothioether-structured nanomaterials are better than those of single-structured materials. Under mild conditions, the hierarchical heterostructures were prepared on $MnMoO_4$ by simple reflux method. And the surface modification is realized. Based on $MnMoO_4/CoMoO_4$ layered heterostructured nanowires, asymmetric supercapacitors have been fabricated. The layered heterostructured nanowires can accommodate lattice expansion and contraction, and the slip at the interface can significantly alleviate the stress concentration in the nano-electrode material. The SSA increased by 18 times compared with before construction, the energy density of the energy storage device increased by one order of magnitude, and the specific capacity retention rate increased from 78% to 98% after 1000 cycles, jointly improving the electrical transport and structural stability of the material (Figure 8.3).

As shown in Figure 8.4, supersaturated solution, including massive $CoMoO_4$ crystal seeds, were formed by inducing cobalt salt. Due to their high surface energy, nanoparticles are prone to growing on the surface of $MnMoO_4$. The substrate, $MnMoO_4$, has similar lattice parameters and can control $CoMoO_4$ nanoparticle self-assembly and oriented crystallization to form hierarchical $MnMoO_4/CoMoO_4$ heterostructure nanowires.

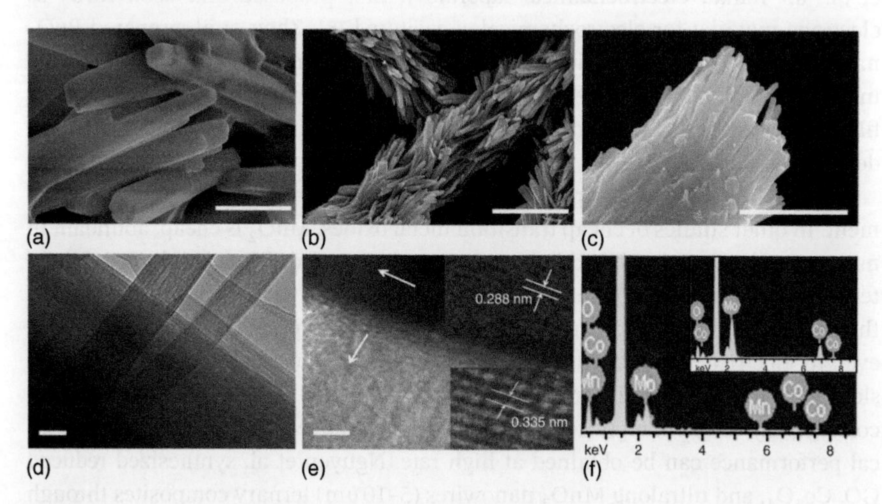

Figure 8.3 Morphology characterization. (a) SEM image of the backbone $MnMoO_4$ nanowires. (b, c) SEM images of $MnMoO_4/CoMoO_4$ nanowires. (d, e) TEM and HRTEM images at the heterojunction of $MnMoO_4/CoMoO_4$ nanowires. (f) The EDS microanalysis of selected areas. Scale bars, 1 μm (a–c); 20 nm (d); 5 nm (e). Source: Mai et al. [29], Republished with permission from Springer Nature.

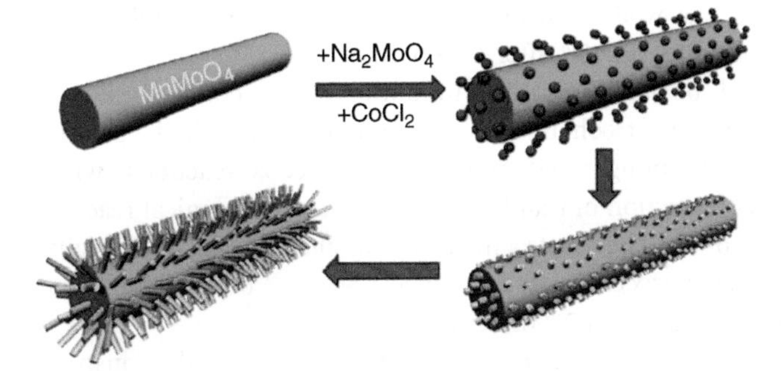

Figure 8.4 The construction of hierarchical $MnMoO_4/CoMoO_4$ nanowires. The green rod represents the backbone $MnMoO_4$ nanowire, and the orange rods are the $CoMoO_4$ nanorods. Red and blue balls are different ions dispersed in the aqueous solution. Source: Mai et al. [29], Republished with permission from Springer Nature.

8.2.2 Conducting Polymer Nanowire Electrode Materials

Conducting polymers are another kind of pseudocapacitive electrode materials. The rapid and reversible redox reaction of N/P- type doping on polymer-conjugated chains enables the storage of large amounts of charge. The P-type doping of conducting polymer refers to the loss of electrons in the polymer skeleton so that the anions in the electrolyte will adsorb and gather in the polymer chain to achieve charge balance. N-type doping, on the other hand, means that the excess charge in the polymer needs to be balanced by cations in the solution. Polymer used in pseudocapacitive electrode materials mainly include polyaniline, polypyrrole, and polythiophene.

Polyaniline is favored by people because of its low price, simple preparation method, and excellent electrochemical performance. Using vanadium pentoxide as the hard template, Bai et al. obtained polyaniline fiber hydrogel [30] by in situ polymerization (Figure 8.5). Based on its ultrafine fiber structure, its mass specific

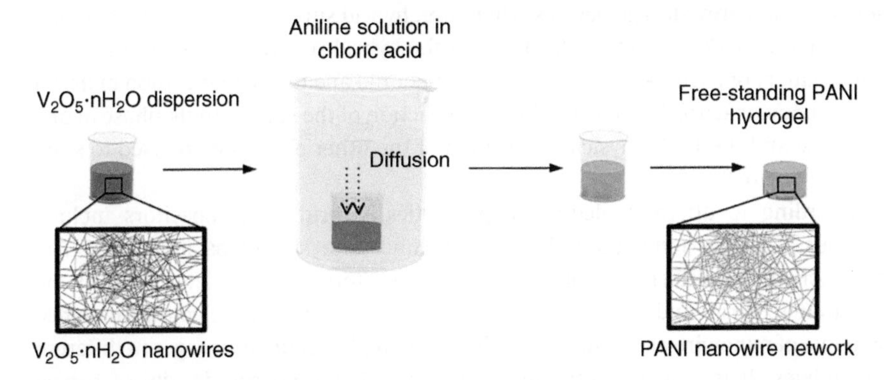

Figure 8.5 Illustrated diagram of synthesis procedure. Source: Zhou et al. [30], Reproduced with permission from ACS.

capacity reached 636 F g^{-1}, as well as excellent rate performance and stability (the specific capacity retention rate was 83% after 10 000 times).

Xie et al. synthesized polyaniline/carbon/titanium nitride composites with ternary components [31]. Electrochemically active polyaniline shells provide pseudocapacitance by performing rapid and reversible redox reactions, while carbon shells avoid the destruction of titanium nitride in electrochemical reaction and improve cycling stability. The electrochemical test showed that the specific capacity of the polyaniline matrix composites reached 1093 F g^{-1} under 1 A g^{-1}, and the specific capacity remained 98% at 2000th cycle.

Polypyrrole has high conductivity and low synthesis cost. Through the construction of polypyrrole nanofiber structure. Moreover, the network is expected to exhibit rapid ion diffusion in bulk materials. Biswas et al. prepared the polypyrrole fiber and graphene nanoflake composites [32], benefited from the polypyrrole fiber network and conductive graphene nanoflake coexisting, the nanostructured electrode showed excellent ability of ions and electrons transfer at 1 A g^{-1} cycle after 1000 times and still has a capacitance of 165 F g^{-1}.

Since Bjornholm et al. first obtained polythiophene nanowires [33], they have been developed as electrode materials for supercapacitors. Polythiophene used as electrode materials for supercapacitors is mainly prepared by modifying thiophene to a certain degree. The monomers in the synthesis process of conducting polymers are generally toxic. Although their supercapacitors have excellent performance, they are difficult to commercially produce in large quantities, which greatly limits their development.

8.3 Nanowire Electrode Materials of Hybrid Supercapacitors

The low energy density of supercapacitors is urged to be improved for commercial applications. Due to the low specific capacitance of EDLC and the low cycle life of pseudocapacitors, conventional supercapacitors can no longer meet the public's demand for energy storage devices. Therefore, hybrid supercapacitors are gradually entering the market. The two electrodes of the hybrid supercapacitor are the traditional battery or pseudocapacitive electrode material and the porous carbon material of the supercapacitor. The electrochemical reaction of the surface/bulk phase of one electrode and the EDL physical adsorption of the other electrode are used to store energy (Figure 8.6) [36].

According to different electrolyte properties, hybrid supercapacitors include aqueous, organic, ionic liquid-based and solid-state supercapacitors [37]. The electrochemical performance of hybrid supercapacitors is mainly determined by the electrode material and electrolyte [38]. In view of the excellent electron transmission rate of nanowire electrode materials, they can play their advantages in different electrolytes. It is worth mentioning that although the energy density of hybrid

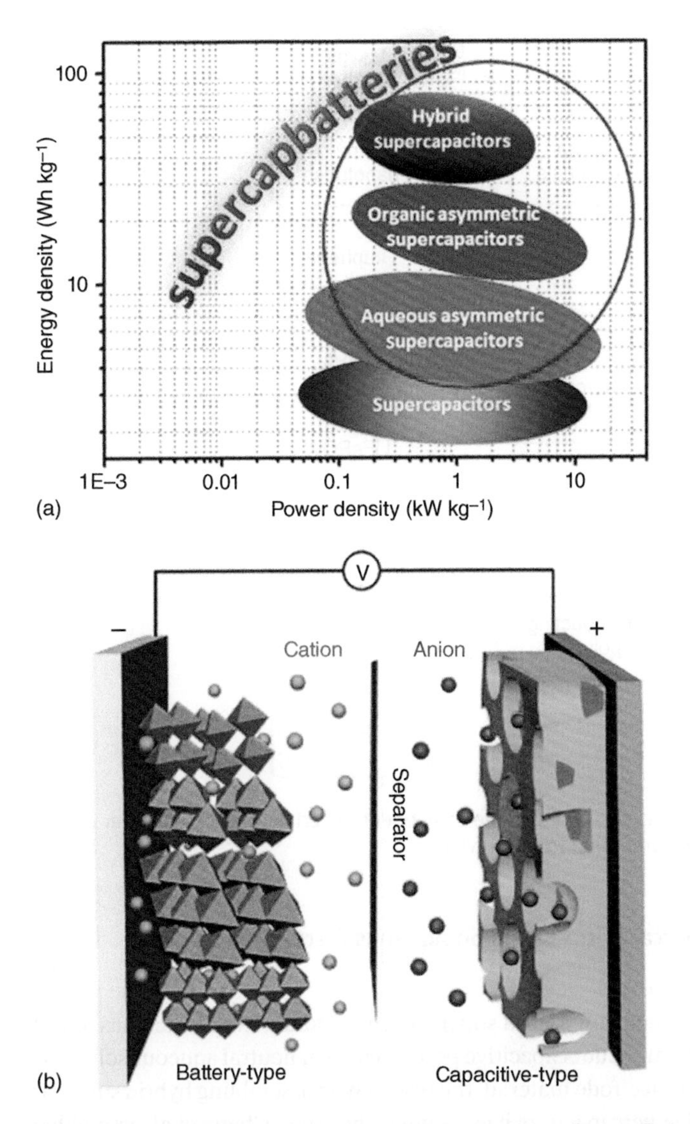

Figure 8.6 (a) Ragone plots for various supercapacitors. Source: Dubal et al. [34], Republished with permission from Elsevier. (b) The fundamental configuration of hybrid supercapacitors. Source: Liu et al. [35], Republished with permission from Elsevier.

supercapacitors has been increased significantly, there is still a certain difference in power density compared with ordinary supercapacitors. Based on the unique advantages of 1D nanowire electrode materials in electric double-layer supercapacitors and pseudocapacitive supercapacitors, their application in hybrid supercapacitors will be more extensive. Appropriate matching of the two one-dimensional nanowire electrode materials can effectively increase its working voltage and thereby increase its energy density [39] (Figure 8.7).

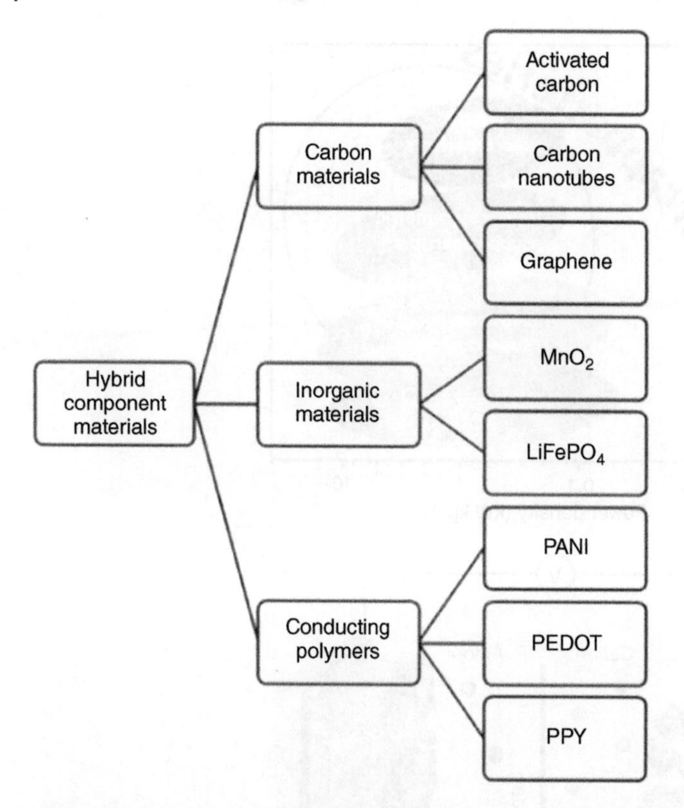

Figure 8.7 Hybrid supercapacitor electrode component materials. Source: Muzaffar et al. [40], Republished with permission from Elsevier.

8.3.1 Hybrid Supercapacitor Based on Aqueous Electrolyte

8.3.1.1 Carbon/Metal Oxide

MnO_2 is a suitable cathode for hybrid supercapacitors, and new research has found that it exhibits excellent pseudocapacitive performance in neutral aqueous solutions [41–44]. As a positive electrode material, the research on assembling hybrid supercapacitors into hybrid supercapacitors has become a hot spot. Cheng et al. assembled graphene anode and graphene/MnO_2 nanowires cathode into an asymmetric hybrid capacitor using neutral Na_2SO_4 electrolyte [41], whose working voltage can reach 2 V and its energy density can reach 30.4 Wh kg^{-1}. Graphene sheets and nanowire composites improve substantially property of hybrid supercapacitors by reducing the resistance of electrode material. Similar to MnO_2, nanowires shorten transportation distance of ions and effectively reduce charge transfer resistance. Yu et al. designed a simple and cheap method to obtain asymmetric supercapacitors without binders [43]. Bacterial cellulose-derived CNFs were used as the carrier load, MnO_2 was used as the positive electrode of the hybrid supercapacitor, and CNFs were used as the negative electrode. In neutral 1 mol l^{-1} Na_2SO_4, the voltage window can be extended to 2 V (Figure 8.8). Nitrogen-doped CNFs have excellent

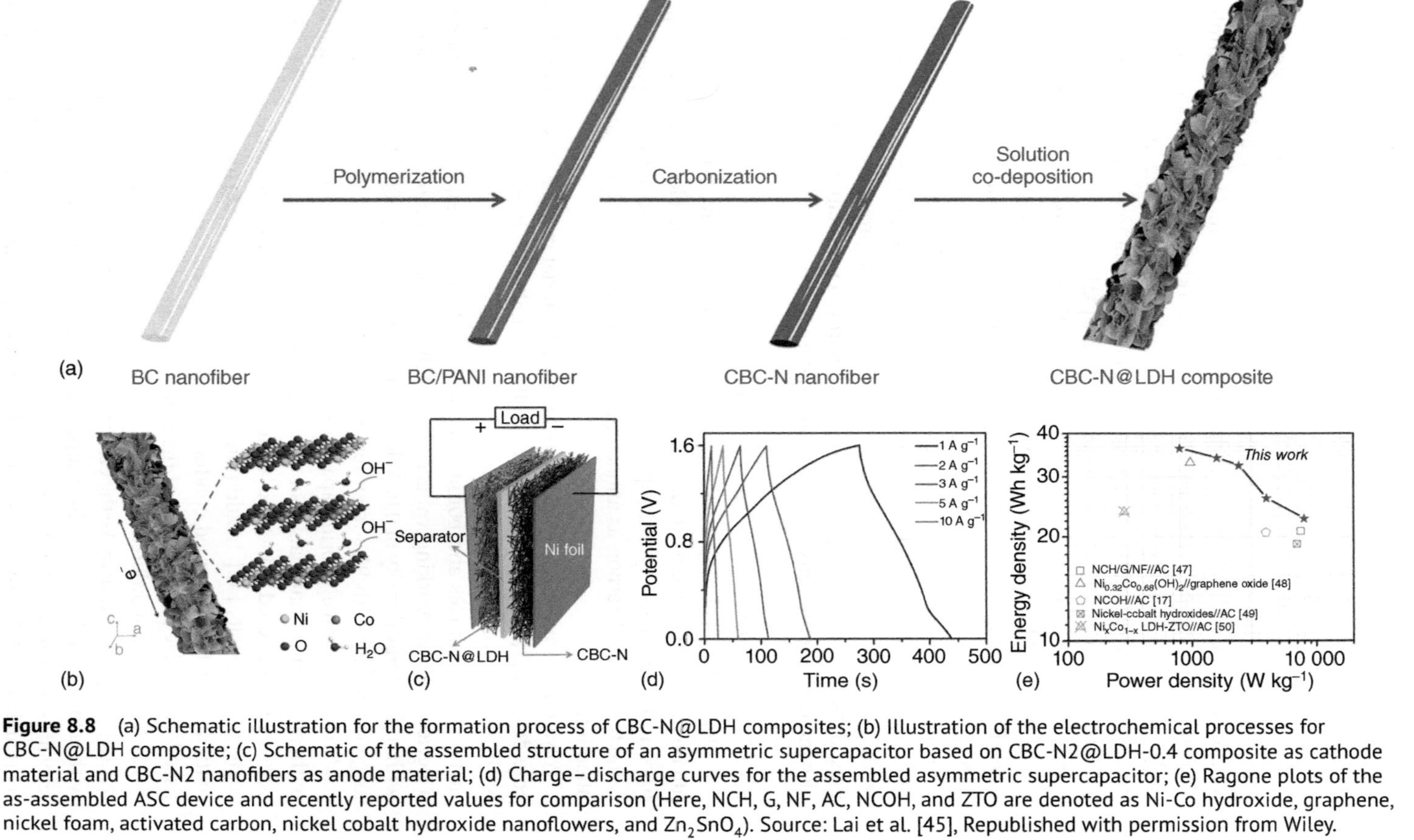

Figure 8.8 (a) Schematic illustration for the formation process of CBC-N@LDH composites; (b) Illustration of the electrochemical processes for CBC-N@LDH composite; (c) Schematic of the assembled structure of an asymmetric supercapacitor based on CBC-N2@LDH-0.4 composite as cathode material and CBC-N2 nanofibers as anode material; (d) Charge–discharge curves for the assembled asymmetric supercapacitor; (e) Ragone plots of the as-assembled ASC device and recently reported values for comparison (Here, NCH, G, NF, AC, NCOH, and ZTO are denoted as Ni-Co hydroxide, graphene, nickel foam, activated carbon, nickel cobalt hydroxide nanoflowers, and Zn_2SnO_4). Source: Lai et al. [45], Republished with permission from Wiley.

structural stability. The nano-sized structure after coating MnO_2 provides a highly electrochemically active area. Its energy density can be as high as 0.033 Wh g^{-1}, its power density can be 0.285 kW g^{-1}, and it has excellent structural cycle stability.

The operating voltage range also affects the property of hybrid supercapacitors. MnO_2 with very low crystallinity is prepared by the sol–gel method. Hybrid capacitor can be formed with activated carbon electrode in 0.65 mol l^{-1} K_2SO_4 electrolyte. The energy density can reach 19 Wh kg^{-1} with working voltage of 2.2 V. However, under such high working voltage, gas will precipitate, resulting in poor cycle performance. By reducing the voltage to 1.5 V, gas precipitation can be avoided. After 23 000 cycles, the energy retention rate is still 76% [42]. In addition, for the same type of hybrid capacitor, at voltage of 2 V, corrosion of the stainless-steel current collector can be inhibited by removing the oxygen in the system, and the cycle performance is significantly improved. After 5000 cycles, the specific capacity attenuation is only 12.5% [44].

In addition, other 1D metal oxides, such as TiO_2, are used in hybrid capacitors. Li et al. assembled a CNT cathode and TO_2-B anode into hybrid capacitor with working voltage of 2.8 V [46]. CNTs can effectively improve the diffusion rate of ions/electrons, while TiO_2 nanowires can improve cycle stability and reversibility, also increase energy density of electrode material.

8.3.1.2 Carbon/Conductive Nanowire Polymer

The electrochemical properties of conductive polymers have received widespread attention since they were discovered and have been used in electrochemical capacitors. In recent years, hybrid supercapacitors composed of conductive polymer as cathode, activated carbon as anode, and water-based electrolyte have gained wide attention.

Polyaniline is a widely researched electrode material in Faraday's pseudoelectric core device [47]. This material has excellent stability in aqueous solutions. Wang et al. used porous graphene oxide/1D carbon nanotubes/polyanilinc composite materials as cathodes and porous graphene oxide as anodes. After assembling as a device [48], its working voltage can reach 1.6 V in a 1 mol l^{-1} H_2SO_4 solution with high energy density of 0.02 Wh g^{-1} and power density of 0.025 kW g^{-1}, as well as good stability with 91% retention through 5000 times. At present, there is also research to improve hybrid capacitors by combining metal oxide nanomaterials with polyaniline materials. For example, Li et al. have prepared Fe_2O_3@PAN core-shell structured nanowire arrays, which show excellent cycle stability and high energy density [49]. In addition, Liu et al. has grown CoO nanowires on nickel foam and further deposited conductive polypyrrole to synergistically improve the pseudocapacitive performance of the material (Figure 8.9) [50].

The larger SSA of ordered mesoporous nanofibers and the excellent electrical conductivity of polypyrrole make the electrolyte ion transfer rate faster. When assembled with activated carbon as a hybrid supercapacitor, its voltage window can reach 1.8 V as well as high energy density of 0.044 Wh g^{-1}. After 20 000 times, the specific capacity remains at 91.5%. In addition to carbon materials, conductive polymers can also

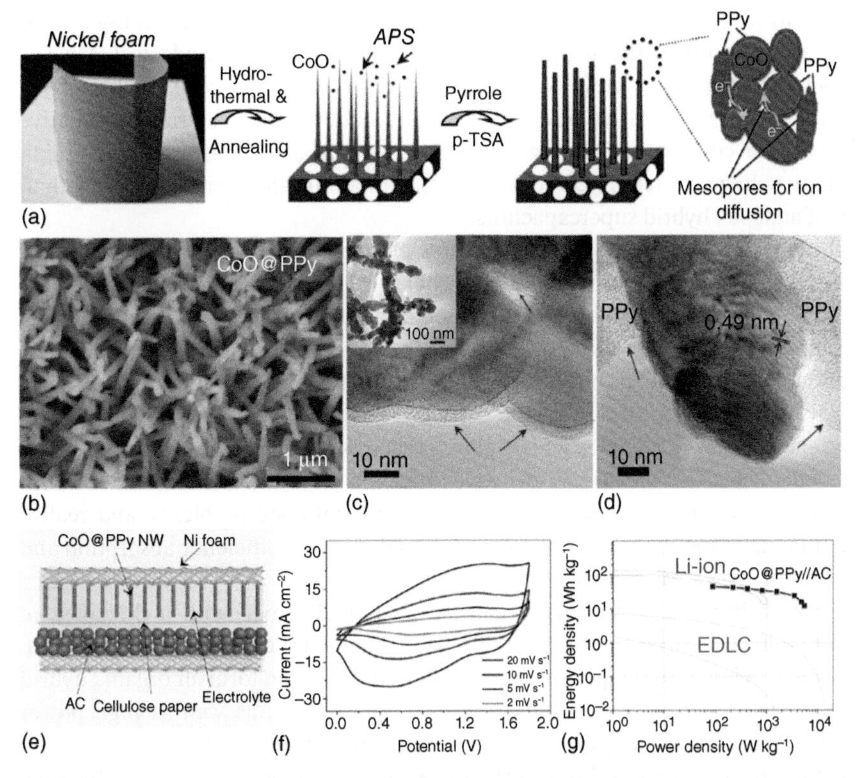

Figure 8.9 (a) The representative synthetic procedure and structure details of the 3D hybrid nanowire electrode. (b) high-magnification SEM image of the hybrid nanowire electrode. (c, d) HRTEM images of surface of individual CoO@PPy hybrid nanowires. Inset shows the general view of several nanowires. (e) Schematic illustration of the asymmetric supercapacitor configuration. (f) CVs. (g) Volumetric energy and powder densities of our supercapacitor compared with other data. Source: Zhou et al. [50], Republished with permission from ACS.

be used as the negative electrode of hybrid supercapacitors. Mousavi used synthetic polypyrrole nanofibers as anode and polyaniline nanofibers as cathode to obtain high-performance hybrid supercapacitors [51], whose working voltage reaches 1.3 V in the aqueous system, the energy density can reach 0.23 Wh g^{-1}, and power density of 0.045 kW g^{-1}.

8.3.2 Other Electrolyte System Hybrid Supercapacitors

8.3.2.1 Organic Electrolyte System

Based on the higher electrochemical window of the organic electrolyte, the assembled hybrid supercapacitor has a higher energy density and is the current market leader. Moreover, mature production processes such as current collectors, separators, and battery cases in batteries can be used in organic hybrid supercapacitors. In addition, the active materials used in the battery can also be used in hybrid supercapacitors (the negative electrode is $Li_4Ti_5O_{12}$, graphite, metal oxide, metal sulfide, metal phosphide, etc.; the positive electrode is $LiMn_2O_4$, $LiCoO_2$, $Na_3V_2(PO_4)_3$, etc.). Such devices generally use electric double- layer capacitor-type material as positive electrode, battery-type material as negative electrode, or conversely. The organic hybrid supercapacitor has a relatively high working voltage (generally Up to 3.8–4.0 V) with an ultrahigh energy density than that of aqueous hybrid supercapacitors.

Although the operating voltage of organic electrolyte is high, current capacity of these hybrid supercapacitors is still low. This is mainly due to the high viscosity of the organic electrolyte, slow ion adsorption and diffusion rate, and high electrical resistance. This characteristic poses a great challenge to electrode materials. It is difficult for traditional bulk materials to make full use of internal active sites, which greatly reduces their specific capacity. The agglomeration phenomenon of nanomaterials is inevitable, which also increases the cost of the preparation of electrode materials. The 1D nanowire electrode materials that are cross-linked with each other can simultaneously solve the above-mentioned problems, and realize the full immersion of the organic electrolyte and the high-efficiency absorption and desorption of electrolyte ions.

Lu et al. have studied a hybrid supercapacitor electrode with a thickness of 100 μm composed of a carbon nanotube/V_2O_5 nanowire composite material [52]. After assembled with the positive active carbon material to form an organic hybrid supercapacitor, its energy density can reach 0.04 Wh g^{-1}, as well as high power density of 0.02 kW g^{-1}. Thanks to the high electron transfer rate of the 1D nanowire structure, the electrode still has excellent capacitance performance under the dual test of thicker size and a high-viscosity organic electrolyte. In addition, Gao et al. has studied flexible electrode materials composed of niobium pentoxide nanowire compounds [53]. The mass fraction of Nb_2O_5 nanowires in the flexible electrode material reached 93.5%, and its energy density reached 106 Wh kg^{-1} after being assembled with activated carbon cathode material to form a lithium-ion hybrid capacitor. After 100 times under 0.5 A g^{-1}, specific capacitance did not decrease significantly. This fully demonstrates the advantages of nanowire electrode materials in organic electrolyte systems.

8.3.2.2 Redox-Active Electrolyte System

Referring to flow battery reaction system, redox reactions can be carried out at electrolyte/electrode interface to increase energy density of the supercapacitor. In this type of supercapacitor, an electrolyte with redox activity is usually added to the electrolyte, and a redox reaction is carried out at electrolyte/electrode interface to provide a large amount of additional specific capacity, while the other electrode has a rapid capacitance reaction to ensure high power density. Mai et al. used the ultrasonic spray drying method to obtain a porous carbon material with plenty of oxygen-containing functional groups [54]. By reacting with $CuCl_2$ additives in the electrolyte, a capacitance of 4700 F g^{-1} was obtained. Moreover, the energy density of the assembled symmetrical porous carbon supercapacitor reached 73 Wh kg^{-1}.

Such supercapacitors require a large number of interfaces for electrochemical reactions, so 1D nanomaterials with high SSA, anti-agglomeration and continuous axial electron conduction characteristics will have significant advantages. At the same time, how to alleviate the problem of self-discharge in the process of using this type of capacitor is a difficult point for research.

8.3.3 Solid Electrolyte or Quasi-Solid-State Hybrid Supercapacitor

In recent years, research on flexible, all-solid-state supercapacitors has received widespread attention. Flexible electrode materials will have a huge advantage in future wearable equipment. The gel electrolyte is mainly acidic PVA/H_3PO_4 and alkaline PVA/KOH. Similar to the electrolyte system, the gel electrolyte is thicker, and the ion transmission rate and the degree of infiltration at electrolyte/electrode interface are reduced. Because nanowire structure is relatively dispersed, combined with its excellent ion transmission path, it can also effectively solve the shortcomings of the gel electrolyte.

In order to give full play to the advantages of the one-dimensional structure of nanowires, Wang et al. improved the electrochemical performance of one-dimensional nano-pseudocapacitive materials [55], using carbon cloth loaded with MnO_2 nanowires as the positive electrode material, and carbon cloth loaded with Fe_2O_3 nanotubes as the negative electrode materials (Figure 8.10). The large interface area and short-ion diffusion path of 1D nanostructures can effectively enhance the property of hybrid supercapacitors. The energy density is 0.55 Wh L^{-1} under a voltage of 1.6 V. Li et al. used carbon aerogel microspheres as the negative electrode, Co_3O_4 nanowires grown on foamed nickel as the positive electrode and PVA/KOH as the gel electrolyte, and assembled them into an all-solid asymmetric hybrid supercapacitor [56]. The voltage window can reach 1.5 V, with a high energy density of 0.18 Wh g^{-1}.

8.4 Summary and Outlook

This chapter summarizes and discusses the functions of different kinds of nanowires in supercapacitors. EDL Carbonaceous nanowires through physical adsorption mechanism are the mainstream in commercial electrode materials due to poor cycling stability of the pseudocapacitive materials. The relationship between SSA as well as pore structure of porous nanowire electrodes and transport kinetics of solvated ions in electrolyte would be studied further. Otherwise, the diffusion process of cations and anions and the selection of the electrolyte window are urged for research. Nanowire electrode materials possess distinct advantage for solving the mentioned issues. Firstly, rate performance would be promised due to the construction of conductivity network from nanowires. Secondly, the unique 1D structure provides strong force endurance, which may be caused during charging and discharging process. Finally, the high L/D ratio makes electrodes have a higher SSA and abundant active sites for redox reaction to obtain good electrochemical performance. Therefore, nanowire electrode materials have good competitiveness and broad research potential in the supercapacitor field and are expected to be applied commercially.

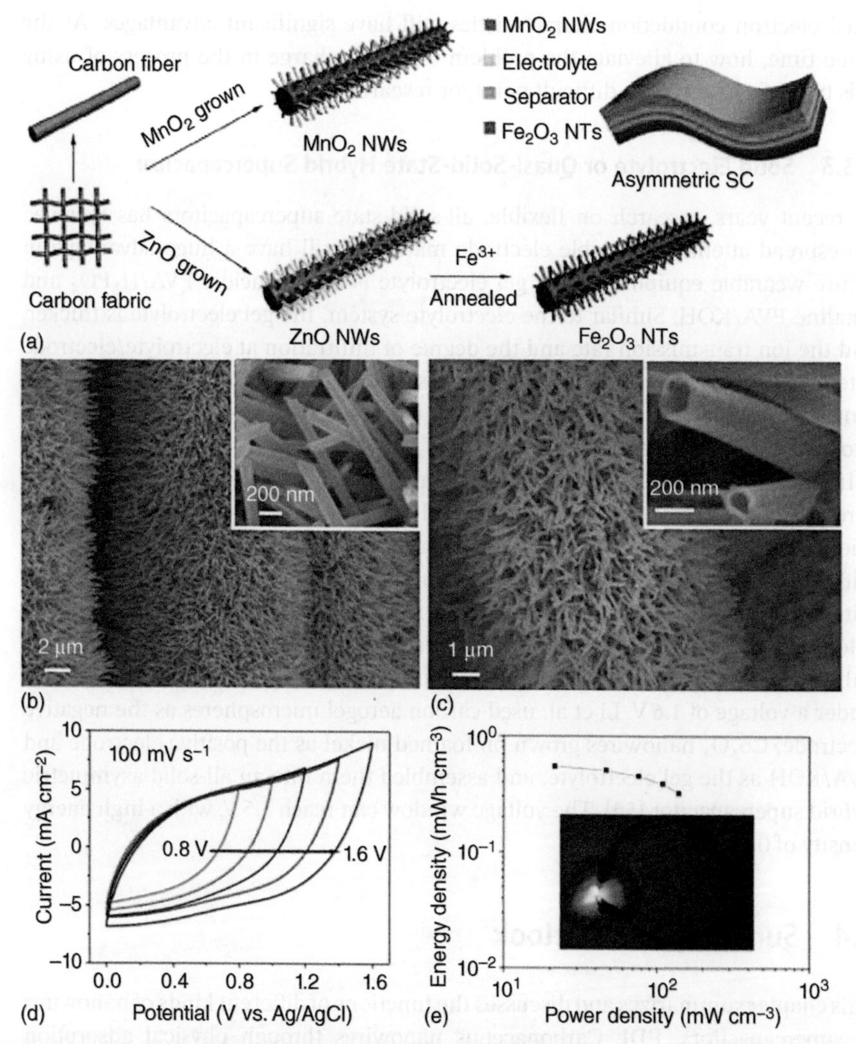

Figure 8.10 (a) Schematic illustration for the formation process of MnO_2 NWs and Fe_2O_3 NTs on carbon cloth and schematic sketch illustrating the designed asymmetric supercapacitor device. SEM image of (b) manganese dioxide nanowires and (c) Fe_2O_3 NTs. (d) CV curves of the assembled solid-state ASC device collected in different scan voltage windows. (e) Ragone plots of the solid-state ASC device. Inset shows a blue LED powered by the tandem ASC devices. Source: Yang et al. [55], Republished with permission from ACS.

References

1 Becker H.I. (1957-7-23). Low voltage electrolytic capacitor. U.S. Patent 2,800,616[P].

2 Wang, X. et al. (2005). Sol–gel template synthesis of highly ordered MnO_2 nanowire arrays. *Journal of Power Sources* 140 (1): 211–215.

3 Wei, Q. et al. (2017). Porous one-dimensional nanomaterials: design, fabrication and applications in electrochemical energy storage. *Advanced Materials* 29 (20): 1602300.

4 Yu, J., Ma, T., and Liu, S. (2011). Enhanced photocatalytic activity of mesoporous TiO_2 aggregates by embedding carbon nanotubes as electron-transfer channel. *Physical Chemistry Chemical Physics* 13 (8): 3491–3501.

5 Yan, J. et al. (2014). Recent advances in design and fabrication of electrochemical supercapacitors with high energy densities. *Advanced Energy Materials* 4 (4): 1300816.

6 Poulin, P.R. (2006). Irreversible organic crystalline chemistry monitored in real time. *Science* 313 (5794): 1756–1760.

7 Baughman, R.H. (2002). Carbon nanotubes–the route toward applications. *Science* 297 (5582): 787–792.

8 Yang, Z. et al. (2019). Carbon nanotube- and graphene-based nanomaterials and applications in high-voltage supercapacitor: a review. *Carbon* 141: 467–480.

9 Huang, J. et al. (2010). Curvature effects in carbon nanomaterials: Exohedral versus endohedral supercapacitors. *Journal of Materials Research* 25 (8): 1525–1531.

10 Hata, K. et al. (2006). Shape-engineerable and highly densely packed single-walled carbon nanotubes and their application as super-capacitor electrodes. *Nature Materials* 5 (12): 987–994.

11 Du, C. and Pan, N. (2006). Supercapacitors using carbon nanotubes films by electrophoretic deposition. *Journal of Power Sources* 160 (2): 1487–1494.

12 Tan, J. et al. (2017). In situ nitrogen-doped mesoporous carbon nanofibers as flexible freestanding electrodes for high-performance supercapacitors. *Journal of Materials Chemistry A* 5 (45): 23620–23627.

13 Zhang, X. et al. (2018). Hollow carbon nanofibers with dynamic adjustable pore sizes and closed ends as hosts for high-rate lithium-sulfur battery cathodes. *Nano Research* 11 (3): 1238–1246.

14 Sun, Q. et al. (2012). Controlled hydrothermal synthesis of 1D nanocarbons by surfactant-templated assembly for use as anodes for rechargeable lithium-ion batteries. *Journal of Materials Chemistry* 22 (33): 1749–1754.

15 Zhang, X. et al. (2013). Synthesis of superior carbon nanofibers with large aspect ratio and tunable porosity for electrochemical energy storage. *Journal of Materials Chemistry A: Materials for Energy and Sustainability* 1 (33): 9449.

16 Bai, Y., Huang, Z., and Kang, F. (2014). Electrospun preparation of microporous carbon ultrafine fibers with tuned diameter, pore structure and hydrophobicity from phenolic resin. *Carbon* 66: 705–712.

17 Teng, M. et al. (2012). Electrospun mesoporous carbon nanofibers produced from phenolic resin and their use in the adsorption of large dye molecules. *Carbon* 50 (8): 2877–2886.

18 Tan, J., Han, Y., He, L. et al. (2017). In aitu nitrogen-doped mesoporous carbon nanofibers as flexible freestanding electrodes for high-performance supercapacitors. *Journal of Materials Chemistry A* 5 (45): 23620–23627.

19 Veron, I., Patrice, S., and Bruce, D. (2014). Pseudocapacitive oxide materials for high-rate electrochemical energy storage. *Energy & Environmental Science* 7 (5): 1597–1614.

20 Wang, G., Zhang, L. et al. (2012). A review of electrode materials for electrochemical supercapacitors. *Chemical Society Reviews* 41 (2): 797–828.

21 Aric, A.S., Bruce, P., Scrosati, B. et al. (2005). Nanostructured materials for advanced energy conversion and storage devices. *Nature Materials* 4 (5): 366–377.

22 Simon, P. and Gogotsi, Y. (2008). Materials for electrochemical capacitors. *Nature Materials* 7 (11): 845–854.

23 Toupin, M., Brousse, T. et al. (2004). Charge storage mechanism of MnO_2 electrode used in aqueous electrochemical capacitor. *Chemistry of Materials* 16 (16): 3184–3190.

24 Wang, D.W., Li, F., and Cheng, H.M. (2008). Hierarchical porous nickel oxide and carbon as electrode materials for asymmetric supercapacitor. *Journal of Power Sources* 185 (2): 1563–1568.

25 Trasatti, S. and Buzzanca, G. (1971). Ruthenium dioxide: a new interesting electrode material. Solid state structure and electrochemical behaviour. *Journal of Electroanalytical Chemistry and Interfacial Electrochemistry* 29 (2): A1–A5.

26 Chen, P., Chen, H., Qiu, J. et al. (2010). Inkjet printing of single-walled carbon nanotube/RuO_2 nanowire supercapacitors on cloth fabrics and flexible substrates. *Nano Research* 3 (8): 594–603.

27 Nguyen, V.H., Tran, V.C., Kharismadewi, D. et al. (2015). Ultralong MnO_2 nanowires intercalated graphene/Co_3O_4 composites for asymmetric supercapacitors. *Materials Letters* 147: 123–127.

28 Zhou, J., Yu, L., Liu, W. et al. (2015). High performance all-solid supercapacitors based on the network of ultralong manganese dioxide/polyaniline coaxial nanowires. *Scientific Reports* 5: 17858.

29 Mai, L.Q., Yang, F., Zhao, Y.L. et al. (2011). Hierarchical $MnMoO_4$/$CoMoO_4$ heterostructured nanowires with enhanced supercapacitor performance. *Nature Communications* 2: 381.

30 Zhou, K., He, Y., Xu, Q. et al. (2018). A hydrogel of ultrathin pure polyaniline nanofibers: oxidant-templating preparation and supercapacitor application. *ACS Nano* 12 (6): 5888–5894.

31 Xie, Y., Xia, C., Du, H. et al. (2015). Enhanced electrochemical performance of polyaniline/carbon/titanium nitride nanowire array for flexible supercapacitor. *Journal of Power Sources* 286 (20): 561–570.

32 Biswas, S. and Drzal, L.T. (2010). Multilayered nanoarchitecture of graphene nanosheets and polypyrrole nanowires for high performance supercapacitor electrodes. *Chemistry of Materials* 22 (20): 5667–5671.

33 Bjørnholm, T., Hassenkam, T., Greve, D.R. et al. (1999). Polythiophene nanowires. *Advanced Materials* 11 (14): 1218–1221.

34 Dubal, D.P., Ayyad, O., Ruiz, V., and Gómez-Romero, P. (2015). Hybrid energy storage: the merging of battery and supercapacitor chemistries. *Chemical Society Reviews* 44: 1777–1790.

35 Liu, H. et al. (2020). Transition metal-based battery-type electrodes in hybrid supercapacitors: a review. *Energy Storage Materials* 28: 122–145.

36 Wang, Y., Song, Y., and Xia, Y. (2016). Electrochemical capacitors: mechanism, materials, systems, characterization and applications. *Chemical Society Reviews* 45 (21): 5925–5950.

37 Conway, B.E. and Pell, W.G. (2002). Power limitations of supercapacitor operation associated with resistance and capacitance distribution in porous electrode devices. *Journal of Power Sources* 105 (2): 169–181.

38 Gualous, H. et al. (2003). Experimental study of supercapacitor serial resistance and capacitance variations with temperature. *Journal of Power Sources* 123 (1): 86–93.

39 Lu, M. (2013). *Supercapacitors: Materials, Systems, and Applications*. Wiley.

40 Muzaffar, A. et al. (2019). A review on recent advances in hybrid supercapacitors: Design, fabrication and applications. *Renewable and Sustainable Energy Reviews* 101: 123–145.

41 Wu, Z. et al. (2010). High-energy MnO_2 nanowire/graphene and graphene asymmetric electrochemical capacitors. *ACS Nano* 4 (10): 5835–5842.

42 Brousse, T., Toupin, M., and BeLanger, D. (2004). A hybrid activated carbon-manganese dioxide capacitor using a mild aqueous electrolyte. *Journal of the Electrochemical Society* 151 (4): A614–A622.

43 Chen, L. et al. (2013). Bacterial-cellulose-derived carbon nanofiber@MnO_2 and nitrogen-doped carbon nanofiber electrode materials: an asymmetric supercapacitor with high energy and power density. *Advanced Materials* 25 (34): 4746–4752.

44 Brousse, T. et al. (2007). Long-term cycling behavior of asymmetric activated carbon/MnO_2 aqueous electrochemical supercapacitor. *Journal of Power Sources* 173 (1): 633–641.

45 Lai, F. et al. (2016). Biomass-derived nitrogen-doped carbon nanofiber network: a facile template for decoration of ultrathin nickel-cobalt layered double hydroxide nanosheets as high-performance asymmetric supercapacitor electrode. *Small* 12 (24): 3235–3244.

46 Wang, Q., Wen, Z.H., and Li, J.H. (2006). A hybrid supercapacitor fabricated with a carbon nanotube cathode and a TiO_2–B nanowire anode. *Advanced Functional Materials* 16 (16): 2141–2146.

47 Park, J.H. and Park, O.O. (2002). Hybrid electrochemical capacitors based on polyaniline and activated carbon electrodes. *Journal of Power Sources* 111 (1): 185–190.

48 Shen, J. et al. (2013). High-performance asymmetric supercapacitor based on nanoarchitectured polyaniline/graphene/carbon nanotube and activated graphene electrodes. *ACS Applied Materials & Interfaces* 5 (17): 8467–8476.

49 Lu, X. et al. (2015). α-Fe_2O_3 @PANI core-shell nanowire arrays as negative electrodes for asymmetric supercapacitors. *ACS Applied Materials & Interfaces* 7 (27): 14843–14850.

50 Zhou, C. et al. (2013). Construction of high-capacitance 3D CoO@Polypyrrole nanowire array electrode for aqueous asymmetric supercapacitor. *Nano Letters* 13 (5): 2078–2085.

51 Ghenaatian, H.R., Mousavi, M.F., and Rahmanifar, M.S. (2012). High performance hybrid supercapacitor based on two nanostructured conducting polymers: self-doped polyaniline and polypyrrole nanofibers. *Electrochimica Acta* 78: 212–222.

52 Chen, Z. et al. (2011). High-performance supercapacitors based on intertwined CNT/V_2O_5 nanowire nanocomposites. *Advanced Materials* 23 (6): 791–795.

53 Song, H. et al. (2016). Flexible Nb_2O_5 nanowires/graphene film electrode for high-performance hybrid Li-ion supercapacitors. *Journal of Power Sources* 328: 599–606.

54 Mai, L.Q., Minhas-Khan, A., Tian, X. et al. (2013). Synergistic interaction between redox-active electrolyte and binder-free functionalized carbon for ultrahigh supercapacitor performance. *Nature Communications* 4: 2923.

55 Yang, P. et al. (2014). Low-cost high-performance solid-state asymmetric supercapacitors based on MnO_2 nanowires and Fe_2O_3 nanotubes. *Nano Letters* 14 (2): 731–736.

56 Liu, W. et al. (2015). High-performance all-solid-state asymmetric supercapacitor based on Co_3O_4 nanowires and carbon aerogel. *Journal of Power Sources* 282: 179–186.

9

Nanowires for Multivalent-ion Batteries

Li-ion battery (LIB) technology has already achieved great success in the application of commercial portable electronic devices, benefiting from its characteristics of high energy density, high power density, and environmental friendliness [1–4]. However, the limited reserve of Li sources, safety issues, and high cost restrict its further applications [5, 6]. Therefore, the exploration of other battery systems based on earth-abundance elements with good safety to replace or complement LIBs is necessary, especially in large-scale applications. Na, Mg, Zn, and Al as anodes in rechargeable batteries have a similar operating mechanism as that of Li. Potential candidates as alternatives to LIBs can be multivalent batteries, including magnesium-ion batteries (MIBs), calcium-ion batteries (CIBs), zinc-ion batteries (ZIBs), and aluminum-ion batteries (AIBs). In respect of multivalent batteries, the Mg, Ca, Zn, and Al metals are abundant in the earth and deliver high specific capacity as well as energy density when employed as the anode owing to the multi-electron reactions [7–12]. In addition, they have many unique properties, such as small ionic radius and reasonably low deposition potential. Certainly, they also have special opportunities and challenges when used as an anode in counterpart batteries. At the present stage, they are uncompetitive with LIBs in many aspects, and methods are being explored to enhance the ions storage performance, such as the optimization of cathode, anode, and electrolyte. There is no denying that multivalent battery systems still require much time and effort before achieving satisfactory results to narrow the gap with LIBs or even surpass them [13].

9.1 Nanowires for Magnesium-Ion Battery

Benefiting from the advantages of magnesium metal anode, i.e. abundant reserves (one of the most abundant elements, ~2% of Earth's crust), high volumetric capacity (3833 mAh cm^{-3}) and low redox potential (−2.37 V versus SHE), rechargeable MIBs were identified as a promising candidate for the next-generation energy storage system with desired cost, safety, life, and energy density [14–18]. Despite these advantages, the development of MIBs is impeded by several key technical issues, such as the incompatibility of Mg metal anode with most of the conventional

Nanowire Energy Storage Devices: Synthesis, Characterization, and Applications, First Edition.
Edited by Liqiang Mai.

electrolytes and the sluggish solid-state diffusion kinetics of divalent Mg^{2+} in host structures [19–21]. Therefore, developing new cathodes for high-performance Mg-based batteries is a challenging, but necessary research direction.

9.1.1 Vanadium-Based Nanowires for MIBs

Poor cycling stability is one of the problems with the application of $V_2O_5 \cdot xH_2O$ in MIBs. An et al. [22] reported graphene-decorated $V_2O_5 \cdot xH_2O$ nanowires (VOG-1) which exhibit excellent cycling stability for Mg^{2+} ion storage. The electrochemical performance of VOG-1 was tested with $Mg(TFSI)_2$/acetonitrile as the electrolyte and activated carbon cloth (ACC) as quasi-reference electrode. A capacity as high as 330 mAh g^{-1} is achieved in the galvanostatic intermittent titration technique (GITT) measurement at 1.6–3.4 V. Besides, the calculated average Mg^{2+} diffusion coefficient of VOG-1 during the intercalation process is about 3×10^{-11} cm^{-2} s^{-1}, which is higher than that of Cheveral Mo_6S_8 ($2–6 \times 10^{-12}$ cm^{-2} s^{-1}). When cycled at 1000 mA g^{-1}, capacity retention of 81% is obtained after 200 cycles. The enhanced Mg^{2+} storage performance of VOG-1 is attributed to the charge shielding effect of crystal water, the fast electronic conduction pathway of graphene, and the short Mg^{2+} ion diffusion channel of nanowires.

Xu et al. [23] fabricated the Mg-ion pre-intercalated bilayered V_2O_5 ($Mg_{0.3}V_2O_5 \cdot 1.1 H_2O$) nanowires (Figure 9.1a) and employed them as cathode materials for MIBs. The pre-intercalated Mg-ions improve the electronic conductivity and stabilize the layer structure (Figure 9.1b). The interlayered water molecules enlarge the interlayer spacing and hold "charge shielding effect," facilitating the diffusion of magnesium-ions. Benefiting from the synergistic effects of interlayered water molecules and pre-intercalated Mg-ions, the $Mg_{0.3}V_2O_5 \cdot 1.1H_2O$ nanowires display improved cycling stability and rate performance (Figure 9.1c,d). Meanwhile, $Mg_{0.3}V_2O_5 \cdot 1.1H_2O$ nanowires exhibit such an ultralong cycling life that a capacity of about 90 mAh g^{-1} is achieved after 10 000 cycles at 2 A g^{-1}, corresponding to 80% of the maximum capacity (Figure 9.1e).

Luo et al. reported the investigation of nanostructured VO_2(B) as a high-voltage and high-capacity cathode material in MIBs [24]. The magnesium storage performance of nanostructured VO_2(B) was measured in a three-electrode system in which carbon rod and Ag/AgCl electrode were used as counter and reference electrode in 1 M $Mg(ClO_4)_2$/AN electrolyte. As a result, the electrochemical performance of VO_2(B) nanorods is much better than that of VO_2 (B) nanosheets. In the first cycle, VO_2(B) nanorods deliver a high capacity of 391 mAh g^{-1} at 25 mA g^{-1} while VO_2(B) nanosheets deliver a capacity of 356 mAh g^{-1}. After 10 cycles, the capacity of VO_2(B) nanorods remains at 94.7%, but that of VO_2(B) nanosheets only remains at 42%. The main reason for the better electrochemical performance of VO_2(B) nanorods is that their special morphology results in a faster ion diffusion rate and reduces the diffusion path [25].

Except for V_2O_5 and VO_2, many non-stoichiometric vanadium oxides have also attracted considerable interest as Mg^{2+} ions intercalation cathodes [26, 27]. Jiao

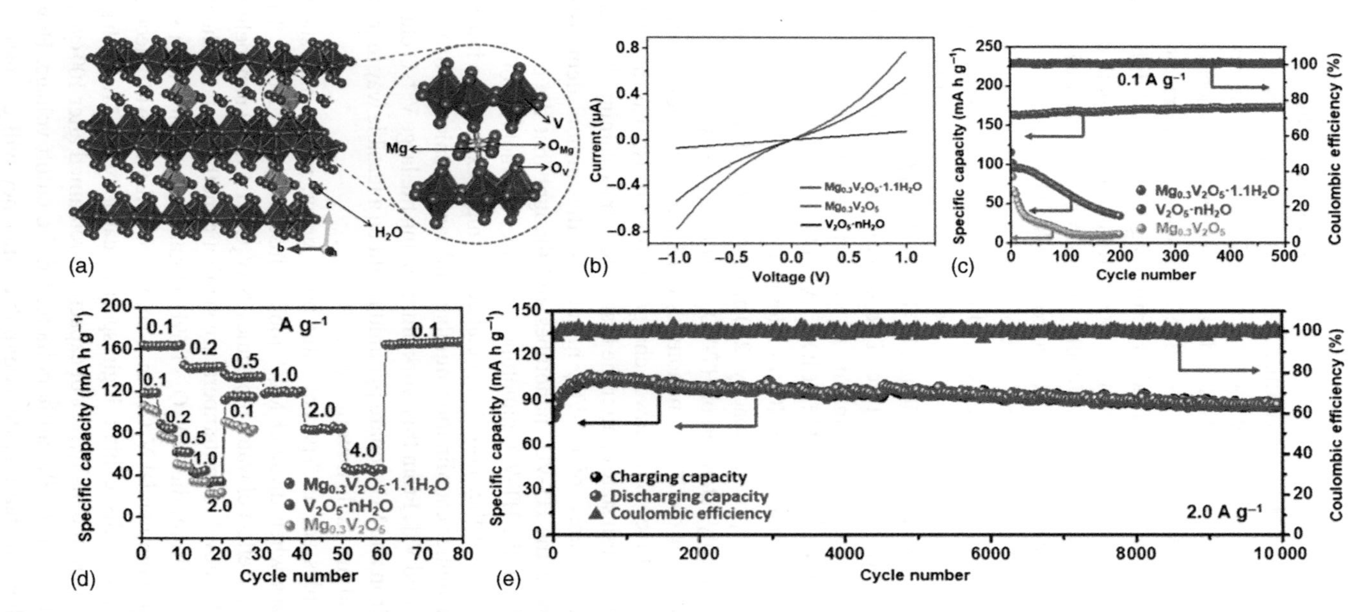

Figure 9.1 (a) The structural illustration of Mg-ion pre-intercalated bilayered V_2O_5 ($Mg_{0.3}V_2O_5 \cdot 1.1H_2O$); (b) I–V plots, (c) cycling stability and (d) rate capability of $Mg_{0.3}V_2O_5 \cdot 1.1H_2O$, $V_2O_5 \cdot 1.1H_2O$ and $Mg_{0.3}V_2O_5$; (e) Long-term cycling performance of $Mg_{0.3}V_2O_5 \cdot 1.1H_2O$. Source: Xu et al. [23], Reproduced with permission from Elsevier.

et al. synthesized the open-ended $VO_{2.37}$ nanotubes by sol–gel reaction followed by hydrothermal treatment [26, 28]. In the 0.25 M $Mg(AlBu_2Cl_2)_2$/THF solution with Mg metal as counter and reference electrode, $VO_{2.37}$ nanotubes exhibit an oxidation peak at 1.16 V and a reduction peak at 0.92 V, which correspond to the (de)intercalation of Mg^{2+} ions. At 1 mA g^{-1}, $VO_{2.37}$ nanotubes deliver a discharge capacity of 76 mAh g^{-1} and Mg^{2+} ions diffuse much faster than polycrystalline V_2O_5 from the electrochemical impedance spectroscopy (EIS) results. The open-ended $VO_{2.37}$ nanotubes with wide inner and outer diameter exhibit better electrochemical performance compared with the polycrystalline V_2O_5, which might result from the specific structure. First, in an open-ended structure, Mg^{2+} ions could diffuse faster compared with the surface-to-bulk structure. Second, the diffusion path of cations is shorter than the ordinary structure. Finally, the open-ended tubes will provide large electrode-electrolyte contact area in electrolyte-filled channels. Similar works were also reported by Ryoung-Hee and co-workers; they synthesized the highly reduced VO_x nanotubes via microwave-assisted hydrothermal method with amine as reducing agent and organic template [27]. Open-ended structures and rolled-tube shapes of VO_x nanotubes are also exhibited in this work, and the roll-tube comes from the octahedral VO_6 (V^{3+}) structure. Two different kinds of VO_x (low and high concentrations of the amine template in VO_x differentiate the LT–VO_x and HT–VO_x) were measured in 0.5 M $Mg(ClO_4)_2$/AN electrolyte with magnesium alloy as counter electrode and $Ag/AgNO_3$ as reference electrode. Although LT-VO_x nanotubes exhibit higher specific capacity (230 mAh g^{-1}) than HT-VO_x nanotubes (218 mAh g^{-1}), the capacity retention of HT-VO_x nanotubes (70.8% after 20 cycles) is much higher than that of LT-VO_x nanotubes (24% after 20 cycles). In the EIS analysis, HT-VO_x nanotubes show lower charge transfer resistance than LT-VO_x nanotubes, whether in fresh state or after the first cycle. For explaining the excellent electrochemical performance of HT-VO_x nanotubes, the authors demonstrated that V^{3+} only exists in the HT-VO_x nanotubes and thought that the existence of octahedral VO_6 (V^{3+}) in the HT-VO_x nanotubes enhances the mobility of Mg^{2+} ions in the crystal structure. Furthermore, VO_6 octahedron might expand the interstitial sites because of the higher coordination number and larger ionic size of V^{3+} ions compared with V^{4+} and V^{5+}. From this work, introducing the large V^{3+}-O cluster to active VO_4 tetrahedron and VO_5 square pyramid might be a feasible way to improve the property of vanadium oxide materials in MIBs.

The water molecules act as "spacers" and bond water, which also stabilizes the lattice of bronzes when divalent Mg^{2+} ions diffuse in the structure. To take advantage of the shield effect of water molecules, $NaV_3O_8 \cdot 1.69H_2O$ nanobelts were synthesized [29]. The novel room-temperature synthetic method retains the water in the structure. In addition, the $NaV_3O_8 \cdot 1.69H_2O$ nanobelts exhibit high surface area compared with commercial V_2O_5. In a coin-type cell with APC electrolyte, $NaV_3O_8 \cdot 1.69H_2O$ nanobelts exhibit high specific capacity of 110 mAh g^{-1} at 10 mA g^{-1}. At high current density, 80% of capacity is remained after 100 cycles. To increase the specific capacity, the authors increased the cutoff voltage. However, with the cutoff voltage increasing, the coulombic efficiency will decrease due to electrolyte decomposition. During cycling performance, no new phase is observed in the materials. But some Mg^{2+} ions will be trapped in the $NaV_3O_8 \cdot 1.69H_2O$ nanobelts. Moreover, these trapped Mg^{2+} ions will affect the reversibility of battery and result in low ion diffusion rate inside $NaV_3O_8 \cdot 1.69H_2O$ nanobelts. In this work, the lateral confinement and high conduction channels of $NaV_3O_8 \cdot 1.69H_2O$ nanobelts facilitate electron movements along the nano-length axis. The fast insertion and extraction of Mg^{2+} ions is attributed to the shielding effect of water. Reducing the internal resistance of trapped Mg^{2+} ions is the next problem to solve in the $NaV_3O_8 \cdot 1.69H_2O$ nanobelts.

9.1.2 Manganese-Based Nanowires for MIBs

The shielding effect of a much lower H_2O content in electrolyte or crystal structure based on V_2O_5 cathode has been experimentally and theoretically investigated, so it provides an alternative strategy to another high performance cathode of MnO_2. Lee et al. firstly demonstrated that the reversible Mg^{2+} insertion into nanostructured MnO_2 can be enhanced in water containing an electrolyte [30]. The electrochemical performance was tested in three-electrode system with the synthesized free-standing MnO_2 nanowire as the working electrode, Ag/AgCl as the reference electrode, and platinum as the counter electrode in $Mg(ClO_4)_2$/PC containing water electrolyte. The authors determined the improved performance by CV curves

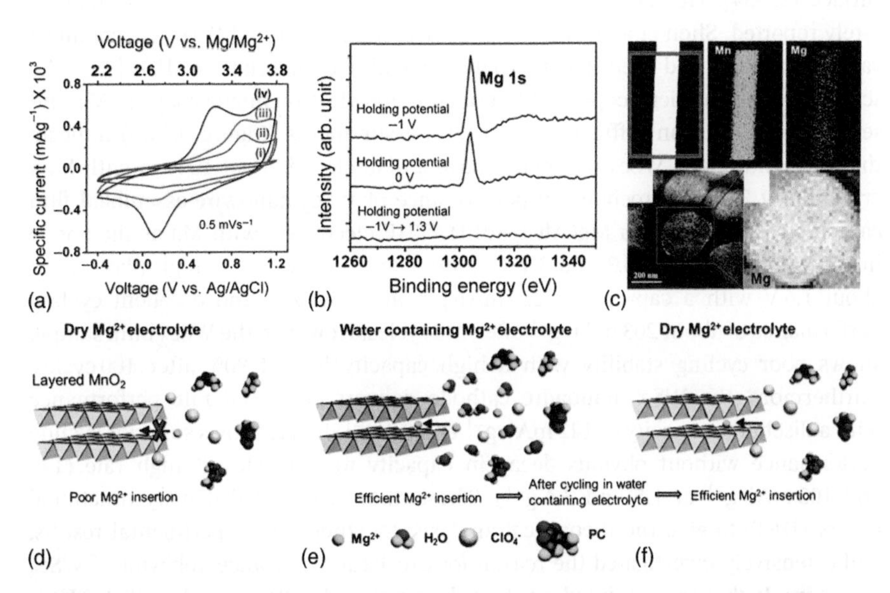

Figure 9.2 (a) CV curves of MnO_2 in various water-containing electrolytes (i) 0.1 M dry $Mg(ClO_4)_2$, (ii) 0.1 M $Mg(ClO_4)_2 \cdot 6H_2O$ with 0.4 M dry $Mg(ClO_4)_2$, (iii) 0.1 M $Mg(ClO_4)_2 \cdot 6H_2O$ with 0.1 M dry $Mg(ClO_4)_2$, (iv) 0.1 M $Mg(ClO_4)_2 \cdot 6H_2O$. (b) Mg 1 s XPS spectra for MnO_2 at different states. (c) STEM mapping of Mn, Mg, and Mg mapping image analyzed by EELS at a fully discharged state. (d) Mg^{2+} insertion process in a dry electrolyte. (e) Mg^{2+} insertion process in a wet electrolyte. (f) Mg^{2+} insertion process in dry electrolyte after cycling in a wet electrolyte. Source: Song et al. [30], Reproduced with permission from the Royal Society of Chemistry.

(Figure 9.2a) in electrolytes with various water contents. Ex situ XPS (Figure 9.2b), STEM mapping, and electron energy loss spectroscopy (EELS) (Figure 9.2c) give direct evidence about insertion of Mg^{2+} into MnO_2 and the uniform distribution. The amount of intercalated Mg^{2+} ions with the co-intercalated water molecules is determined via the combination of inductively coupled plasma-atomic emission spectroscopy (ICP-AES) and electrochemical quartz crystal microbalance (EQCM) techniques. The molar ratio of co-inserted H_2O and Mg^{2+} is in agreement with that of V_2O_5 (about 3 : 1) [31]. The difference in MnO_2 system is that the improved Mg^{2+} insertion behavior can be maintained in the dry electrolyte after the initial cycling in water containing electrolyte, while the enhanced capacity is no longer observed in V_2O_5 system. The schematic of Mg^{2+} insertion mechanism into MnO_2 is presented in Figure 9.2d–f. In terms of cycling performance, it displays a high specific capacity of about 160 mAh g^{-1} in the first cycle, and the capacity retention is 67% after 200 cycles at 1.6 C (1 C = 0.616 A g^{-1}) in 0.1 M $Mg(ClO_4)_2 \cdot 6H_2O$/PC electrolyte, much better than previously reported results.

9.1.3 Other Nanowires for MIBs

WSe_2 has attracted much attention for its extraordinary properties of low thermal conductivity, efficient p-type field effect performance, and high hydrophobic sticky surface [32–34]. However, the application of WSe_2 in energy storage systems is rarely reported. Shen et al. synthesized a novel nanostructured WSe_2 via chemical vapor deposition and used it as the potential cathode material for MIBs [35]. The scanning electron microscopy (SEM), transmission electron microscopy (TEM), and selected-area electron diffraction (SAED) measurements (Figure 9.3a,b) indicate that the synthesized WSe_2 nanowires have a diameter of about 100 nm with high crystallinity. The electrochemical performance of WSe_2 nanowire-assembled film cathode was evaluated in $Mg(AlCl_2BuEt)_2$/THF electrolyte with Mg as the anode in the voltage range of 0.3 and 3 V (Figure 9.3c–f). It exhibits a high platform at about 1.6 V with a capacity of 220 mAh g^{-1} at 0.05 A g^{-1}, and excellent cycling performance of about 203 mAh g^{-1} after 160 cycles. However, the WSe_2 bulk sample shows poor cycling stability with a high capacity loss of 90% after 100 cycles. Furthermore, the WSe_2 nanowire cathode delivers enhanced rate performance with a discharge capacity of 142 mAh g^{-1} at 0.8 A g^{-1}. It also shows superior cycling performance without obvious decay in capacity for 50 cycles at high rate (120 and 103 mAh g^{-1} at 1.5 and 3 A g^{-1}). The authors conducted density functional theory (DFT) to give the theoretical analysis, together with experimental results, and extensively investigated the reason for excellent Mg-storage behavior of WSe_2 nanowire. It should be pointed out that the employed voltage window (0.3–3 V) is beyond the anodic limit of electrolyte (about 2.5 V), thus promoting the undesirable side reactions of electrolyte.

Up to now, some cathode materials have demonstrated excellent Mg-storage performance; however, most of the electrochemical performances are performed in conventional electrolytes, which are not compatible with Mg anodes. In order to make practical use of these cathodes, it is necessary to investigate other

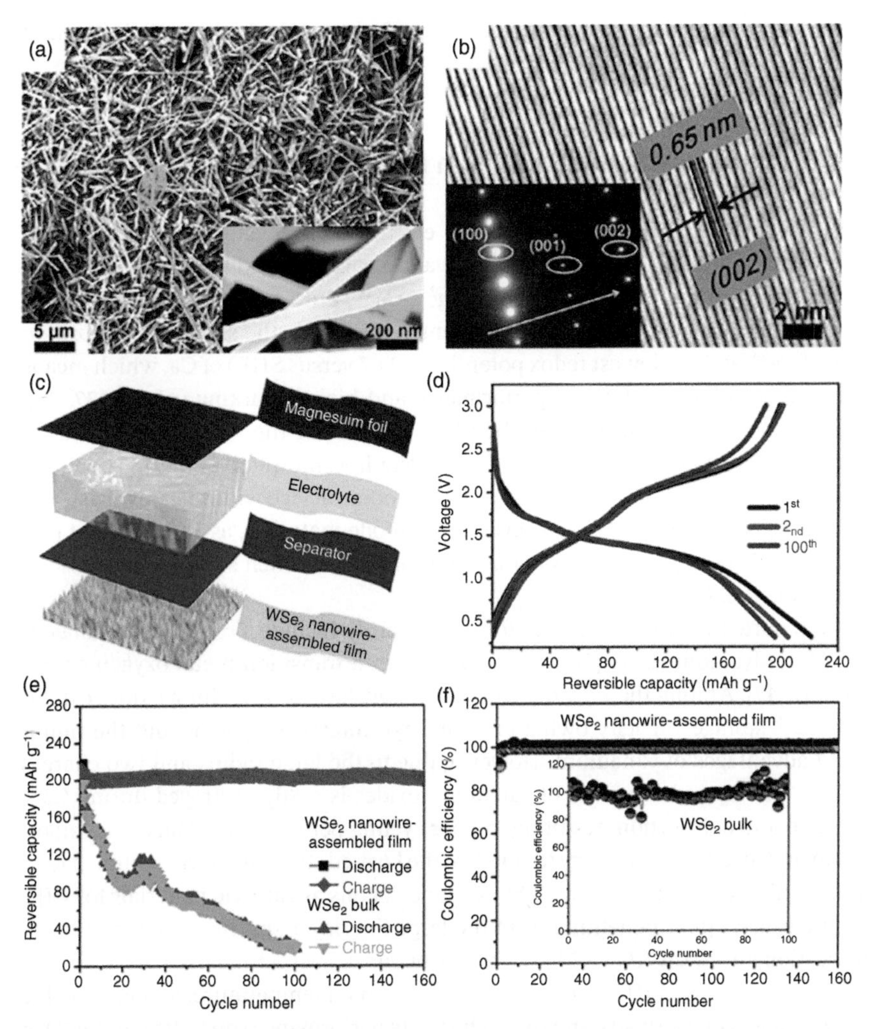

Figure 9.3 (a) SEM images of the synthesized WSe_2 nanowires. (b) TEM image and SAED pattern. (c) Schematic illustration of the assembled cell. (d) The discharge/charge profiles of WSe_2 nanowire-assembled film at the current density of $0.05\,A\,g^{-1}$. (e) The cycling performance and (f) corresponding Coulombic efficiency of WSe_2 nanowire-assembled film and WSe_2 bulk. Source: Liu et al. [35], Reproduced with permission from the American Chemical Society.

anode systems. Liu et al. developed the nanostructured Bi anode for MIBs [36]. The Bi nanotubes with the diameter of around 8 nm are synthesized through a hydrothermal method. For comparison, Bi microparticles (about 100 μm) and Bi nanoparticles (30–50 nm) were also prepared. The electrochemical performance was tested in the $Mg(BH_4)_2$ + $LiBH_4$/diglyme electrolytes against the Mg cathode. As expected, the Bi nanotubes demonstrate much better cycling performance of 92.3% capacity retention ($303\,mAh\,g^{-1}$) after 200 cycles and the best rate performance of 350 and $216\,mAh\,g^{-1}$ at 0.05 C and 5 C, respectively.

The authors concluded that the Mg-storage performance is significantly influenced by the morphology and size of Bi.

9.2 Nanowires for Calcium-Ion Batteries

Since each carrier can transfer multiple electrons, multivalent-ion batteries are expected to achieve higher energy density and are considered potential candidates for post-LIBs. Among multivalent-ion (Mg^{2+}, Ca^{2+}, Zn^{2+}, and Al^{3+}) batteries, CIBs have received increasing attention in recent years due to the smallest polarization force of Ca^{2+} and the lowest redox potential (−2.87 versus SHE) of Ca, which means that CIBs may have better rate performance and higher working voltage [37, 38]. In addition, Ca is the fifth most abundant element in the earth's crust, and Ca metal tends to plating/stripping with dendrite-free morphology [39]. Therefore, CIBs have the potential to become an energy storage device with safe, low-cost, and high-energy-density [40, 41]. However, the cathode materials, as an important part of CIBs, still show poor electrochemical performance, which severely impedes the development of CIBs.

The reported cathode materials for CIBs mainly include Prussian blue analogs [42, 43], polyanionic salts [38, 44–46], and layered transition metal oxygen or sulfides [47–49]. Among these cathode materials, the layered vanadium oxides exhibit high Ca^{2+} storage capacity owing to their large interlayer spacing and the multivalent advantages of vanadium. However, due to the large radius and two charges of Ca^{2+}, the layered structure of vanadium oxides is easily destroyed during Ca^{2+} insertion and deinsertion, resulting in poor cycling performance. Thus, it is important to find an effective strategy to stabilize the layered structure of vanadium oxides. Wang et al. presented $CaV_6O_{16} \cdot 2.8H_2O$ nanoribbons as cathode materials for CIBs (Figure 9.4a–c) [50]. Benefiting from the large interlayer spacing, pillar, and water lubrication effect, $CaV_6O_{16} \cdot 2.8H_2O$ nanoribbons deliver a high discharge capacity of 175.2 mAh g^{-1} at 50 °C and 131.7 mAh g^{-1} at room temperature, a long cycle life of 1000 cycles (Figure 9.4d), and the highest rate performance (up to 1000 mA g^{-1}) in organic electrolyte. Furthermore, a single-phase Ca^{2+} insertion and extraction reaction is revealed by in situ X-ray diffraction and in situ Fourier transform infrared spectroscopy. DFT computations indicate that Ca^{2+} tends to diffuse along the b direction with a low energy barrier of 0.36 eV.

Besides, Dong et al. investigated the electrochemical performance of $K_2V_6O_{16}$ nanowires as cathode materials for CIBs [51]. The $K_2V_6O_{16}$ nanowires exhibit a high initial capacity of 113.9 mAh g^{-1} at 20 mA g^{-1} and high capacity retention of 78.30% at 50 mA g^{-1} after 100 cycles. Adil et al. investigate zirconium-doped $NH_4V_4O_{10}$ nanobelts as a cathode for CIBs at room temperature [52]. The zirconium-doped $NH_4V_4O_{10}$ nanobelts exhibited an initial discharge capacity of 78 mAh g^{-1} with an average discharge voltage of ~3.0 V versus Ca^{2+}/Ca and presented a high rate capability, excellent cycling stability for 500 cycles with ~89% capacity retention.

Except for vanadium-based nanowires, manganese-based nanowires have also been employed as cathode materials for CIBs. Zuo et al. investigated the calcium

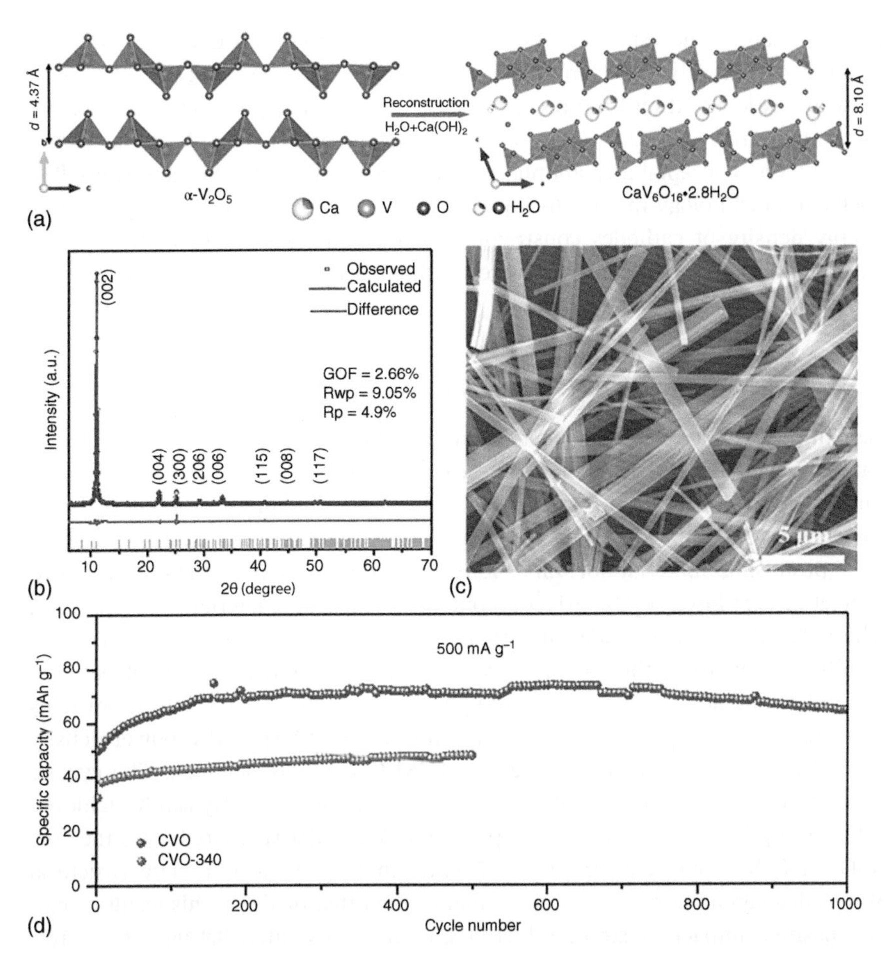

Figure 9.4 (a) Schematic diagram of the formation and crystal structure of $CaV_6O_{16} \cdot 2.8$ H_2O. (b) XRD pattern and Rietveld refinement of $CaV_6O_{16} \cdot 2.8H_2O$ nanoribbons and (c) SEM image of $CaV_6O_{16} \cdot 2.8H_2O$ nanoribbons. (d) Long cycling performance of $CaV_6O_{16} \cdot 2.8H_2O$ (CVO) and CVO-340 at 500 mA g^{-1} and 50 °C. Source: Wang et al. [50], Reproduced with permission from Wiley-VCH GmbH.

storage performance of four MnO_2 nanowires or nanowire assemblies with different crystal phases (α, β, γ, and δ-phase) [53]. Among these, the δ-MnO_2 nanowire assemblies exhibit superior electrochemical performance in terms of rate capability (capacities of 125, 118, 105, and 87 mA g^{-1} at 0.1, 0.2, 0.5, and 1 A g^{-1}) and a reversible capacity of 125 mAh g^{-1} at 100 mA g^{-1} because of high Ca-ion diffusion coefficient and capacitive capacity contribution.

9.3 Nanowires for Zinc-Ion Batteries

The development and application of LIBs in large-scale energy storage are limited by the inevitable challenges, such as increasingly scarce lithium resources, high cost, low safety, and especially the utilization of noxious and flammable organic electrolytes [54, 55]. As a promising alternative candidate, aqueous ZIBs have attracted wide attention from researchers owing to their high theoretical capacity ($5854\,mAh\,cm^{-3}$ and $820\,mAh\,g^{-1}$), low redox potential (−0.762 V versus SHE), environmental acceptability, abundance, intrinsic nonflammability, and simplified packaging technology in air [56–58]. However, the poor cycling stability and low energy density of cathodes constrained the current application of ZIBs. Thus, developing high performance cathodes with high capacity and cycle stability is necessary.

9.3.1 Vanadium-Based Nanowires for ZIBs

Benefiting from the wide valence variation from V^{5+} to V^{3+}, vanadium-based cathodes usually have a high specific capacity. In particular, the layered vanadium oxides with a large interlayer spacing could accommodate cathodic ions' insertion/extraction, conveniently. Bilayer V_2O_5 with larger interlayer spacing is a promising material for Zn^{2+} ions intercalation/deintercalation. Recently, Yan et al. explored $V_2O_5 \cdot nH_2O$/graphene as cathode material for ZIBs [59]. The reduced graphene oxide-supported $V_2O_5 \cdot nH_2O$ nanowire framework could significantly increase the conductivity of cathode and enhance contact between electrolyte and active material, resulting in a higher ion diffusion coefficient. This aqueous Zn/V_2O_5 system displays a capacity of $295\,mAh\,g^{-1}$ at a power density of $6.4\,kW\,kg^{-1}$, and an initial capacity of $381\,mAh\,g^{-1}$ is obtained. The cycling performance test at $6\,A\,g^{-1}$ shows that 71% of maximum capacity can be achieved after 900 cycles. For the control experiment, VOG-350 (prepared by annealing $V_2O_5 \cdot nH_2O$/graphene at 350 °C for 2 hours in vacuum to eliminate structural water) displays a poorer cycling performance than that of VOG. This result reveals the positive impact of structural H_2O on Zn^{2+} ions intercalating into bilayer $V_2O_5 \cdot nH_2O$. They further demonstrated it with an XRD study and NMR analysis and proposed that the effective charge of Zn^{2+} could be reduced by water solvation, leading to a reduction of electrostatic interactions between Zn^{2+} ions and V_2O_5 framework, which can effectively promote ion diffusion. He et al. [60] utilized $Na_{0.33}V_2O_5$ nanowires as cathode material for ZIBs, which display greatly enhanced cycling stability compared to undoped V_2O_5 cathode. After 1000 cycles at $1\,A\,g^{-1}$, $Na_{0.33}V_2O_5$ still delivers a capacity of $218.4\,mAh\,g^{-1}$, which is over 93% of the highest capacity in the 2–1000th cycles. Based on the ex situ XRD and TEM technologies, the improved structural stability of $Na_{0.33}V_2O_5$ compared to V_2O_5 during cycling was demonstrated, which benefit from the "pillar effect" of sodium ions in interlayer.

Nazar et al. [61] reported the electrochemical performance of $Zn_{0.25}V_2O_5 \cdot nH_2O$ nanobelts as cathode material for ZIBs. The pre-intercalated Zn-ions stabilize the layered structure and thus ensure cycling stability. Consequently, the capacity retention of $Zn_{0.25}V_2O_5 \cdot nH_2O$ cathode is over 80% after cycling at $2400\,mA\,g^{-1}$ for 1000 cycles. After that, another two zinc vanadates for aqueous ZIBs were reported: $Zn_3V_2O_7(OH)_2{\cdot}2H_2O$ nanowires [62] and $Zn_2V_2O_7$ nanowires [63], respectively. $Zn_3V_2O_7(OH)_2 \cdot 2H_2O$ nanowires were found to display a reversible capacity of 213 mAh g^{-1} at 50 mA g^{-1} and cycling stability up to 300 cycles at 200 mA g^{-1}. $Zn_2V_2O_7$ nanowires exhibit a reversible capacity of ~250 mAh g^{-1} at 50 mA g^{-1} and ~ 150 mAh g^{-1} at 4000 mA g^{-1}. Good cycling stability for 1000 cycles at 4000 mA g^{-1} is obtained. The Zn storge mechanism for these phases was all demonstrated to be intercalation/deintercalation reaction. Considering the low cost of zinc, vanadium, and aqueous electrolyte, combined with the superior reported electrochemical performance, the aqueous Zn//Zn-V-O batteries are very promising for large-scale energy storage.

Xia et al. [64] demonstrated that layered $Ca_{0.25}V_2O_5 \cdot nH_2O$ nanobelts are promising cathode materials for aqueous ZIB cathode. A high capacity of 340 mAh g^{-1} at 0.2 C and capacity retention of 96% after 3000 cycles at 80 C are achieved. When compared with the reported $Zn_{0.25}V_2O_5 \cdot nH_2O$ nanobelts at earlier stage, $Ca_{0.25}V_2O_5 \cdot nH_2O$ nanobelts show higher electrical conductivity (185 S m^{-1} compared to 46 S m^{-1}) and larger layer spacing (10.6 Å compared to 10.2 Å) due to the larger CaO_7 polyhedra than the ZnO_6 octahedra in the layers. Besides, the lower molecular weight can result in higher gravimetric capacity.

The zinc storage performance of $H_2V_3O_8$ nanowires was investigated by He et al. [65] and Pang et al. [66] with $Zn(CF_3SO_3)_2$ aqueous electrolyte. The high capacity of about 400 mAh g^{-1} is achieved with two slope plateaus at about 0.8 and 0.5 V, respectively. Stable cycling over 1000 cycles is obtained at high rates. In addition, $Na_{1.1}V_3O_{7.9}$ nanoribbons [67], $NaV_3O_8 \cdot 1.5H_2O$ nanobelts [68], $Na_2V_6O_{16} \cdot 1.63H_2O$ nanowires [69], and $Na_2V_6O_{16} \cdot 3H_2O$ nanorods [70] have been demonstrated as exciting high-performance ZIB cathodes with high capacity (over 300 mAh g^{-1}), excellent cycling stability and rate capability. For example, $NaV_3O_8 \cdot 1.5H_2O$ nanobelts were reported to realize a high reversible capacity of 380 mAh g^{-1} and capacity retention of 82% after 1000 cycles, while $Na_2V_6O_{16} \cdot 1.63H_2O$ nanowires were found to exhibit a capacity retention of 90% over 6000 cycles at 5000 mA g^{-1}. These exciting results suggest that appropriate vanadate nanowires as cathodes are one of the good choices for aqueous ZIBs in practical applications.

9.3.2 Manganese-Based Nanowires for ZIBs

The development of the Zn//α-MnO_2 batteries is hindered by their restricted cycle life and inferior rate capability in the primary period. The electrochemical performance of α-MnO_2 cathode can be significantly improved by adjusting the concentration and component of electrolyte. Pan et al. realized the high reversible capacity and excellent cycling stability of α-MnO_2 nanowires by restraining the dissolution of Mn^{2+} by pre-added Mn^{2+} ions ($MnSO_4$) into the electrolyte [71]. Although the $MnSO_4$ addition has no impact on redox processes, it can considerably increase the MnO_2 cathodes' cycling stability (Figure 9.4b). Subsequently, the capacity retention of α-MnO_2 nanowires using electrolyte with the $MnSO_4$ additive is 92% after 5000 cycles at a rate of 5 C. In addition, Wu et al. coated α-MnO_2 nanowires with graphene to solve the problems of rapid capacity decay and material dissolution [72].

Graphene scroll-coated α-MnO_2 nanowires exhibit higher capacity, increased conductivity, and enhanced rate performance compared to bare α-MnO_2 nanowires. The graphene scroll-coated α-MnO_2 nanowires deliver excellent long-term cycle stability; their specific capacity reaches 145.3 mAh g^{-1} at 3 A g^{-1} with a capacity retention of 94% after 3000 cycles. This strategy provides new ideas for the development of aqueous ZIBs and even other types of electrode material optimization. In addition, Wang et al. presented superfine α-MnO_2 nanowires (~10 nm) with rich crystal defects (oxygen vacancies and cavities) as cathode materials for ZIBs [73]. The defects facilitate the adsorption and diffusion of hydrogen/zinc for fast ion transportation and the buildup of a local electric field for improved electron migration. In addition, the superfine nanostructure could provide sufficient active sites and short diffusion pathways for further promotion of capacity and reaction kinetics of MnO_2. Remarkably, the defect-enriched MnO_2 nanowires manifest an energy density as high as 406 Wh kg^{-1} and excellent durability over 1000 cycles without noticeable capacity degradation.

Except for α-MnO_2 nanowires, δ-MnO_2 nanowires have also been employed as cathode materials for ZIBs. For example, Liu et al. reported the δ-MnO_2 nanowire networks as ZIB cathode materials [74]. A distinctive three-dimensional (3D) nanonetwork structure results in enhancement of electrolyte osmosis and a significant increase in contact between electrode and electrolyte, and also provided more active sites and convenient rapid ion transport routes. Moreover, the fine nanowire structure and the optimum layer spacing resulted in easier insertion/deinsertion of ions in the active material. Taking advantage of this feature, the δ-MnO_2 cathode provides high reversible capacity of 342 mAh g^{-1}, fast rate capability of 150 mAh g^{-1} at 6.494 C (1 C = 308 mA g^{-1}), and good cycling stability of 148 mAh g^{-1} retained after 400 cycles at 1.623 C.

9.4 Nanowires for Aluminum Ion Batteries

Along with the increasing world energy consumption and exhaustion of fossil fuel, exploring sustainable energy sources to meet drastic demand is urgently needed [75–77]. Developing large-scale energy storage systems is crucial to transformation of sustainable energy into electric energy infrastructure [78–82]. New types of multivalent ion batteries, such as MIBs [17, 83], ZIBs [60, 84], and AIBs [7, 10, 85], have received considerable attention derived from the abundance and safety of corresponding metal anodes. Aluminum is the cheapest metal (~1.4 USD kg^{-1}) due to the largest reserve (82 000 ppm in earth's crust) when compared with other multivalent metal anodes [10]. Moreover, the trivalent charge carriers can provide high volumetric capacity (8046 mAh cm^{-3}) and gravimetric capacity (2980 mAh g^{-1}) [86]. Despite the above advantages, long-standing challenges still limit the development of AIBs. The major obstacle comes from inherently strong polarization effect between Al^{3+} and framework lattice of cathode material, which increases the diffusion resistance and limits the specific capacity [82, 87]. Thus, finding suitable cathode materials becomes a formidable challenge for AIBs [88].

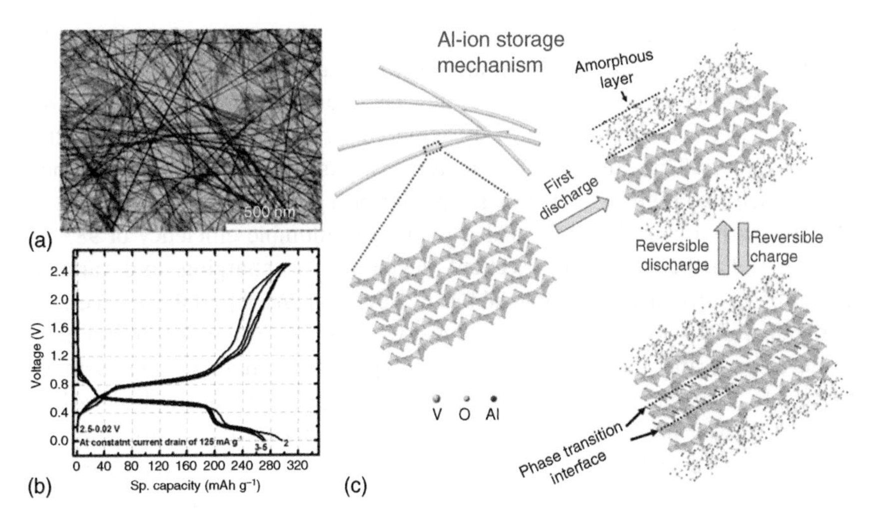

Figure 9.5 (a) TEM image of V_2O_5 nanowires, (b) charge/discharge curves of V_2O_5 nanowires in AIBs. Source: Jayaprakash et al. [86], Reproduced with permission from the Royal Society of Chemistry. (c) Illustration of the structural evolution of V_2O_5 nanowires during Al-ion storage. Source: Gu et al. [89], Reproduced with permission from Elsevier.

The first stable AIB based on V_2O_5 with extended cycle life was reported by Jayaprakash et al. [86] in 2011. In this AIB, the V_2O_5 nanowires are used as the cathode material, with ionic liquid as electrolyte and aluminum metal as anode (Figure 9.5a). The V_2O_5 nanowires exhibit high discharge capacities of 305 mAh g^{-1} in the 1st cycle and 273 mAh g^{-1} in the 20th cycle at 125 mA g^{-1} with a discharge plateau at about 0.55 V (Figure 9.5b). However, the cathode reaction in this system is controversial. Reed et al. [90] reported that the V_2O_5 nanowires are electrochemically inactive in above AIB system, and the capacity is attributed to the reactions of ionic liquid electrolyte with the iron and chromium metal in the stainless steel current collector. Gu et al. [89] demonstrated the reversible insertion/extraction of aluminum-ions into V_2O_5 nanowires. The surface amorphization of V_2O_5 nanowires arises during the intercalation of Al^{3+} ions (Figure 9.5c). However, V_2O_5 nanowires with Ni current collector display a lower discharge capacity (53.4 mAh g^{-1} in 2nd cycle at 10 mA g^{-1}) compared to V_2O_5 nanowires with stainless steel current collector. Therefore, it can be concluded that both the reactions of ionic liquid electrolyte with stainless steel current collectors and reversible insertion/extraction of aluminum-ion into V_2O_5 nanowires exist.

In addition, Yu et al. presented hexagonal NiS nanobelts as cathode materials for AIBs [91]. The Al/NiS battery keeps a capacity of over 100 mAh g^{-1} after 100 cycles at a current density of 200 mA g^{-1}. Significantly, good cycle stability with a 90% capacity retention over 100 cycles is also obtained. Zhang et al. reported unzipped carbon nanotubes as cathode materials for AIBs [92]. A flexible unzipped multi-walled carbon nanotube (UCNT) film consisting of graphene nanoribbons and a carbon nanotube backbone is prepared via a simple, low-cost, and scalable method for high-performance aluminum battery positive electrode. In the nanostructure, the neonatal graphene nanoribbons provide numerous active intercalation sites to store $AlCl_4^-$ anions, while the internal core CNTs are responsible for the rapid transportation of electrons to the active sites. Fast transportation of electrons and anions in UCNT film is simultaneously achieved, promoting the kinetics in the aluminum battery. Thus, the UCNT-Al battery exhibits a high storage performance with a capacity as high as 100 mAh g^{-1} at 2000 mA g^{-1} and a long cycle life of 5600 cycles with a capacity of 75 mAh g^{-1} and a Coulombic efficiency of 98% at 5000 mA g^{-1}. Furthermore, the excellent flexibility, fast charging, and variable discharging performance are beneficial for the development of portable devices.

To obtain AIBs with higher energy density, Al–S, Al–Se and Al–Te batteries were developed, and nanowires were also employed in these battery systems. For example, Bian et al. reported that Al–S batteries with multiwalled carbon nanotube (MWCNT)/S composites as the cathode materials [93]. This Al–S cell delivers an initial specific capacity of ~740 mAh ${g_S}^{-1}$ and retains a specific capacity of up to ~520 mAh ${g_S}^{-1}$ after 100 charge/discharge cycles. In addition, the Al–S cell exhibits a high discharge voltage plateau (1.6–2.0 V). Lei et al. presented graphene-encapsulated selenium@polyaniline nanowires as cathode materials for Al–Se batteries [94]. Se nanowires are well sealed in a PANI layer with a thickness of 15 nm, and then the constructed Se@PANI core–shell nanowires are uniformly encapsulated in graphene nanosheets to form a 3D hierarchical architecture of Se@PANI@G. As a result, the AIBs with the as-prepared Se@PANI@G composite positive electrode can deliver a high specific capacity of ~445.5 mAh g^{-1} during the second cycle at a current density of 200 mA g^{-1}, with a retained discharge specific capacity of 164.0 mAh g^{-1} after 160 cycles. The improved cycling performance can be associated with the high electrical conductivity of graphene sheets and the unique PANI shell, together with the 3D hierarchical architecture of Se@PANI@G. In addition, Zhang et al. proposed the Al–Te batteries with Te nanowires as cathode materials [95]. The Te nanowires deliver an ultrahigh discharge capacity ~1026 mA h g^{-1} at 0.5 A g^{-1} with an initial 1.4 V discharge voltage, which is competitive with the record-setting energy density of the documented AIBs.

9.5 Summary and Outlook

Scientific breakthroughs about multivalent-ion batteries have been reported at an increasing rate in the last ten years. Although many problems remain to be solved and the large gap in electrochemical performance between multivalent batteries and Li-ion batteries still exists, the unique advantage of multivalent metal and the resource scarcity problem of lithium stimulate the researchers to constantly lucubrate the new types of high-performance multivalent-ion batteries. A large amount of effort is still needed to achieve the prospective target. Effective strategies should be proposed to optimize the crystal structure and promote the electrochemical intercalated/deintercalated process in the future. The conventional optimization strategies borrowed from Li-ion battery, such as nanocrystallization treatment to improve active surface area, ion doping, carbon coatings, and preparation of porous structures, have improved the performance of nanowires for multivalent-ion batteries. Moreover, other specific optimized methods, such as co-intercalation of water molecules or halogen, pre-intercalation stratagem, amorphization treatment, and controlling the multivalent cation intercalation extent, are also important to improve the electrochemical performance of nanowires. On the other hand, advanced characterization methods are important for material development. With the rapid progress of detection method, the intrinsic reason for many troublesome issues in nanowires for multivalent-ion batteries will be revealed, which will guide the development of optimizing approach. After continuous exploration and innovation, we optimistically anticipate that nanowires for multivalent-ion batteries will break through bottleneck in the near future.

References

1 Grey, C. and Tarascon, J. (2017). Sustainability and in situ monitoring in battery development. *Nature Materials* 16 (1): 45–56.

2 Ohmer, N., Fenk, B., Samuelis, D. et al. (2015). Phase evolution in single-crystalline $LiFePO_4$ followed by in situ scanning X-ray microscopy of a micrometre-sized battery. *Nature Communications* 6 (1): 1–7.

3 Bonnick, P. and Muldoon, J. (2020). A trip to oz and a peak behind the curtain of magnesium batteries. *Advanced Functional Materials* 30 (21): 1910510.

4 Sun, R., Pei, C., Sheng, J. et al. (2018). High-rate and long-life VS2 cathodes for hybrid magnesium-based battery. *Energy Storage Materials* 12: 61–68.

5 Mei, L., Xu, J., Wei, Z. et al. (2017). Chevrel phase Mo6T8 (T = S, Se) as electrodes for advanced energy storage. *Small* 13 (34): 1701441.

6 Tang, H., Xu, N., Pei, C. et al. (2017). $H_2V_3O_8$ nanowires as high-capacity cathode materials for magnesium-based battery. *ACS Applied Materials & Interfaces* 9 (34): 28667–28673.

7 Lin, M.-C., Gong, M., Lu, B. et al. (2015). An ultrafast rechargeable aluminium-ion battery. *Nature* 520 (7547): 324–328.

8 Huang, J., Guo, Z., Ma, Y. et al. (2019). Recent progress of rechargeable batteries using mild aqueous electrolytes. *Small Methods* 3 (1): 1800272.

9 Ponrouch, A., Bitenc, J., Dominko, R. et al. (2019). Multivalent rechargeable batteries. *Energy Storage Materials* 20: 253–262.

10 Ambroz, F., Macdonald, T.J., and Nann, T. (2017). Trends in aluminium-based intercalation batteries. *Advanced Energy Materials* 7 (15): 1602093.

11 Li, Z., Vinayan, B.P., Jankowski, P. et al. (2020). Multi-electron reactions enabled by anion-based redox chemistry for high-energy multivalent rechargeable batteries. *Angewandte Chemie, International Edition* 132 (28): 11580–11587.

12 Zhao, X., Zhao-Karger, Z., Fichtner, M. et al. (2020). Halide-based materials and chemistry for rechargeable batteries. *Angewandte Chemie, International Edition* 59 (15): 5902–5949.

13 Erickson, E.M., Markevich, E., Salitra, G. et al. (2015). Development of advanced rechargeable batteries: a continuous challenge in the choice of suitable electrolyte solutions. *Journal of the Electrochemical Society* 162 (14): A2424.

14 Mao, M., Gao, T., Hou, S. et al. (2018). A critical review of cathodes for rechargeable Mg batteries. *Chemical Society Reviews* 47 (23): 8804–8841.

15 Muldoon, J., Bucur, C.B., and Gregory, T. (2014). Quest for nonaqueous multivalent secondary batteries: magnesium and beyond. *Chemical Reviews* 114 (23): 11683–11720.

16 Tan, S., Xiong, F., Wang, J. et al. (2020). Crystal regulation towards rechargeable magnesium battery cathode materials. *Materials Horizons* 7 (8): 1971–1995.

17 Xiong, F., Fan, Y., Tan, S. et al. (2018). Magnesium storage performance and mechanism of CuS cathode. *Nano Energy* 47: 210–216.

18 Pei, C., Xiong, F., Yin, Y. et al. (2021). Recent progress and challenges in the optimization of electrode materials for rechargeable magnesium batteries. *Small* 17 (3): 2004108.

19 Li, B., Masse, R., Liu, C. et al. (2019). Kinetic surface control for improved magnesium-electrolyte interfaces for magnesium ion batteries. *Energy Storage Materials* 22: 96–104.

20 Zhang, Q., Takeuchi, E.S., Takeuchi, K.J. et al. (2015). High energy density electrode materials for rechargeable magnesium batteries. *ECS Transactions* 66 (9): 171.

21 Cheng, X., Zhang, Z., Kong, Q. et al. (2020). Highly reversible cuprous mediated cathode chemistry for magnesium batteries. *Angewandte Chemie, International Edition* 132 (28): 11574–11579.

22 An, Q., Li, Y., Yoo, H.D. et al. (2015). Graphene decorated vanadium oxide nanowire aerogel for long-cycle-life magnesium battery cathodes. *Nano Energy* 18: 265–272.

23 Xu, Y., Deng, X., Li, Q. et al. (2019). Vanadium oxide pillared by interlayer Mg^{2+} ions and water as ultralong-life cathodes for magnesium-ion batteries. *Chem* 5 (5): 1194–1209.

24 Luo, T., Liu, Y., Su, H. et al. (2018). Nanostructured-VO_2(B): a high-capacity magnesium-ion cathode and its electrochemical reaction mechanism. *Electrochimica Acta* 260: 805–813.

25 Won, J.M., You, N.K., Lee, J.K. et al. (2015). Superior electrochemical properties of rutile VO_2 -carbon composite microspheres as a promising anode material for lithium ion batteries. *Electrochimica Acta* 156: 179–187.

26 Jiao, L., Yuan, H., Si, Y. et al. (2006). Electrochemical insertion of magnesium in open-ended vanadium oxide nanotubes. *Journal of Power Sources* 156 (2): 673–676.

27 Kim, R.-H., Kim, J.-S., Kim, H.-J. et al. (2014). Highly reduced VO_x nanotube cathode materials with ultra-high capacity for magnesium ion batteries. *Journal of Materials Chemistry A* 2 (48): 20636–20641.

28 Jiao, L., Yuan, H., Wang, Y. et al. (2005). Mg intercalation properties into open-ended vanadium oxide nanotubes. *Electrochemistry Communications* 7 (4): 431–436.

29 Rashad, M., Zhang, H., Asif, M. et al. (2018). Low-cost room-temperature synthesis of $NaV_3O_8 \cdot 1.69H_2O$ nanobelts for Mg batteries. *ACS Applied Materials & Interfaces* 10 (5): 4757–4766.

30 Song, J., Noked, M., Gillette, E. et al. (2015). Activation of a MnO_2 cathode by water-stimulated Mg^{2+} insertion for a magnesium ion battery. *Physical Chemistry Chemical Physics* 17 (7): 5256–5264.

31 Nam, K.W., Kim, S., Lee, S. et al. (2015). The high performance of crystal water containing manganese birnessite cathodes for magnesium batteries. *Nano Letters* 15 (6): 4071–4079.

32 Chiritescu, C., Cahill, D.G., Nguyen, N. et al. (2007). Ultralow thermal conductivity in disordered, layered WSe_2 crystals. *Science* 315 (5810): 351–353.

33 Fang, H., Chuang, S., Chang, T.C. et al. (2012). High-performance single layered WSe_2 p-FETs with chemically doped contacts. *Nano Letters* 12 (7): 3788–3792.

34 Boscher, N.D., Carmalt, C.J., and Parkin, I.P. (2006). Atmospheric pressure chemical vapor deposition of WSe_2 thin films on glass-highly hydrophobic sticky surfaces. *Journal of Materials Chemistry* 16 (1): 122–127.

35 Liu, B., Luo, T., Mu, G. et al. (2013). Rechargeable Mg-ion batteries based on WSe_2 nanowire cathodes. *ACS Nano* 7 (9): 8051–8058.

36 Shao, Y., Gu, M., Li, X. et al. (2014). Highly reversible Mg insertion in nanostructured Bi for Mg ion batteries. *Nano Letters* 14 (1): 255–260.

37 Arroyo-de Dompablo, M.E., Ponrouch, A., Johansson, P. et al. (2019). Achievements, challenges, and prospects of calcium batteries. *Chemical Reviews* 120 (14): 6331–6357.

38 Xu, Z.-L., Park, J., Wang, J. et al. (2021). A new high-voltage calcium intercalation host for ultra-stable and high-power calcium rechargeable batteries. *Nature Communications* 12 (1): 1–9.

39 Pu, S.D., Gong, C., Gao, X. et al. (2020). Current-density-dependent electroplating in Ca electrolytes: from globules to dendrites. *ACS Energy Letters* 5 (7): 2283–2290.

40 Hosein, I.D. (2021). The promise of calcium batteries: open perspectives and fair comparisons. *ACS Energy Letters* 6 (4): 1560–1565.

41 Ji, B., He, H., Yao, W. et al. (2021). Recent advances and perspectives on calcium-ion storage: key materials and devices. *Advanced Materials* 33 (2): 2005501.

42 Tojo, T., Sugiura, Y., Inada, R. et al. (2016). Reversible calcium ion batteries using a dehydrated prussian blue analogue cathode. *Electrochimica Acta* 207: 22–27.

43 Shiga, T., Kondo, H., Kato, Y. et al. (2015). Insertion of calcium ion into prussian blue analogue in nonaqueous solutions and its application to a rechargeable battery with dual carriers. *Journal of Physical Chemistry C* 119 (50): 27946–27953.

44 Kim, S., Yin, L., Lee, M.H. et al. (2020). High-voltage phosphate cathodes for rechargeable Ca-ion batteries. *ACS Energy Letters* 5 (10): 3203–3211.

45 Wang, J., Tan, S., Xiong, F. et al. (2020). $VOPO_4 \cdot 2H_2O$ as a new cathode material for rechargeable Ca-ion batteries. *Chemical Communications* 56 (26): 3805–3808.

46 Jeon, B., Heo, J.W., Hyoung, J. et al. (2020). Reversible calcium-ion insertion in NASICON-type $NaV_2(PO_4)_3$. *Chemistry of Materials* 32 (20): 8772–8780.

47 Chae, M.S., Kwak, H.H., and Hong, S.-T. (2020). Calcium molybdenum bronze as a stable high-capacity cathode material for calcium-ion batteries. *ACS Applied Energy Materials* 3 (6): 5107–5112.

48 Ren, W., Xiong, F., Fan, Y. et al. (2020). Hierarchical copper sulfide porous nanocages for rechargeable multivalent-ion batteries. *ACS Applied Materials & Interfaces* 12 (9): 10471–10478.

49 Murata, Y., Takada, S., Obata, T. et al. (2019). Effect of water in electrolyte on the Ca^{2+} insertion/extraction properties of V_2O_5. *Electrochimica Acta* 294: 210–216.

50 Wang, J., Wang, J., Jiang, Y. et al. (2022). $CaV_6O_{16} \cdot 2.8H_2O$ with Ca^{2+} pillar and water lubrication as a high-rate and long-life cathode material for ca-ion batteries. *Advanced Functional Materials* 32 (25): 2113030.

51 Dong, L.-W., Xu, R.-G., Wang, P.-P. et al. (2020). Layered potassium vanadate $K_2V_6O_{16}$ nanowires: a stable and high capacity cathode material for calcium-ion batteries. *Journal of Power Sources* 479: 228793.

52 Adil, M., Sarkar, A., Sau, S. et al. (2022). Non-aqueous rechargeable calcium-ion batteries based on high voltage zirconium-doped ammonium vanadium oxide cathode. *Journal of Power Sources* 541: 231669.

53 Zuo, C., Xiong, F., Wang, J. et al. (2022). MnO_2 polymorphs as cathode materials for rechargeable ca-ion batteries. *Advanced Functional Materials* 2202975.

54 Liu, K., Liu, Y., Lin, D. et al. (2018). Materials for lithium-ion battery safety. *Science Advances* 4 (6): eaas9820.

55 Tarascon, J.-M. and Armand, M. (2001). Issues and challenges facing rechargeable lithium batteries. *Nature* 414 (6861): 359–367.

56 Liu, C., Xie, X., Lu, B. et al. (2021). Electrolyte strategies toward better zinc-ion batteries. *ACS Energy Letters* 6 (3): 1015–1033.

57 Liang, Y., Tao, Z., and Chen, J. (2012). Organic electrode materials for rechargeable lithium batteries. *Advanced Energy Materials* 2 (7): 742–769.

58 Liu, Y., Huang, M., Xiong, F. et al. (2022). Improved zinc-ion storage performance of the metal-free organic anode by the effect of binder. *Chemical Engineering Journal* 428: 131092.

59 Yan, M., He, P., Chen, Y. et al. (2018). Water-lubricated intercalation in $V_2O_5 \cdot nH_2O$ for high-capacity and high-rate aqueous rechargeable zinc batteries. *Advanced Materials* 30 (1): 1703725.

60 He, P., Zhang, G., Liao, X. et al. (2018). Sodium ion stabilized vanadium oxide nanowire cathode for high-performance zinc-ion batteries. *Advanced Energy Materials* 8 (10): 1702463.

61 Kundu, D., Adams, B.D., Duffort, V. et al. (2016). A high-capacity and long-life aqueous rechargeable zinc battery using a metal oxide intercalation cathode. *Nature Energy* 1 (10): 1–8.

62 Xia, C., Guo, J., Lei, Y. et al. (2018). Rechargeable aqueous zinc-ion battery based on porous framework zinc pyrovanadate intercalation cathode. *Advanced Materials* 30 (5): 1705580.

63 Sambandam, B., Soundharrajan, V., Kim, S. et al. (2018). Aqueous rechargeable Zn-ion batteries: an imperishable and high-energy $Zn_2V_2O_7$ nanowire cathode through intercalation regulation. *Journal of Materials Chemistry A* 6 (9): 3850–3856.

64 Xia, C., Guo, J., Li, P. et al. (2018). Highly stable aqueous zinc-ion storage using a layered calcium vanadium oxide bronze cathode. *Angewandte Chemie, International Edition* 57 (15): 3943–3948.

65 He, P., Quan, Y., Xu, X. et al. (2017). High-performance aqueous zinc–ion battery based on layered $H_2V_3O_8$ nanowire cathode. *Small* 13 (47): 1702551.

66 Pang, Q., Sun, C., Yu, Y. et al. (2018). $H_2V_3O_8$ nanowire/graphene electrodes for aqueous rechargeable zinc ion batteries with high rate capability and large capacity. *Advanced Energy Materials* 8 (19): 1800144.

67 Cai, Y., Liu, F., Luo, Z. et al. (2018). Pilotaxitic $Na_{1.1}V_3O_{7.9}$ nanoribbons/graphene as high-performance sodium ion battery and aqueous zinc ion battery cathode. *Energy Storage Materials* 13: 168–174.

68 Wan, F., Zhang, L., Dai, X. et al. (2018). Aqueous rechargeable zinc/sodium vanadate batteries with enhanced performance from simultaneous insertion of dual carriers. *Nature Communications* 9 (1): 1–11.

69 Hu, P., Zhu, T., Wang, X. et al. (2018). Highly durable $Na_2V_6O_{16} \cdot 1.63\ H_2O$ nanowire cathode for aqueous zinc-ion battery. *Nano Letters* 18 (3): 1758–1763.

70 Soundharrajan, V., Sambandam, B., Kim, S. et al. (2018). $Na_2V_6O_{16} \cdot 3H_2O$ barnesite nanorod: an open door to display a stable and high energy for aqueous rechargeable Zn-ion batteries as cathodes. *Nano Letters* 18 (4): 2402–2410.

71 Pan, H., Shao, Y., Yan, P. et al. (2016). Reversible aqueous zinc/manganese oxide energy storage from conversion reactions. *Nature Energy* 1 (5): 1–7.

72 Wu, B., Zhang, G., Yan, M. et al. (2018). Graphene scroll-coated α-MnO_2 nanowires as high-performance cathode materials for aqueous Zn-ion battery. *Small* 14 (13): 1703850.

73 Wang, J., Wang, J.-G., Qin, X. et al. (2020). Superfine MnO_2 nanowires with rich defects toward boosted zinc ion storage performance. *ACS Applied Materials & Interfaces* 12 (31): 34949–34958.

74 Liu, D.-S., Mai, Y., Chen, S. et al. (2021). A 1D–3D interconnected δ-MnO_2 nanowires network as high-performance and high energy efficiency cathode material for aqueous zinc-ion batteries. *Electrochimica Acta* 370: 137740.

75 Chu, S. and Majumdar, A. (2012). Opportunities and challenges for a sustainable energy future. *Nature* 488 (7411): 294–303.

76 Etacheri, V., Marom, R., Elazari, R. et al. (2011). Challenges in the development of advanced Li-ion batteries: a review. *Energy & Environmental Science* 4 (9): 3243–3262.

77 Larcher, D. and Tarascon, J.-M. (2015). Towards greener and more sustainable batteries for electrical energy storage. *Nature Chemistry* 7 (1): 19–29.

78 Bonaccorso, F., Colombo, L., Yu, G. et al. (2015). Graphene, related two-dimensional crystals, and hybrid systems for energy conversion and storage. *Science* 347 (6217): 1246501.

79 Tepavcevic, S., Xiong, H., Stamenkovic, V.R. et al. (2012). Nanostructured bilayered vanadium oxide electrodes for rechargeable sodium-ion batteries. *ACS Nano* 6 (1): 530–538.

80 Sun, Y.-K., Chen, Z., Noh, H.-J. et al. (2012). Nanostructured high-energy cathode materials for advanced lithium batteries. *Nature Materials* 11 (11): 942–947.

81 Kennedy, T., Brandon, M., and Ryan, K.M. (2016). Advances in the application of silicon and germanium nanowires for high-performance lithium-ion batteries. *Advanced Materials* 28 (27): 5696–5704.

82 Sathiya, M., Abakumov, A.M., Foix, D. et al. (2015). Origin of voltage decay in high-capacity layered oxide electrodes. *Nature Materials* 14 (2): 230–238.

83 Levi, E., Gofer, Y., and Aurbach, D. (2010). On the way to rechargeable Mg batteries: the challenge of new cathode materials. *Chemistry of Materials* 22 (3): 860–868.

84 Xu, C., Li, B., Du, H. et al. (2012). Energetic zinc ion chemistry: the rechargeable zinc ion battery. *Angewandte Chemie* 124 (4): 957–959.

85 Elia, G.A., Marquardt, K., Hoeppner, K. et al. (2016). An overview and future perspectives of aluminum batteries. *Advanced Materials* 28 (35): 7564–7579.

86 Jayaprakash, N., Das, S., and Archer, L. (2011). The rechargeable aluminum-ion battery. *Chemical Communications* 47 (47): 12610–12612.

87 Ling, C. and Mizuno, F. (2013). Phase stability of post-spinel compound AMn_2O_4 (A= Li, Na, or Mg) and its application as a rechargeable battery cathode. *Chemistry of Materials* 25 (15): 3062–3071.

88 Zafar, Z.A., Imtiaz, S., Razaq, R. et al. (2017). Cathode materials for rechargeable aluminum batteries: current status and progress. *Journal of Materials Chemistry A* 5 (12): 5646–5660.

89 Gu, S., Wang, H., Wu, C. et al. (2017). Confirming reversible Al^{3+} storage mechanism through intercalation of Al^{3+} into V_2O_5 nanowires in a rechargeable aluminum battery. *Energy Storage Materials* 6: 9–17.

90 Reed, L.D. and Menke, E. (2013). The roles of V_2O_5 and stainless steel in rechargeable Al–ion batteries. *Journal of the Electrochemical Society* 160 (6): A915.

91 Yu, Z., Kang, Z., Hu, Z. et al. (2016). Hexagonal NiS nanobelts as advanced cathode materials for rechargeable Al-ion batteries. *Chemical Communications* 52 (68): 10427–10430.

92 Zhang, E., Wang, J., Wang, B. et al. (2019). Unzipped carbon nanotubes for aluminum battery. *Energy Storage Materials* 23: 72–78.

93 Bian, Y., Li, Y., Yu, Z. et al. (2018). Using an $AlCl_3$/urea ionic liquid analog electrolyte for improving the lifetime of aluminum-sulfur batteries. *ChemElectroChem* 5 (23): 3607–3611.

94 Lei, H., Tu, J., Li, S. et al. (2022). Graphene-encapsulated selenium@polyaniline nanowires with three-dimensional hierarchical architecture for high-capacity aluminum–selenium batteries. *Journal of Materials Chemistry A* 10 (28): 15146–15154.

95 Zhang, X., Jiao, S., Tu, J. et al. (2019). Rechargeable ultrahigh-capacity tellurium–aluminum batteries. *Energy & Environmental Science* 12 (6): 1918–1927.

10

Conclusion and Outlook

Electrochemical energy storage has the characteristics of good energy density, high power density, and stable cycle life and has thus become a widely studied energy storage form. Recently, in order to obtain higher-performance electrochemical energy storage devices, various nanomaterials have been studied and explored. Among them, nanowire materials have unique advantages compared with other simple-structured nanomaterials. The unique anisotropy, large specific surface area, excellent tension adaptability, fast axial electron transport, and radial-ion diffusion of nanowire materials play a key role in shortening the ion transport path and suppressing the volume expansion of electrode materials. Research and applications of high-performance nanowire electrode materials have become hotspots in the field of electrochemical energy storage devices.

10.1 Structure Design and Performance Optimization of 1D Nanomaterials

In order to explore the application potential of nanowire materials, some nanowire materials with new structures have been prepared one after another. For example, the aforementioned porous nanowires [1–3], hierarchical nanowires [4], heterogeneous nanowires [5, 6], hollow nanowires [7], and nanowire arrays [8] all extend the application of nanowire materials. These are actually breakthroughs in the synthesis strategy of nanowires. Specifically, conventional nanowire preparation processes generally include hydrothermal/solvothermal methods, sol–gel methods, electrospinning methods, and high-temperature solid-phase methods.

The hydrothermal/solvothermal method utilizing the dissolution-recrystallization reaction mechanism is one of the most reliable nanowire preparation methods. This method requires steps such as centrifugation and sample washing, and the solvent is not easily removed. The product purity is general, which can be affected by many factors such as solvent, temperature, or pressure. However, the equipment used is simple, easy to operate, and it can also achieve a high yield with good crystallinity. Homogeneous conventional nanowires can be prepared simply and quickly

Nanowire Energy Storage Devices: Synthesis, Characterization, and Applications, First Edition.
Edited by Liqiang Mai.

by this method. In addition, heterogenous nanowires can also be prepared through the heterogeneous nucleation process of the secondary hydrothermal method [9]. Heterogeneous nanowire arrays can be prepared by a hydrothermal-chemical bath deposition method [10]. The advantages of the hydrothermal/solvothermal method in the preparation of nanowire materials provide the possibility for the large-scale preparation of many nanowire materials.

The sol–gel method can effectively control the size and uniformity of nanowires by using the properties of colloids combined with calcination methods, and the obtained phase is purer, which is one of the fine synthesis methods of nanowires. In this method, compounds with high chemical activity components are skillfully mixed under liquid phase conditions to undergo hydrolysis and polymerization reactions, and sol and gel are obtained in turn, and then fine nanowires are obtained by drying and calcining. At present, the research on sol–gel technology has explored the corresponding theoretical basis in the polymer discipline and many chemical disciplines (such as physical chemistry, colloidal chemistry, coordination chemistry, and metal organic chemistry). Its application technology has also been gradually improved and matured and has been widely used [11–13]. However, the severe requirements of this method on calcination conditions and reaction time, as well as its cumbersome reaction steps, long preparation cycle, and other disadvantages also need to be fully considered.

Electrospinning is a powerful method for preparing nanowire materials. The microscopic morphology of the prepared nanowire materials is not limited by the crystal growth orientation of the material itself and can effectively realize single-component nanofibers such as polymers and inorganics, as well as polymer/polymer, polymer/inorganic and inorganic/inorganic composite nanofibers. This technology greatly enriches the variety of one-dimensional nanomaterials [14–17]. In addition, the equipment required by the electrospinning technology is simple, the cost of raw materials is low, the selection of raw materials is easy, and the process is controllable. Therefore, electrospinning is also one of the important ways to efficiently prepare conventional nanowires.

The high-temperature solid-phase method prepares the target product by solid-phase reaction at high temperature. This method has the advantages of low cost, large output, simple preparation process, no agglomeration of the prepared powder particles and good filling property [18]. Therefore, high-temperature solid-phase method is also one of the potential methods to realize the industrialization of nanowires.

In addition to the above conventional nanowire preparation methods, methods such as co-precipitation method [19], ultrasonic spray pyrolysis method [20], chemical vapor deposition method [21], physical vapor deposition method [22], template method [23], self-assembly orientation overlap method [24], and coupled plasma etching method [25] are getting more and more attention. These novel methods have been widely used in the preparation of unconventional nanowire materials such as graded nanowires, heterogeneous nanowires, hollow nanowires, and nanowire arrays.

Ultrasonic spray pyrolysis can directly obtain submicron powder with uniform distribution composed of nano-scale primary particles [26]. Co-precipitation method is often used to prepare insoluble inorganic salt nanowires [19]. The chemical vapor deposition method can induce the directional growth of nanowires to obtain high-purity single nanowires [21]. The physical vapor deposition method can deposit nanowires on the surface that there is the substrate with a small amount of consumables under the premise of no pollution to the environment [22]. The template method and self-assembly method can synthesize porous nanowires with large specific surface area [23], and the material and structure of the template can be designed according to the size, function, and structural requirements of the desired material to meet the actual needs.

Generally speaking, with the endless emergence of nanowire materials, nanowire materials have also developed from the initial simple conventional nanowires to porous, hollow, heterogeneous, and other nanowires. Different synthesis methods and growth mechanisms have their own advantages, disadvantages, and influencing factors. Among them, the synthesis methods of conventional nanowires are mainly hydrothermal/solvothermal method, sol–gel methods, electrospinning methods, and solid-phase sintering method. With the in-depth research of nanowire materials, novel synthesis methods such as ultrasonic spray pyrolysis, co-precipitation, and vapor deposition have also expanded the application scope of nanowires. They are all synthesis methods of nanowires derived from the synthesis methods of conventional nanomaterials based on the two basic ideas of controlling the growth of crystal anisotropy by dominant crystal planes and artificially induced forced directional growth.

Each synthetic method has its own unique advantages and limitations. In order to prepare ideal nanowire products that can be practically applied, various methods are usually combined to synthesize or modify nanowires with special structures. Researchers should combine knowledge from chemistry, physics, and materials science to delve into microscopic processes and internal synthesis mechanisms to achieve precise designs. At the same time, in order to achieve breakthroughs in synthetic methods, it is recommended that researchers effectively combine new ideas with traditional synthetic methods so that synthetic nanowires have higher feasibility, suitable cost, and good performance.

This book combines the latest research progress in nanowire materials, focuses on the synthesis methods of nanowire electrode materials, starts with morphology control, and focuses on the structural differences caused by different synthesis methods. The typical research works are selectively listed to briefly describe the preparation methods, morphologies, and corresponding characterizations of nanowires. The common synthesis methods and growth mechanisms of different types of nanowires (conventional, porous, hierarchical, heterogeneous, hollow, and array) are introduced in detail, and their respective advantages, disadvantages, and influencing factors are shown. The different structures of the nanowires determine the specific properties of the nanowires. There are many typical nanomaterial preparation methods that can also be used to prepare nanowire materials. The advantages and disadvantages of different preparation methods can better realize the combination of nanowire materials and production practice, and can be better applied in various aspects.

10.2 Advanced Characterization Methods for 1D Nanomaterials

The unique properties of nanomaterials often require special detection methods to observe. Common material characterization methods include X-ray diffraction, X-ray photoelectron spectroscopy, Raman spectroscopy, scanning electron microscopy, and transmission electron microscopy. These characterization methods have strong analytical and detection capabilities. However, these characterization methods are usually interfered by factors such as changes in the material after it is separated from the electrochemical environment, making it difficult to accurately interpret the experimental results. In situ characterization technology can directly observe the information changes of the electrochemical reaction of the materials of nanowire electrode [27]. It can reveal the nature of the electrochemical reaction of nanowire electrode materials through real-time monitoring of the change in phase, morphology, and atomic valence bond connection state of the material. It provides a direction to optimize the performance of nanowire electrode materials and shows good scientific value and application potential in exploring the intrinsic mechanism.

Special structures produce special properties, and one-dimensional nanowire materials have special structures that reach the nanometer level in two dimensions, thus exhibiting special properties that bulk materials do not possess. For example, nanowire electrode materials possess unique anisotropy, which is combined with continuous axial electron transport, large specific surface area, and fast ion diffusion. In the research field of micro-nano energy storage devices', micro-nano devices assembled based on a single nanomaterial play an important role, such as single nanowires and single nanosheet devices, the unique anisotropy of nanowire materials provides great convenience for the construction of single nanodevices.

In order to achieve accurate characterization of nanowire electrode materials, by constructing a single-nanowire energy storage device, the electrode/current collector interface and electrode/electrolyte interface can be subtly extracted from the mixed and complex electrochemical system for research. In turn, the connection between the structure of individual nanowires, electrical transport, electrode/current collector contact, electrode charge–discharge state, and capacity fading can be investigated. Researchers can also intuitively present principles that are difficult to reveal by conventional means through ingenious device design, achieve accurate measurement of basic physical property parameters such as electricity, heat, and magnetism, and provide an effective method to study the evolution of material properties at the nanoscale. Like the first single nanowire solid-state electrochemical device in the world, the material structure information is detected by a monochromatic laser, and the multifield real-time monitoring of electricity/electrochemistry/spectroscopy is realized [28]. At the same time, it also realizes the precise integration of micro-nano circuits and energy storage devices, and has high detection accuracy and strong collection signals, which provides great convenience for studying the basic reaction principles of electrode materials in electrochemical reactions. With the continuous maturity of semiconductor controllable precision processing technology, research on single nanowire devices has gradually expanded from the frontier of basic physics to the application of electrochemical energy storage. And the combination of multifield real-time monitoring will definitely help us understand the reaction mechanism.

In situ TEM has important applications in material synthesis, chemical catalysis, life sciences, and energy storage materials. It can observe and control the progress of gas-phase reactions and liquid-phase reactions in real time at the atomic scale, so as to study scientific issues such as the essential mechanism of the reaction. The fourth chapter of this book lists the application of in situ spectroscopy technology in nanowire energy storage devices, which truly realizes the precise measurement of the properties of nanowires. In situ electron microscopy technology for electrode material morphology plays a key role in studying its electrochemical performance and electrochemical reaction mechanism. The in situ characterization of the material morphology can intuitively reflect the changes in the microscopic morphology of the material during the electrochemical reaction. These microscopic changes can often provide evidence for the hypothesis of the electrochemical mechanism, so that the hypothesis can be confirmed accordingly.

In situ TEM characterization of secondary batteries is relatively mature, and there are many specific sample rods that can achieve targeted in situ characterization of the electrochemical electrodes and solid and liquid electrolytes of the battery. For example, the solid-state open in situ battery sample holder can allow in situ high spatial resolution imaging; the sealed liquid in situ battery sample holder can simulate the real battery liquid environment to study the evolution of SEI and Li dendrites. The in situ heating sample holder can analyze the thermal stability of metal oxide-based cathodes by realizing the heating function. In situ sample rods based on ionic liquid open cells enable the investigation of phase transitions in metal oxide-based cathode materials. The nanofilm battery sample rod also enables in situ observations of changes in solid-state electrolytes. These accessories have greatly advanced the process of in situ characterization of nanowire materials.

In situ TEM can deeply explore the electrochemical reaction mechanism of nanomaterials by exploring the changes in morphology, organization, and structure of nanomaterials during electrochemical processes. In addition, by adding accessories such as energy dispersive spectrometers to the transmission electron microscope, it is also possible to characterize the changes in element species and distributions of nanowire materials during the electrochemical reaction process to assist in the analysis of the electrochemical mechanism of materials. In Chapter 4 of this book, the choice of nanowires and the choice of counter electrodes are very flexible, which also proves that the observation of nanowires by in situ TEM is universal. In addition, the citation of cryo-EM can also greatly enrich the application of electrode materials, SEI, electrolyte observation, etc.

In addition to in situ TEM, in situ SEM also enables fine characterization of material surfaces and material cross-sections. It also enables in situ characterization of changes in element species and content in conjunction with energy spectrometers. Electrochemical auxiliary information, such as reaction sites on the electrode surface, can be obtained by observing the real-time changes of the sample surface or cross section.

Through more than 100 years of human research in the field of spectroscopy, the use of spectroscopy to detect the internal information of materials has become one of the most routine testing methods for materials testing. Electrochemical reactions involve multiple systems that are complex and changeable, and it is difficult to characterize them clearly by ordinary testing methods. If the electrochemical reaction process of the material cannot be accurately characterized, it is difficult to point out the direction for structural optimization and performance improvement of the material. Some ex situ tests can damage the external environment of the battery material, thereby affecting the test results. The emergence of in situ X-ray diffraction technology has met people's needs for basic scientific research. Compared to traditional ex-situ testing techniques, in situ X-ray diffraction (XRD) requires the test object to be in a reactive state. It can closely combine the in-situ concept with the X-ray diffractometer to track and react on microscopic dynamic information in real time and detect intermediate products and phase changes in the reaction process. In the existing in situ electrochemical testing technology, since this technology can directly observe the change of the diffraction peak generated by the crystal plane of the material, the change of the crystallographic parameter of the material can be obtained through analysis. So it is also widely recognized and used.

As mentioned in Chapter 4, by introducing structures similar to ordinary button batteries, such as battery shell, positive electrode material, negative electrode material, battery separator, electrolyte, and current collector, in situ measurement of electrode material XRD can be achieved under working conditions. In 2001, McBreen et al. of Brookhaven National Laboratory in the United States first used organic films as window materials for in-situ battery molds [29]. Later, the team of Mai Liqiang of Wuhan University of Technology used beryllium sheet as the window material for mold X-ray. The researchers proposed and optimized the in situ XRD cell mold in a relatively short period of time. On this basis, the in situ mold has undergone some improvements. For example, put the in situ battery in the matching box use resistance wire heating and low-temperature alcohol circulation to create high- and low-temperature environment of the mold, respectively. This method can complete the test at a temperature of −20 to 60 °C, making up for the problem of in situ reaction at a single temperature [30].

In situ electrochemical X-ray technique can be an effective technique to study the electrochemical reaction mechanism. However, limited by the complex equipment design and harsh test conditions, this technology still has a lot of room for development. Its development direction mainly has the following three points: (i) Design of the in-situ battery device. At present, there is a big problem with the packaging of in situ batteries, and the sealing performance is not very good. As a result, components such as water and oxygen in the air can easily enter the battery, thereby affecting the reaction in the electrodes. This phenomenon is more obvious when the test time is longer. In addition, when beryllium sheets are used as X-ray window materials, some electrode materials or electrolytes are prone to react with beryllium sheets, which will affect the test results and even fail the experiment. At the same time, the high price of beryllium flakes and their toxicity and easy oxidation prompt people to search for better window materials. In addition, expanding the test angle range through a reasonable design of the device is also one of the directions for device optimization. (ii) Accuracy and sensitivity of in situ testing. In situ XRD is generally designed to capture the intermediate products of the reaction, and these intermediate products may appear in a very short time, which requires the diffraction equipment to have sufficient data acquisition speed. (iii) Combined use of in situ characterization techniques. A single in situ test method is usually qualitative and can generally only reflect information about a certain aspect of the material, but a reasonable combination of these methods can give the whole picture of the entire electrochemical reaction. For example, in situ XRD is used in conjunction with in situ SEM, in situ TEM, in situ Raman spectroscopy, in situ Fourier infrared spectroscopy, etc. The evolutionary links between phase, morphology, and chemical bonds can be obtained simultaneously. The combined use of these technologies will certainly contribute to the further understanding, design, and control of secondary ion electrode materials and battery systems.

Raman spectroscopy can detect information about molecular vibrations and rotations, so it can be used to characterize the valence bonds and functional groups present in the molecules in the sample. In situ Raman spectroscopy combines the in situ testing concept with a Raman spectrometer. By monitoring the position of the Raman absorption peak and the peak intensity change of the electrode material in real time during the electrochemical process, we can understand the structural change of the material during the electrochemical process. Inspired by microfluidic research, Liqiang Mai's team successfully characterized the structural evolution of a single nanowire in an electrochemical reaction by combining a micro-reaction cell designed with a single nanodevice platform. The electrochemical processes of energy storage materials were explored at the atomic scale by combining in situ Raman spectroscopy with in situ atomic force microscopy. Through in situ Raman characterization, they successfully explained the high capacity exhibited by transition metal-layered oxides [31]. The electrochemical reaction model for co-intercalation of anions and cations is demonstrated. It opens up new ideas for the study of the electrochemical mechanism of cathode.

As a common characterization method, X-ray photoelectron spectroscopy plays a key role in the characterization of element types and contents of materials. The reference point for the binding energy of a solid sample is the Fermi level, which is used to detect the energy required for the electron to transition to the Fermi level. The X-rays interact with the sample, and the energy of the X-rays is absorbed by the solid. The kinetic energy of the photoelectrons can be analyzed by the photoelectron energy analyzer in the photoelectron spectrometer, and then the value of the binding energy can be obtained. The number of detected photoelectrons is distributed according to the correspondingly measured binding energy, and the spectrum of the X-ray photoelectron spectrum can be obtained. Photoelectron spectroscopy can be used for elemental composition analysis and quantitative analysis, which has important research value in the field of electrochemistry.

Synchrotron radiation is electromagnetic radiation emitted when charged particles travel along curved orbits under the influence of an electromagnetic field. With the construction of the synchrotron radiation source, information such as elemental composition, electronic state, and microstructure of materials can be analyzed by using the signal changes before and after X-ray incidence. Therefore, X-ray absorption spectroscopy has been unprecedentedly developed. Synchrotron radiation light source has been developed for three generations. There are currently four synchrotron radiation facilities in China, located in Beijing, Hefei, Shanghai, and Taiwan. XAS has some discrete peaks or undulations near the absorption edge and its high-energy end, which are called fine structure. Synchrotron X-ray absorption fine structure (XAFS) technology is widely used to study the chemical environment around atoms of various elements in condensed matter by adjusting the energy of X-rays. And obtain important information such as the local structure, oxidation state, and neighboring coordination atoms of the constituent elements in the material. In addition, through the in situ synchrotron radiation technology that combines the in situ concept with the synchrotron radiation technology, information such as the structural evolution, local structural changes, and transition metal ion oxidation states of electrode materials during electrochemical energy storage can be obtained. This has important guiding significance for accurately revealing the battery reaction mechanism, optimizing the material composition and structure, and developing high-performance new materials.

In the application of nanowire energy storage materials, the problem of electrode attenuation is very serious. Researchers need to track the electrochemical process in real time and study the mechanism of attenuation. In addition to the in situ characterization of nanowire energy storage devices and the commonly used electron microscopy and spectroscopic characterizations described above, researchers can also take advantage of some unique advantages of other testing methods as an auxiliary or supplement. In situ characterization and research of materials, such as scanning probe microscopy, including atomic force microscopy and scanning tunneling microscopy. With very high resolution, atoms can be easily "see", which is difficult to achieve with ordinary electron microscopes. Different from some analytical instruments that estimate the surface structure of the sample by indirect or computational methods, scanning probe microscope (SPM) obtains a real-time, real high-resolution image of the sample surface. In addition, a huge advantage of SPM scanning imaging compared to other electron microscopes is that it enables 3D imaging of the sample surface. Similar to SPM, in situ testing methods that have not been widely used in the characterization of nanoscale energy storage materials but have great prospects include electron paramagnetic resonance imaging, neutron diffraction, and nuclear magnetic resonance. Combining these methods with single nanowire energy storage devices, electron microscopy, and spectroscopy, it is possible to more accurately detect the electrochemical process of electrode materials in situ, deeply study the reaction mechanism and propose optimization strategies to improve their electrochemical performance.

When the material is characterized by AFM, the slight change of the microcantilever can be detected by the sensor, and then the surface information of the material can be obtained. Therefore, in situ AFM can be used to probe the surface changes of materials. This has important implications for some hard-to-probe material interfaces, such as the change of SEI of nanowire materials during electrochemical processes. Nuclear magnetic resonance (NMR) technology has a wide range of applications in physics, medicine, chemistry, biology, food and other fields, it can provide information on molecular dynamics and molecular chemical structure. At present, it has become a conventional technical means for analyzing the molecular structure and characterizing the physical and chemical properties of substances.

The principle of neutron diffraction is similar to that of XRD. It was mainly used to study the crystal structure of matter, and Bragg formula is also applicable. It is complementary to X-ray diffraction in application. When X-rays or a stream of electrons encounter matter and scatter, the scattering centers are mainly electrons in atoms. The scattering power increases with the increase of the atomic number of the substance, and decreases with the increase of the diffraction angle 2θ. In neutron diffraction, the neutron stream is uncharged, and when interacting with matter, it mainly interacts with atomic nuclei, resulting in isotropic scattering. This scattering power has no certain relationship with the atomic number of the substance, so the diffraction spectrum avoids the influence of the atomic number of the element. Therefore, neutron diffraction is more reliable for the characterization of some substances containing elements with relatively large atomic numbers.

Time-of-flight mass spectrometry is also a commonly used test method to characterize elements. The ions generated by the ion source are first collected by the collector. The ions with zero initial velocity are accelerated by a pulsed electric field, then enter the field-free drift tube, and fly to the ion receiver at a constant speed. The higher the ion mass, the longer it takes to reach the receiver; the lower the ion mass, the shorter the time it takes to reach the receiver. According to this principle, ions of different masses can be separated according to their mass-to-charge ratio (the ratio of mass to charge). With lateral resolution down to below 50 nm, SIMS uses a focused beam of primary ions to bombard and ionize the sample surface, and then generate characteristic secondary ions directed toward the mass spectrometer. Due to the high sensitivity of the surface and the high resolution capability of the depth profile, time of flight secondary ion mass spectrometry (TOF SIMS) can be used to study the related problems of the SEI in the battery system, such as the composition difference of the related interface before and after the electrochemical reaction occurs.

The research of modern science is inseparable from the development of modern testing technology, and the in situ characterization of the unique 1D structure of nanowires is a hot issue in current research, and related research results have been reported many times in journals such as *Nature* and *Science*. Among the many characterization methods, the most common are in situ TEM and in situ XRD. These two characterization techniques play a key role in studying the morphology and phase changes of nanowires in electrochemical reactions, and can often give researchers convincing evidence. However, for some characterization methods with harsh test conditions, such as in situ synchrotron radiation and in situ neutron diffraction, since the bombarded particles require enormous energy, a huge acceleration field or light source must be constructed. For synchrotron radiation, only Beijing, Shanghai, Hefei, and Taiwan have light sources for synchrotron radiation testing in China, and relevant testing resources are still lacking. There is still a lot of room for development in the characterization of single nanowire energy storage devices. The optical field generated by the femtosecond laser also has a relatively large influence on a single device, and can also provide some information for the characterization of a single nanowire.

In the process of studying the reaction mechanism of nanowire materials, it is often necessary to explain various test results so that the conclusions drawn are more convincing. By infiltrating more electrochemical in situ techniques into characterization techniques and monitoring the changes of nanowire materials in the entire electrochemical reaction in real time, it will undoubtedly enable people to more accurately understand the electrochemical energy storage mechanism of nanowires and explore new theory.

10.3 Applications and Challenges of Nanowire Energy Storage Devices

In the previous chapters, we discussed the application of nanowire materials in the field of energy storage in detail. Here, we will summarize the applications and challenges of nanowires in different energy storage systems. As an important energy storage device, ion batteries have been widely studied and applied. Lithium-ion batteries have gradually replaced lead-acid batteries and become mainstream batteries due to their high specific energy, long cycle performance, low battery self-discharge, and light weight. However, due to the scarcity of lithium resources, research on other types of ion batteries is also making breakthroughs.

Other types of ion batteries can be divided into monovalent-ion battery, multivalent-ion battery, and mixed-ion battery according to the amount of charge carried by the carriers. The bottleneck in the development of ion batteries is mainly limited by the development of high-performance electrode materials. Nanowire materials can enhance the diffusion kinetics of carriers and improve the stability of materials. The advantages of in-situ characterization of electrochemical processes and the construction of flexible devices, which are beneficial to the construction of in situ devices, have made nanowire electrode materials of interest not only in lithium-ion batteries, but also in other types of ion batteries.

10.3.1 Application of Nanowire Structures in Lithium-ion Batteries

The rapid development of new electric vehicles and large-scale energy storage systems has put forward higher requirements for the energy density of lithium-ion battery. As for the anode, it is necessary to introduce higher-capacity anode materials by solving the problem of volume expansion. In terms of positive electrode materials, it is necessary to design high-voltage positive electrode materials. The most important advantage of nanowire electrode materials is that they can allow faster migration and transport of lithium ions, and more efficient intercalation and deintercalation reactions occur at the interface between electrode and electrolyte, which is beneficial to the fast charge and discharge performance of the battery.

The cathode materials of traditional lithium-ion battery mainly include $LiCoO_2$, $LiMn_2O_4$ and $LiFePO_4$. Chapter 5 introduces the modification of the cathode by nanowires in detail. The rate performance of nanowire-structured $LiCoO_2$ synthesized by hydrothermal method is significantly improved, and it has great application potential in the consumer electronics market that pursues fast charging [32]. $LiMn_2O_4$ nanowires synthesized by low-pressure solid-phase reactions also exhibited better kinetic performance, high reversibility, and cycling stability [33]. In addition, both the high working-voltage Ni-doped LiNiMnO and Li-rich manganese-based layered oxides have improved electrochemical performance after obtaining the nanowire structure [34, 35]. Since the 1D nanostructure can effectively relieve the internal stress and shorten the ion diffusion distance, transition metal oxides, and polyanionic lithium-ion battery cathode materials have also been significantly improved in rate and cycle performance.

According to the different reaction mechanisms, lithium battery anode materials can be divided into three categories: intercalation reaction type, alloying reaction type, and conversion reaction type. Intercalation reaction-type lithium battery anode materials are diverse, abundant, and environmentally friendly. Intercalation-reactive lithium-ion nanowire anode materials can promote the rapid migration of lithium ions and have broad application prospects. Alloying reaction-type materials have high specific capacity and low operating voltage, which can overcome the safety problems of traditional lithium-ion battery electrode materials, but the violent expansion of the volume will generate huge stress and cause the electrode to pulverize. The unique 1D structure can buffer volume expansion both laterally and along the nanowire growth direction without causing pulverization of the electrode material, providing the possibility for the large-scale use of alloy-based anodes. Most of the anode materials in Li-ion batteries have conversion reaction mechanisms, such as single metal oxides, sulfides, phosphides, and nitrides, as well as their complexes. The nanowire structure can provide shorter electron and ion transport paths, larger contact area, and space for effective buffering of internal stress for electrochemical reactions.

10.3.2 Applications of Nanowire Structures in Na-ion Battery

Sodium and lithium have many similar chemical properties, and current research on cathode materials for sodium-ion batteries mainly focuses on layered transition metal oxides, Prussian blue analogs and polyanionic compounds. The electrochemical performance of layered transition metal oxide nanowires and graded nanofibers synthesized by electrospinning is significantly better than that of nanoparticles. Moreover, the encapsulation of the nanowires by the conductive carbon layer is more beneficial to improving the electronic conductivity. Among the polyanionic compound nanowire materials, the low electrical conductivity of NASICON-type sodium superionic conductors limits their application. The nanowire structure formed by NASICON particles combined with mesoporous nanowires can significantly increase its electrical conductivity and improve its electrochemical performance [36]. There are few reports on the synthesis of Prussian blue analogs with nanowire morphology; however, according to the existing nanowire synthesis strategy, the Prussian blue nanowire structure is still very promising. Using the unique structural advantages of nanowires to further improve the performance of Prussian blue analogs as cathode materials for new sodium-ion battery still has great scientific value.

According to the reaction mechanism, the anode materials of sodium-ion batteries currently studied can be divided into three categories: intercalation type, alloying type, and reactive type. As mentioned above, the hollow carbon nanowires obtained by pyrolysis of hollow polyaniline nanowires have excellent structural stability and electrical conductivity. The $Na_2Ti_3O_7$ nanotube arrays grown on the metallic titanium substrate in the titanium-based anode also showed excellent electrochemical performance [37]. In the negative electrode of the alloying reaction, the electrochemical performance of the carbon-encapsulated Sn/Ni nanorod composites is also significantly improved relative to the pure Sn material [38]. The carbon-wrapped $Ni_{0.85}Se$/C hollow nanowires exhibit excellent rate capability in switching reaction electrode materials [39]. The unique advantages of the nanowire structure can improve the electrochemical performance of these three types of anodes.

10.3.3 Applications of Nanowire Structures in Other Monovalent-ion Batteries

Potassium ions can reversibly and stably intercalate into commercial graphite anodes like lithium ions, which has the potential to be applied in large-scale energy storage grids. However, the radius of potassium ions is significantly larger than that lithium, and potassium ion batteries generally have the problems of slow diffusion kinetics of ions in electrode materials and poor structural stability of electrode materials. By constructing nanowire materials, it is expected to significantly improve the performance of potassium-ion batteries by taking advantage of the short radial diffusion distance and good stress release characteristics of nanowires. The Fe/Mn-based layered oxide $K_{0.7}Fe_{0.5}Mn_{0.5}O_2$ nanowire cathode and the thorn spherical $K_2Ti_8O_{17}$ anode composed of nanorods have also been reported to have good potassium storage properties [40].

In addition to the commonly studied alkali metal-ion batteries, nonmetal carrier batteries such as ammonium ions and protons have also been reported. Although the research on the non-metallic carrier battery system is still in the preliminary stage, there are only a few reports on the application of nanowires uesd in it, but the unique structure of nanowires makes it have great application potential in the field of nonmetallic carrier batteries.

10.3.4 Application of Nanowires in Lithium–Sulfur Batteries

Lithium–sulfur batteries have higher energy density and better battery safety, and they are expected to be adapt to more severe environments. Compared with nanoparticle, nanosheet, bulk, and other structure, the nanowire structure has the advantages of more continuous electron transport path and a woven structure. It can overcome the low electronic conductivity of sulfur and lithium sulfide, which leads to the slow kinetics of the electrochemical process. Although sulfur itself is difficult to form a nanowire structure, it can be effectively loaded on 1D nanomaterials to form nanocomposites. 1D carbon nanostructures, conductive polymer nanowires and metal compound nanowires have been reported and successfully applied to lithium–sulfur batteries.

At present, providing sufficient space to buffer the volume expansion of sulfur by using porous materials. Adding conductive agent to provide fast ion and electron transfer rate and using special structure to short transfer path. Efficiently adsorbing polysulfides. These three methods are used to solve the problems of volume expansion, difficult oxidation of lithium sulfide, and shuttle effect caused by the dissolution of polysulfides in the electrolyte.

Up to now, sulfur cathode materials based on nanowires can be divided into the following types according to the type of composite materials: ① carbon/sulfur nanowire composite material; ② conductive polymer/sulfur nanowire composite material; and ③ metal compound/sulfur nanowire composite material.

1D carbon materials (carbon nanotubes, carbon nanofibers, etc.) are ideal carriers for lithium–sulfur batteries due to their high electrical conductivity and porosity. In the research on the preparation of carbon/sulfur nanowire composite structures, high temperature melting, or chemical synthesis can be used to uniformly load sulfur in the pores of carbon nanotubes. Effective contact between sulfur and carbon nanotubes can improve the specific capacity of lithium–sulfur batteries. Unique carbon nanowire structures with more pores or smaller confinement channels can simultaneously combine the confinement effect of multi-channel hollow nanofibers and the weavable properties. The sulfur loading can be increased to build a 3D continuous conductive framework to better confine the dissolution shuttle of polysulfides. In addition, the surfaces of CNFs and CNTs doped with N, O, S, etc. will have C—N, C—O, C—S bonds hanging on the surface. These polar bonds can form Li—N, Li—O, Li—S bonds with polysulfides that have the same polarity and have the effect of chemical adsorption.

Conductive polymer materials will form covalent bonds with polysulfides through abundant polar bonds on the surface to achieve chemical sulfur fixation while taking electrical conductivity into account. In addition, conductive polymers can also form a variety of copolymers through high-temperature vulcanization reaction with sulfur to construct high-performance composite electrodes. Although the cost of polymer materials is low and the synthesis and processing process is simple, the research on conductive polymer/sulfur composite nano-cathode materials at high loading still needs to be strengthened.

Titanium oxide, manganese dioxide, and other metal oxides have stronger chemical adsorption with polysulfides, which can increase the sulfur loading. The sulfide-solid sulfur matrix can prevent the dissolution of polysulfides through strong polar adsorption of polysulfides through metal sulfur bonds on the surface or edges and can also promote the transformation of lithium polysulfides through the mediation of thiosulfate. Promote kinetics and reduce the dissolution of polysulfides. In addition, some traditional metal electrocatalysts such as Pt, Co, and Ni have also been studied as solid sulfur substrates.

1D carbon nanostructures have unique advantages in the application of highly loaded sulfur cathodes. Their high electrical conductivity can provide efficient electron transport for insulating sulfur and lithium sulfide, and can also greatly reduce the introduction of conductive additives, thereby significantly increasing the sulfur content. However, it cannot perform good chemical adsorption on polysulfides, which limits its further development. Nanowire materials still need more in-depth structural design and experimental exploration in lithium–sulfur batteries.

10.3.5 Application of 1D Nanomaterials in Supercapacitors

Supercapacitors have the characteristics of high power density and good cycle life, but their low energy density limits their development. The surface state, ionic diffusion, and electronic conductivity of supercapacitor electrode materials are critical to their performance. When increasing the materials' specific surface area, attention should also be paid to the ease of diffusion of ions in the electrode from the bulk electrolyte to the energy storage interface. Taken together, the applications of nanowire electrode materials in supercapacitors and secondary batteries are not strictly differentiated. The 1D staggered network is still beneficial for ion diffusion and so on in supercapacitors. The difference is that, due to the characteristics of surface energy storage, there is no ion diffusion in the bulk phase.

The electrode material of the electric double-layer supercapacitor is mainly porous carbon. The surface charge of porous carbon is mainly generated by the adsorption of ions in the electrolyte. Therefore, no electrochemical reaction is involved in the cycle process, and the energy storage is a real capacitive effect. Pore structure is the core of research on electrode materials for electric double-layer supercapacitors. The porous nanowire structure constructed by nanowire electrode materials can accelerate the diffusion of ions in the pore structure and increase the diversity of pore structure design of carbon-based electrode materials. In addition, the nanowire cross-linked network can realize the continuity of the macroporous structure, which further promotes the slow transport process of electrolyte ions between the interface and the bulk electrolyte during the charging and discharging process.

Carbon nanotubes have high crystallinity and electrical conductivity, a unique 1D hollow cavity structure, and the utilization rate of the interconnected pore structure. It can adsorb more electrolyte ions, accelerate the diffusion rate of ions, and reduce their internal transmission resistance. It is very beneficial to the transport of electrons and the formation of electric double layers, and it improves the rate performance of supercapacitors. However, its high synthesis cost limits its widespread use, and the development of a simple synthesis process for carbon nanotubes is an important research direction.

Defects on the surface of carbon fiber can adsorb a large number of electrolyte ions, and at the same time, its unique 1D structure can accelerate the transport of ions. Moderate activation can further enhance the specific surface area of carbon nanofibers and increase electrochemically active sites. At present, the synthesis method of carbon nanofibers is simpler than that of carbon nanotubes, but its commercial application is still difficult. Finding a facile synthetic method for one-step synthesis of carbon nanofibers and applying it to the field of supercapacitors is a major challenge for researchers.

Pseudocapacitor electrode materials are mainly transition metal compounds, conductive polymers, and so on. The key factors affecting its performance mainly include specific surface area, conductivity, structural stability, etc. The most widely studied 1D transition metal oxide is RuO_2 [41]. In the research of other inexpensive transition metal oxides, MnO_2 is inexpensive, exhibits good pseudocapacitive properties in neutral electrolytes, and has a wide voltage window, making it an inexpensive and effective novel capacitor material [42, 43]. However, its conductivity is low and the electrochemical reaction rate is slow, so it needs to be combined with a conductive agent.

Conductive polymers can achieve massive charge storage through fast and reversible n/p-type doping redox reactions on conjugated chains. The polymers used for pseudocapacitive electrode materials are mainly polyaniline, polypyrrole, and polythiophene. Polypyrrole has higher conductivity and a lower synthesis cost. It is also one of the most effective ways to improve pseudocapacitors by constructing a conductive polymer fiber structure network.

The low energy density restricts the commercial application of supercapacitors. Limited by the low specific capacity of electric double-layer supercapacitors and the low cycle life of pseudocapacitors, hybrid supercapacitors with both high energy density and good power density are gradually entering the market. Although the energy density of hybrid supercapacitors has increased significantly, its power density is still different from that of ordinary supercapacitors. Nanowire is a potential electrode materials for new hybrid supercapacitors with both high energy density and good power density due to the unique properties. Whether in aqueous electrolytes or organic electrolytes, the unique structure of nanowire electrode materials significantly increases the charge transfer rate, showing cycling stability, high specific capacity, and good rate performance. Its uniform dispersion structure can effectively solve the shortcomings of slow diffusion rate of organic electrolytes and high viscosity, and it has great development potential and broad application prospects in the current research on supercapacitor electrode materials.

10.3.6 Nanowires for Other Energy Storage Devices

10.3.6.1 Metal Air Batteries

Metal-air batteries are rich in raw materials, cost-effective, and completely pollution-free, and are considered a good choice for next-generation new energy vehicle power battery. At present, metal-air batteries that have made research progress mainly include zinc-air batteries, aluminum-air batteries, lithium-air batteries, and magnesium-air batteries. Zinc-air batteries have been successfully commercialized and widely used in hearing aids and beacon lights, and aluminum-air batteries have also entered the pilot stage. In terms of specific energy, lithium-air batteries have the greatest development potential. However, the formation of metal dendrites, loss of electrolyte, and clogging of air electrodes severely restrict the application of Li-air batteries. The slow decomposition of peroxides and nondecomposable by-products in the charging reaction lead to large charging polarization voltages, thus requiring excellent oxygen reduction/oxidation catalysts. In addition, the air cathode hinders the diffusion of oxygen due to the blockage of the deposition of discharge products. Therefore, cathode materials must have porous structure and high specific surface area to provide more deposition sites. Nanowire materials have a large number of active sites and high specific surface area. Therefore, loading highly active nanowire electrocatalysts on porous materials is a very effective strategy, which can improve the energy density, specific capacity, efficiency, and lifetime of batteries.

10.3.6.2 Multivalent-ion Battery

Multivalent ions, such as magnesium ions, calcium ions, zinc ions, and aluminum ions, have received more and more attention in recent years. In multivalent ion batteries, related metals are usually used directly as negative electrodes. Due to the high charge density of multivalent ions, polarization is easily generated during the intercalation process, and even strong bonding occurs with the host material, which is difficult to be released. Therefore, the main research bottleneck of multivalent-ion batteries lies in the electrolyte and cathode materials. Most cathode materials used in monovalent ion battery systems are difficult to work in multivalent ion batteries. Therefore, the research of cathode materials is a hot and difficult point. Nanoization can shorten the ion diffusion path of the host material, which is conducive to the reversible desorption of multivalent ions. In recent years, the advantageous properties of nanowires have also been well exploited in multivalent ion battery systems.

Tungsten diselenide nanowires, titanium dioxide nanowires, and manganese dioxide nanowires can be used as cathodes for magnesium ion battery, and vanadium oxide nanowires can be used as cathodes for calcium-ion battery in addition to being used as cathodes for magnesium ion battery. Electrode materials with nanowire structures contribute to the reversible de-intercalation of zinc ions, so they have been reported many times in zinc-ion battery systems. At present, the cathode materials of zinc-ion battery are mainly manganese oxide, vanadium oxide, metal ferricyanide, and so on. The anode of aluminum ion battery uses metal aluminum, and the cathode is mainly divided into two types: intercalation/extraction reaction and conversion reaction. Since the diffusion of trivalent aluminum ions in the cathode material is affected by strong lattice repulsion, compared with the traditional monovalent metal cathode materials, the anode materials of aluminum-ion battery require more active sites and larger contact area of the electrolyte. Nanoscale materials are an effective way to improve the electrochemical performance of materials. By reducing the particle size, the ion transport distance can be shortened and the number of active sites near the surface of the material can be increased. Typical nanomaterials such as TiO_2 nanowires, graphene-supported SnO_2 nanoparticles and VS_4 nanosheets, are used in aluminum-ion batteries for the first time. Among them, nanowireization, as a typical way of nanometerization, has a relatively common application in aluminum ion battery system.

10.3.6.3 Metal Sulfur Batteries

Room temperature sodium–sulfur battery is still in the early stages of research. Although available electrolytes and electrode materials have been developed, it will take a long time to achieve high energy density at low cost. How to better solve the problem of slow kinetics and dissolution of sodium polysulfide is a topic that current researchers should pay attention to. Similar to lithium–sulfur batteries, some matrix materials with catalytic effect are introduced into the cathode material to improve the kinetics, while adsorbing sodium polysulfide. Or add organic small molecules to the composite electrode to chemically connect with sodium polysulfide molecules to prevent dissolution. These related studies can further improve the energy density of sodium–sulfur batteries, and then the problems of high load and commercial mass production still need to be solved. At the same time, sodium metal is more active than lithium metal, and the guarantee of safety is also a major problem for sodium–sulfur batteries. It can be seen that the commercial production of room-temperature sodium–sulfur batteries is still difficult.

Compared the high activity of lithium and sodium metal negative electrodes, magnesium metal negative electrodes show good stability. Using magnesium as the negative electrode of metal chalcogenide battery system can greatly improve the safety of the battery. In addition, compared lithium and sodium metals, magnesium metal has become a promising high-energy-density battery system due to its higher volumetric energy density, no obvious dendrite behavior, and abundant crustal content. However, the development of magnesium–sulfur batteries is relatively late, and there are still many unsolved or even unknown problems. Among them, the slow reaction kinetics of magnesium ions and sulfur element can lead to great battery polarization. In addition, nonnucleophilic magnesium electrolytes compatible with elemental sulfur also need to be developed. Magnesium–sulfur batteries also suffer from the problem that magnesium polysulfides are easily dissolved in organic electrolytes. Nano-films prepared from nanowire materials can play a good role in fixing sulfur. Magnesium–sulfur batteries also suffer from the problem that magnesium polysulfides are easily dissolved in organic electrolytes. The surface of ordinary glass fiber film is covered with a layer of carbon fiber interwoven film, which is used to prevent the dissolution and loss of magnesium polysulfide. The specific capacity of the magnesium–sulfur batteries with carbon fiber coating is significantly better than that of the magnesium–sulfur batteries without carbon fiber coating, and it also exhibits better coulombic efficiency. Overall, the industrial application of nanowire materials still faces enormous challenges. Current synthesis methods are not suitable for industrial production due to low yield, the properties of nanowires are directly affected by their structure and size, so the cost of synthesising nanowires with special structures is still high, and there is still an urgent need for reliable synthesis methods that are cost-effective and can accurately control the morphology and size of nanowires. In addition to applications in electrode materials, nanowire materials also have great application prospects in collectors and battery separators, etc. The development of high-performance collectors and special functional separators can effectively improve the service life of energy storage devices. By providing an excellent platform to investigate fundamental scientific questions, nanowires can also greatly broaden the application of advanced characterisation techniques, based on the development of in situ or ex situ characterisation techniques capable of revealing the reaction mechanisms of materials at the atomic level on both spatial and temporal scales. A combination of theoretical computational simulations and experiments is a good choice in developing new nanowire materials. Currently developed machine learning capability has pointed out a new direction for the discovery of new nanowires materials as well as prediction of the possible properties. Despite the aforementioned challenges, the unique physical and chemical properties of nanowire materials, as well as various synthesis methods for controlling the morphology and size of nanowires, provide exciting opportunities for constructing advanced electrochemical energy storage devices for practical applications.

References

1 Li, F., He, J., Zhou, W.L., and Wiley, J.B. (2003). Synthesis of porous wires from directed assemblies of nanospheres. *Journal of the American Chemical Society* 125 (52): 16166–16167.
2 Bechelany, M., Abou Chaaya, A., Frances, F. et al. (2013). Nanowires with controlled porosity for hydrogen production. *Journal of Materials Chemistry A* 1 (6): 2133–2138.
3 Yuan, C., Zhang, X., Hou, L. et al. (2010). Lysine-assisted hydrothermal synthesis of urchin-like ordered arrays of mesoporous $CO(OH)_2$ nanowires and their application in electrochemical capacitors. *Journal of Materials Chemistry* 20 (48): 10809–10816.
4 Liu, J., Jiang, J., Cheng, C. et al. (2011). Co_3O_4 nanowire@MnO_2 ultrathin nanosheet core/shell arrays: a new class of high-performance pseudocapacitive materials. *Advanced Materials* 23 (18): 2076–2081.
5 Yu, L., Zhang, G., Yuan, C., and Lou, X.W. (2013). Hierarchical $NiCo_2O_4$@MnO_2 core–shell heterostructured nanowire arrays on Ni foam as high-performance supercapacitor electrodes. *Chemical Communications* 49 (2): 137–139.
6 Mai, L.-Q., Yang, F., Zhao, Y.-L. et al. (2011). Hierarchical $MnMoO_4/CoMoO_4$ heterostructured nanowires with enhanced supercapacitor performance. *Nature Communications* 2 (1): 381.
7 Jiang, Y., Wu, Y., Zhang, S. et al. (2000). A catalytic-assembly solvothermal route to multiwall carbon nanotubes at a moderate temperature. *Journal of the American Chemical Society* 122 (49): 12383–12384.
8 Liu, H., Hu, L., Meng, Y.S., and Li, Q. (2013). Electrodeposited three-dimensional Ni–Si nanocable arrays as high performance anodes for lithium ion batteries. *Nanoscale* 5 (21): 10376–10383.
9 Wu, H., Xu, M., Wang, Y., and Zheng, G. (2013). Branched Co_3O_4/Fe_2O_3 nanowires as high capacity lithium-ion battery anodes. *Nano Research* 6 (3): 167–173.
10 Xia, X., Tu, J., Zhang, Y. et al. (2012). High-quality metal oxide core/shell nanowire arrays on conductive substrates for electrochemical energy storage. *ACS Nano* 6 (6): 5531–5538.

11 Wang, X., Wang, X., Huang, W. et al. (2005). Sol–gel template synthesis of highly ordered MnO_2 nanowire arrays. *Journal of Power Sources* 140 (1): 211–215.

12 Cao, H.Q., Xu, Y., Hong, J.M. et al. (2001). Sol–gel template synthesis of an array of single crystal CdS nanowires on a porous alumina template. *Advanced Materials* 13 (18): 1393–1394.

13 Jo, Y.R., Myeong, S.H., and Kim, B.J. (2018). Role of annealing temperature on the sol–gel synthesis of VO_2 nanowires with in situ characterization of their metal–insulator transition. *RSC Advances* 8 (10): 5158–5165.

14 Hsu, P.-C., Kong, D., Wang, S. et al. (2014). Electrolessly deposited electrospun metal nanowire transparent electrodes. *Journal of the American Chemical Society* 136 (30): 10593–10596.

15 Ren, W., Zheng, Z., Luo, Y. et al. (2015). An electrospun hierarchical LiV_3O_8 nanowire-in-network for high-rate and long-life lithium batteries. *Journal of Materials Chemistry A* 3 (39): 19850–19856.

16 Lim, J.-M., Moon, J.H., Yi, G.-R. et al. (2006). Fabrication of one-dimensional colloidal assemblies from electrospun nanofibers. *Langmuir* 22 (8): 3445–3449.

17 Jin, Y., Yang, D., Kang, D., and Jiang, X. (2010). Fabrication of necklace-like structures via electrospinning. *Langmuir* 26 (2): 1186–1190.

18 Zhang, J., Li, W., Jia, Q. et al. (2015). Molten salt assisted synthesis of 3C–SiC nanowire and its photoluminescence properties. *Ceramics International* 41 (10, Part A): 12614–12620.

19 Yan, A., Liu, Y., Liu, Y. et al. (2012). A NaAc-assisted large-scale coprecipitation synthesis and microwave absorption efficiency of Fe_3O_4 nanowires. *Materials Letters* 68: 402–405.

20 Vivekchand, S.R.C., Gundiah, G., Govindaraj, A., and Rao, C.N.R. (2004). A new method for the preparation of metal nanowires by the nebulized spray pyrolysis of precursors. *Advanced Materials* 16 (20): 1842–1845.

21 Fu, Q.-G., Li, H.-J., Shi, X.-H. et al. (2006). Synthesis of silicon carbide nanowires by CVD without using a metallic catalyst. *Materials Chemistry and Physics* 100 (1): 108–111.

22 Chengshan, X., Hong, L., Huizhao, Z. et al. (2009). Synthesis of GaN nanowires with tantalum catalyst by magnetron sputtering. *Rare Metal Materials and Engineering* 38 (7): 1129–1131.

23 Liu, R. and Lee, S.B. (2008). MnO_2/poly(3,4-ethylenedioxythiophene) coaxial nanowires by one-step coelectrodeposition for electrochemical energy storage. *Journal of the American Chemical Society* 130 (10): 2942–2943.

24 Xu, L., Jiang, Z., Qing, Q. et al. (2013). Design and synthesis of diverse functional kinked nanowire structures for nanoelectronic bioprobes. *Nano Letters* 13 (2): 746–751.

25 Morber, J.R., Wang, X., Liu, J. et al. (2009). Wafer-level patterned and aligned polymer nanowire/micro- and nanotube arrays on any substrate. *Advanced Materials* 21 (20): 2072–2076.

26 Htay, M.T. and Hashimoto, Y. (2015). Field emission property of ZnO nanowires prepared by ultrasonic spray pyrolysis. *Superlattices and Microstructures* 84: 144–153.

27 Ma, X., Luo, W., Yan, M. et al. (2016). In situ characterization of electrochemical processes in one dimensional nanomaterials for energy storages devices. *Nano Energy* 24: 165–188.

28 Hu, P., Yan, M., Wang, X. et al. (2016). Single-nanowire electrochemical probe detection for internally optimized mechanism of porous graphene in electrochemical devices. *Nano Letters* 16 (3): 1523–1529.

29 Balasubramanian, M., Sun, X., Yang, X.Q., and McBreen, J. (2001). In situ X-ray diffraction and X-ray absorption studies of high-rate lithium-ion batteries. *Journal of Power Sources* 92 (1): 1–8.

30 Yan, M., Zhang, G., Wei, Q. et al. (2016). In operando observation of temperature-dependent phase evolution in lithium-incorporation olivine cathode. *Nano Energy* 22: 406–413.

31 Liu, Q., Hao, Z., Liao, X. et al. (2019). Langmuir–Blodgett nanowire devices for in situ probing of zinc-ion batteries. *Small* 15 (30): 1902141.

32 Xiao, X., Yang, L., Zhao, H. et al. (2012). Facile synthesis of $LiCoO_2$ nanowires with high electrochemical performance. *Nano Research* 5 (1): 27–32.

33 Lee, H.-W., Muralidharan, P., Ruffo, R. et al. (2010). Ultrathin spinel $LiMn_2O_4$ nanowires as high power cathode materials for Li-ion batteries. *Nano Letters* 10 (10): 3852–3856.

34 Badami, P.P. (2016). *Performance Evaluation and Characterization of Lithium-Ion Cells Under Simulated PHEVs Drive Cycles [M.S.]*. Ann Arbor: Arizona State University.

35 Zhang, X., Cheng, F., Yang, J., and Chen, J. (2013). $LiNi_{0.5}Mn_{1.5}O_4$ porous nanorods as high-rate and long-life cathodes for Li-ion batteries. *Nano Letters* 13 (6): 2822–2825.

36 Jiang, Y., Yao, Y., Shi, J. et al. (2016). One-dimensional $Na_3V_2(PO_4)_3$/C nanowires as cathode materials for long-life and high rate Na-ion batteries. *ChemNanoMat* 2 (7): 726–731.

37 Ni, J., Fu, S., Wu, C. et al. (2016). Superior sodium storage in $Na_2Ti_3O_7$ nanotube arrays through surface engineering. *Advanced Energy Materials* 6 (11): 1502568.

38 Liu, Y., Xu, Y., Zhu, Y. et al. (2013). Tin-coated viral nanoforests as sodium-ion battery anodes. *ACS Nano* 7 (4): 3627–3634.

39 Yang, X., Zhang, J., Wang, Z. et al. (2018). Carbon-supported nickel selenide hollow nanowires as advanced anode materials for sodium-ion batteries. *Small* 14 (7): 1702669.

40 Wang, X., Xu, X., Niu, C. et al. (2017). Earth abundant Fe/Mn-based layered oxide interconnected nanowires for advanced K-ion full batteries. *Nano Letters* 17 (1): 544–550.

41 Han, J., Xu, M., Niu, Y. et al. (2016). Exploration of $K_2Ti_8O_{17}$ as an anode material for potassium-ion batteries. *Chemical Communications* 52 (75): 11274–11276.

42 Kharismadewi, D. and Shim, J.-J. (2015). Ultralong MnO_2 nanowires intercalated graphene/Co_3O_4 composites for asymmetric supercapacitors. *Materials Letters* 147: 123–127.

43 Zhou, J., Yu, L., Liu, W. et al. (2015). High performance all-solid supercapacitors based on the network of ultralong manganese dioxide/polyaniline coaxial nanowires. *Scientific Reports* 5 (1): 17858.

Index

Nanowire Energy Storage Devices: Synthesis, Characterization, and Applications, First Edition.
Edited by Liqiang Mai.